普通高等教育"十一五"国家级规划教材

全国高等医药院校药学类专业第三轮实验双语教材

基础化学实验与指导

Experiment and Guidance for Fundamental Chemistry

主　编　王　越　叶宝芬

副主编　洪　瑾　熊晔蓉

徐光富　郝利君

中国健康传媒集团

中国医药科技出版社 · 北京

内 容 提 要

　　本教材是"全国高等医药院校药学类专业第三轮实验双语教材"之一，是根据基础化学的教学大纲及要求，结合专业培养目标配套编写的实验与指导。全书分为绪论、基础化学实验基本操作、基础性实验、元素的化学性质实验、无机化合物综合性制备实验、化学分析实验等六个章节。同时本教材为书网融合教材，即纸质教材有机融合电子教材、教学配套资源（PPT、微课等），使教学资源更加多样化、立体化。

　　本教材主要供高等医药院校药学及相关专业使用。

图书在版编目（CIP）数据

基础化学实验与指导：汉英对照/王越，叶宝芬主编. —北京：中国医药科技出版社，2024.5（2025.7重印）.

全国高等医药院校药学类专业第三轮实验双语教材

ISBN 978 - 7 - 5214 - 4690 - 6

Ⅰ.①基⋯　　Ⅱ.①王⋯ ②叶⋯　　Ⅲ.①化学实验 - 双语教学 - 医学院校 - 教学参考资料 - 汉、英　　Ⅳ.①O6 - 3

中国国家版本馆 CIP 数据核字（2024）第 110199 号

美术编辑　　陈君杞

版式设计　　友全图文

出版　**中国健康传媒集团** | 中国医药科技出版社

地址　北京市海淀区文慧园北路甲 22 号

邮编　100082

电话　发行：010 - 62227427　　邮购：010 - 62236938

网址　www. cmstp. com

规格　889 × 1194mm $^1/_{16}$

印张　15 $^1/_4$

字数　435 千字

版次　2024 年 5 月第 1 版

印次　2025 年 7 月第 2 次印刷

印刷　北京盛通印刷股份有限公司

经销　全国各地新华书店

书号　ISBN 978 - 7 - 5214 - 4690 - 6

定价　**49.00 元**

获取新书信息、投稿、
为图书纠错，请扫码
联系我们。

前 言

　　《基础化学实验与指导》双语教材改版于原《无机化学实验与指导》第三版。自 2006 年出版以来，被国内众多医药院校采用。期间，我们不断收集一线教师和学生的反馈，并在多次的教研活动中反复交流以达到共识。在此基础之上，为推动高等药学教育改革创新，重构学生知识、能力、素质体系，满足行业对药学专业人才日益提高的新需求，本版进行了进一步修订。丰富了本版修订教材的素材，为了适应高校教学课程改革，本版更名为《基础化学实验与指导》。在此，编者对给本教材提出各类宝贵意见和分享教学体验的专家及师生们表示衷心的感谢！

　　化学实验教学在巩固学生的基础知识、培养实践能力和提升科学素质等各方面起着不可替代的作用。根据高等医药院校对化学基础实验教学发展的新要求，在本版修订中，我们扩充了全书内容，对编排体系进行了适当调整，并增加了相关的数字化资源，包括每个实验项目的 PPT、同步习题及解答，还针对个别实验的关键技术或实验步骤制作了微视频，融入优质教学资源，势必会有助于提升教学效果。同时也对前版教材中不妥之处进行了修正。

　　在教学实践过程中，有关书中具体实验项目的选用可根据实际的教学计划，配合主干教材理论课程的进度和安排，但不完全依附于理论课教学的框架，由教师灵活决定实验内容和顺序。

　　尽管在修订过程中我们力求做到选材更恰当，翻译更准确，但由于受限于编者学识水平所限，教材中仍有欠妥之处，恳请同行专家及读者批评指正。

编 者

2024 年 3 月

目 录

第一部分　基础化学实验的基本原理、基本方法与基本操作

第一章　绪　论

一、化学实验的目的与任务

化学是一门实验性科学，化学实验是化学教学不可缺少的重要组成部分。通过实验，既能发现和发展理论，又能检验和评价理论。化学实验的目的是开拓学生智能，培养学生严肃、严密、严格的科学态度和良好的科学素养，提高学生的动手能力和独立工作能力，并为将来从事科学研究奠定坚定的基础。因而，化学实验的作用不是验证学生所学的化学理论知识，而是要通过实验，训练学生进行科学实验的方法和技能，进而使学生进一步学会对实验现象进行观察、分析、归纳、总结，培养学生独立工作、分析问题和解决问题的能力。

二、化学实验的学习方法与要求

要很好地完成实验的各个环节，除具有坚实的理论基础外，还要有正确的学习方法。

1. 认真预习　实验前应认真阅读实验教材，明确实验的目的；了解实验内容、原理和方法；清楚所用药品或试剂的等级、理化性质（熔点、沸点、密度、毒性与安全等数据）；熟悉所用仪器；设计实验装置、实验步骤；估计实验中可能发生的现象和预期结果；明确实验数据处理方法和有关计算公式。在此基础上写好实验预习报告。

2. 认真实验　依据实验的内容、方法、步骤及要求进行实验，做到遵守实验操作规程，仔细观察实验现象，结合理论认真分析实验结果，如实而详细地记录实验现象和原始数据。

3. 认真书写实验报告　实验报告不仅是概括和总结实验过程的文献性质资料，而且是学生通过实验过程获取化学知识的重要途径。

因而，书写实验报告同样是化学实验课程的基本训练内容。实验报告能从一定的角度反映科学工作者的科学态度、实际水平与能力。实验报告的格式与要求基本包括：实验名称、实验目的、实验原理、实验仪器（厂家、型号、测量精度等）、药品与试剂（纯度等级）、实验装置（流程图或表格等）、实验现象及观测数据、实验结果（包括数据处理）、讨论。

实验结果的讨论是实验报告的重要组成部分，它包括实验工作者学术性的体会（并非感性的表达），实验结果的可靠性与合理性评价，实验现象的分析及解释等。

基础化学实验大致分为三种类型：一是验证性实验；二是测定性实验；三是无机制备实验。验证性实验主要是物质性质的验证，可加深对反应原理和物质性质的理解。测定性实验主要是测定数据及数据处理过程。制备实验要写出物质制备原理、流程、原料量、产量、产率、产品质量与性质等。

三、实验误差及有效数字

化学实验过程中经常使用仪器对一些物理量进行测量，从而对体系中的一些化学性质和物理性质

作出定量描述，揭示事物的客观规律。但事实上，任何测量的结果（数据）只能是相对准确的，或者说是存在某种程度上的不确定（不可靠）性，这种不确定（不可靠）被称为实验误差。产生这种误差的原因，是因为测量仪器、方法、实验条件以及实验工作者本人不可避免地存在一定局限性等。

对于不可避免的实验误差，实验者须了解其产生的原因、性质及有关规律，从而在实验中设法控制和减少误差，并对测量的结果进行适当处理，以达到可信的程度。

1. 绝对误差与相对误差 测量中的误差，主要有两种表示方法：绝对误差与相对误差。

（1）绝对误差 测量值与真值（真实值）之差称为绝对误差。若以 χ 代表测量值，以 μ 代表真实值，则绝对误差 δ

$$\delta = \chi - \mu$$

绝对误差是以测量值的单位为单位，可以是正值，也可以是负值，即测量值可能大于或小于其真值。测量值越接近真值，绝对误差越小；反之，越大。

实际上绝对准确的实验结果是无法得到的。化学研究中所谓的真值是指有经验的研究人员用可靠（相对而言）的测定方法多次平行测定得到的平均值。常用的真值一般有三类：理论真值、约定真值及相对真值。

（a）理论真值 如三角形的内角和为 180° 等。

（b）约定真值 如国际计量大会定义的单位（国际单位）及我国的法定计量单位。

（c）相对真值 对科技工作者而言，由于没有绝对纯的化学试剂，因而常用标准参考物质的证书上所给示的含量作为相对真值。

（2）相对误差 绝对误差与真值的比值称为相对误差。

$$\frac{\delta}{\mu} = \frac{\chi - \mu}{\mu}$$

相对误差反映测量误差在测量结果中所占的比例，它没有单位，通常以 %、‰ 表示。

例如，若测定纯 NaCl 中 Cl 的百分含量为 60.52%，而其真值应为 60.66%，则

绝对误差 = 60.52% - 60.66% = -0.14%

$$相对误差 = \frac{60.52\% - 60.66\%}{60.66\%} \times 1000‰ = -2.3‰$$

2. 准确度与精密度 准确度表示测定结果与真值接近程度。测量值与真值越接近，就越准确。准确度的大小用绝对误差或相对误差表示。误差越大，准确度越低；反之，准确度越高。例如一物体的真实质量是 10.000g，甲的测量结果为 10.001g，乙的测量结果为 10.008g。前者的绝对误差是 0.001g，后者的绝对误差是 0.008g。10.001g 比 10.008g 的绝对误差小，所以前者比后者测量得更准确，或者说前一结果比后一结果的准确度高。

精密度表示平行测量的各测量值（实验值）之间相互接近的程度。各测量值间越接近，精密度就越高；反之，精密度低。精密度可用偏差、相对平均偏差、标准偏差与相对标准偏差表示。

（1）偏差 测量值与平均值之差称为偏差。偏差越大，精密度越低。若令 $\bar{x}$ 代表一组平行测量值的平均值，则单个测量值 x_i 的偏差 d_i 为

$$d_i = x_i - \bar{x}$$

因此，d 有正、负之分。各个偏差绝对值的平均值称为平均偏差

$$\bar{d} = \frac{\sum_{i=1}^{n} |x_i - \bar{x}|}{n}$$

式中，n 表示测量次数。

（2）相对平均偏差，定义如下

$$\frac{\bar{d}}{\bar{x}} \times 100\% = \frac{\sum_{i=1}^{n} |x_i - \bar{x}|/n}{\bar{x}} \times 100\%$$

（3）标准偏差或标准差，定义如下：

$$S = \sqrt{\frac{\sum_{i=1}^{n}(x_i - \bar{x})^2}{n-1}} \quad 或 \quad S = \sqrt{\frac{\sum_{i=1}^{n} x_i^2 - \frac{1}{n}(\sum_{i=1}^{n} x_i)^2}{n-1}}$$

使用标准偏差是为了突出较大偏差的影响。

（4）相对标准差（变异系数），定义如下：

$$RSD = \frac{S}{\bar{x}} \times 100\% = \frac{\sqrt{\frac{\sum_{i=1}^{n}(x_i - \bar{x})^2}{n-1}}}{\bar{x}} \times 100\%$$

实际工作中，往往用 RSD 表示测定结果的精密度。

例如，四次测定某溶液的浓度，结果为 0.2041、0.2049、0.2039 和 0.2043mol/L。则平均值（$\bar{x}$），平均偏差（$\bar{d}$），相对平均偏差$\left(\frac{\bar{d}}{\bar{x}}\right)$，标准偏差（$S$）及相对标准偏差（RSD）为：

平均值$(\bar{x}) = (0.2041 + 0.2049 + 0.2039 + 0.2043)/4 = 0.2043$mol/L

平均偏差$(\bar{d}) = (0.0002 + 0.0006 + 0.0001 + 0.0000)/4 = 0.0003$mol/L

相对平均偏差$\left(\frac{\bar{d}}{\bar{x}}\right) = \frac{0.0003}{0.2043} \times 1000\text{‰} = 1.5\text{‰}$

标准偏差$(S) = \sqrt{\frac{0.0002^2 + 0.0006^2 + 0.0004^2 + 0.0000^2}{4-1}} = 0.0004$mol/L

相对标准偏差 $RSD = \frac{0.0005}{0.2043} \times 100\% = 0.3\%$

精密度是保证准确度的前提条件，没有好的精密度就不可能有好的准确度。因为事实上，准确度是在一定的精密度下，多次测量的平均值与真值相互接近的程度。测量值的准确度表示测量结果的正确性，测量值的精密度表示测量结果的重复性或再现性。

3. 系统误差和随机误差 依据误差产生的原因及性质，误差可分为系统误差与随机误差。

（1）系统误差 系统误差是由某些固定的原因造成的，使得测量结果总是偏高或偏低。实验方法不够完善、仪器不够精确、试剂不够纯以及测量者本人的习惯、仪器使用的理想环境达不到要求等因素都有可能产生系统误差。系统误差的特征是：①单向性，即误差的符号及大小恒定或按一定规律变化；②系统性，即在相同条件下重复测量时，误差会再现，因此系统误差可用校正等方法予以减免。常见的系统误差大致分为：仪器误差、方法误差和个人误差。

（a）仪器误差 所有的测量仪器都可能产生系统误差。例如天平失于校准（如不等臂）；磨损或腐蚀的砝码；移液管、滴定管、容量瓶等玻璃仪器的实际容积和标示容积不符；电池电压下降，接触不良造成电路电阻增加等都会造成系统误差。

（b）方法误差 这是由于测试方法不完善造成的。其中有化学和物理化学方面的原因，常常难以发现。因此，这是一种影响最为严重的系统误差。例如某些反应速率很慢或未定量地完成，干扰离子的影响，沉淀溶解、共沉淀和后沉淀等都会系统地导致测定结果偏高或偏低。

（c）个人误差　是一种由操作者本身的一些主观因素造成的误差。例如在读取刻度值时，总是偏高或总是偏低。

（2）偶然误差　偶然误差又称随机误差。它指同一操作者在同一条件下对同一量进行多次测定，而结果不尽相同，以一种不可预测的方式变化着的误差。它产生的直接原因往往难于发现和控制。偶然误差有时正、有时负，数值有时大、有时小，因而具有一定的不确定性。在各种测量中，随机误差总是不可避免地存在，并且不可能加以消除，它构成了测量的最终精密度。偶然误差对测定结果的影响通常服从统计规律，因而，可以采用在相同条件下多次测定同一量，再求其算术平均值的方法来减免。

（3）过失误差　由于操作者的疏忽大意，没有完全按照操作规程实验等原因造成的误差称为过失误差，这种误差使测量结果与事实明显不合，有较大的偏离且无规律可循。含有过失误差的测量值，不能作为一次实验值引入数据处理。这种过失误差，需要通过加强责任心，仔细工作来避免。判断是否发生过失误差必须慎重，应有充分的依据，最好重复这个实验来检查，如果经过细致实验后仍然出现这个数，要依据已有的科学知识判断是否有新的问题，甚至有新的发现，这在实践中是常有的事。

4. 有效数字及运算法则

（1）有效数字　在科学研究过程中，各种物理量的测量值（观测值）的记录必须与测试仪器的精度相一致。通常情况下，任何一种仪器标尺读数的最低一位应该用内插法估计到两条刻度线间距的1/10。因而，任何一个测量值的最后一位数字应是有一定误差的。这种误差来自于估计的不可靠性，有时称为**不确定度**，一般为 ±0.1 分度。最后这位数字是可疑的，但绝非臆造，应该是可信的，因而是"有效"的。记录时应保留这位数字才能正确地反映出测量的精确程度。这种在不丧失测量准确度的情况下，表示某个测量值所需要的最小位数的数字称为**有效数字**。也就是说，有效数字就是实际能够测量到的数字，它总是和测量或测定联系在一起。有效数字的构成包括若干位确定的数字和一位不确定的数字。例 253.8 这个数有 4 位有效数字，用科学表示法写成 2.538×10^2，若写成 2.5380×10^2，就意味着它有 5 位有效数字。"0 是一个特殊的数字。当它出现在两个非零数字之间或小数点右方的非零数字之后时都是有效的。如 10.0500g，其中每个 0 都是有效的，它有 6 位有效数字。而 0.0290 中，2 之前的两个 0 都是无效的，因这两个 0 只是用来决定小数点的位置，取决于所用的单位，当用毫克计量时，可写成 28.0mg。最后一个 0 仍是有效的。

但是，像 83600 这类数字的有效数字却含混不清，可能意味着下列情况之一：

8.36×10^4	3 位有效数
8.360×10^4	4 位有效数
8.3600×10^4	5 位有效数

因此，像 83600 这类数值最好用上述科学表示法之一书写，以便准确地表示出它究竟有几位有效数字。

（2）数的修约　当计算涉及到几个测量值，而它们的有效数字的位数不相同时，便要舍去多余的数字，称之为数的**修约**（rounding）。过去通常采用"四舍五入"或"四舍六入五成双"规则。现介绍国家标准新的修约规则。

（a）在数据处理中，常遇到一些准确度不相等的数值，此时如果按一定规则对数值进行修约，既可节省计算时间，又可减少错误。

（b）修约的含义是用一修约数代替一已知数，修约数来自选定的修约区间的整数倍。

例：

　　　　修约区间：0.1

整数倍：12.1，12.2，12.4 等。

修约区间：10

整数倍：1 210，1 220，1 230，1 240 等。

（c）如果只有一个整数倍最接近已知数，则此整数倍就认为是修约数。

例：

修约区间：0.1

已知数	修约数
12.223	12.2
12.251	12.3
12.275	12.3

（d）如果有两个连续的整数倍同等地接近已知数，则有两种不同的规则可选用。

规则 A：选取偶数整数倍作为修约数。此规则广泛应用于处理测量数据。

例：

修约区间：0.1

已知数	修约数
12.25	12.2
12.35	12.4

规则 B：选取较大的整数倍作为修约后的数。此规则广泛应用于计算机。

例：

修约区间：0.1

已知数	修约数
12.25	12.3
12.35	12.4

（e）用上述规则作多次修约时，可能会产生误差。因此推荐一次完成修约。

例：12.251 应修约成 12.3，而不是第一次修约成 12.25，然后修约为 12.2。

（f）上述规则只用在对选择修约数没有特别规定的情况。

（3）运算规则

（a）加减法 和或差的有效位数按照各原始数据中小数点后位数最少的数据确定。用科学表示法表示的数据，若指数不同，应先化成相同的指数，然后才能加减。

例如，将下列 Cl^- 浓度相加减：3.00×10^{-2} mol/L、5.55×10^{-3} mol/L 和 1.00×10^{-5} mol/L。可用两种方法计算：（a）将各个加数直接相加，然后修约结果；（b）先修约各个加数，然后再相加。

先加后修约	先修约再相加
3.00×10^{-2}	3.00×10^{-2}
0.555×10^{-2}	0.56×10^{-2}
+）0.001×10^{-2}	+）0.00×10^{-2}（可忽略不计）
3.556×10^{-2}（mol/L）	3.56×10^{-2}（mol/L）
↓	
3.56×10^{-2}（mol/L）	

为减小舍入误差的累积，有时在修约各个加数时，比小数位数最小的数多保留一位有效位数。例如：

先加后对 结果修约	修约成小数点后 2 位再相加	修约成小数点后 3 位再相加
2.25	2.25	2.25
3.4375	3.44	3.438
+）4.27502	+）4.28	+）4.275
9.96252	9.97	9.963
↓		↓
9.96		9.96

（b）乘除法　积与商的有效位数按照原始数据中有效数字位数最少的数确定。例如：

$$1.262 \times 10^{-5}$$
$$\times）4.78$$

$$6.03236 \times 10^{-5}$$
$$\downarrow$$
$$6.03 \times 10^{-5}$$

在计算过程中，为防止修约造成误差的累积，可多保留一位有效数字进行计算，最后将计算结果按修约规则修约。例如：

$$5.3179 \times 10^{12} \qquad\qquad 5.32 \times 10^{12}$$
$$\times）3.6 \times 10^{-19} \qquad\qquad \times）3.6 \times 10^{-19}$$

$$\longrightarrow$$

$$1.9152 \times 10^{-6} \longrightarrow 1.9 \times 10^{-6}$$

（c）对数与反对数　对数尾数的有效位数应与真数的有效位数相同。例如：

$$\lg \underset{\text{真数}}{345} = \underset{\text{首数}}{2} . \underset{\text{尾数}}{538}$$

因此 345 可写成 3.45×10^2，它的对数的首数相应于 3.39×10^2 中 10 的幂，起决定小数点位置的作用。

再进一步看一下对数尾数最后一位有效数字（第 3 位）与真数的关系：

$$10^{2.671} = 469 \quad (468.8)$$
$$10^{2.670} = 468 \quad (467.7) \qquad \left.\begin{array}{l}\text{括号中的数值}\\\text{为修约成 3 位}\\\text{数以前的结果}\end{array}\right]$$
$$10^{2.669} = 467 \quad (466.6)$$

可见指数的小数点第 3 位改变 ±1 时，结果 468（即对数的真数）的最后一位数字改变 ±1。

将对数转换成反对数时，有效位数则应与尾数的位数相同，例如 $\lg 10^{-3.42}$ 的反对数为

$$10^{-3.42} = 3.8 \times 10^{-4}$$

又如，$[H^+] = 6.6 \times 10^{-10}$ mol/L 的溶液，pH 应为 9.18，不是 9 或 9.2。

四、化学实验室规则与事故处理

为了保证实验的顺利进行及实验室安全，进入实验室的所有工作人员必须遵守实验室规则和安全守则，懂得常见事故的处理方法。

1. 化学实验室规则

（1）在实验室工作的人员必须遵守纪律，保持肃静，集中思想，认真操作，仔细观察，积极思考，真实记录。

（2）正确使用实验仪器、设备。

（3）药品应按规定的量取用，已取出的试剂不能再放回原试剂瓶中以免污染试剂。取用药品的用具应保持清洁、干燥，以保持试剂不被污染及浓度一定。取用药品后应立即盖上瓶盖，以免放错瓶塞，污染药品。

（4）实验前要检查所需仪器是否齐全，是否破损，以便及时补齐、更换。实验过程中应保持器皿清洁，保持实验台面清洁整齐，实验结束后仪器、药品放回原处。

（5）废的固体、纸、玻璃渣、火柴梗等应倒入废品篮内；废液应倒入指定的废液回收桶，不得倒入水槽流入下水道，剧毒废液由实验室统一处理。

（6）实验完成后应保持实验室清洁，检查水、电、气安全，关好门窗。

（7）实验室一切物品不得私自带出室外。

2. 化学实验室安全规则　化学实验室中有易燃、易爆、有毒或腐蚀性的药品，化学实验过程中使用水、电、气，如果使用不当，则存在不安全因素。凡进入实验室的人员必须重视安全问题，遵守操作规程，绝不可麻痹大意，严格遵守实验室安全守则，以免事故的发生。

（1）易燃的试剂如乙醇、丙酮、乙醚等，使用时应远离火源，用完后立即塞紧瓶塞。

（2）加热、浓缩液体时要防止液体冲击容器，试管口要朝向无人处。

（3）产生有刺激性气味、有毒气体的实验要在通风橱中进行，嗅气体的气味时，只能用手轻轻地煽动空气，使少量气体进入鼻孔。

（4）使用有毒试剂如砷化物、铬盐、氰化物、汞及其化合物等，要严格防止进入口内和伤口内，废液严禁排入下水道。

（5）防止浓酸、浓碱液溅在皮肤或衣物上，尤其不能溅入眼内。

（6）湿手不要接触电器插头，人体不能与导电物体接触。实验结束后应切断电源。

（7）严禁随意混合各种化学试剂，以避免发生意外事故。

（8）严禁在实验室内饮食、吸烟、不得将食物或餐饮具带入实验室，实验后要清洗双手。

3. 常见事故的简单处理　实验室发生事故后，应冷静沉着，立刻采取有有效措施处理事故。

（1）触电　不慎触电时，立即切断电源，或尽快地用绝缘物（干燥的木棒、竹竿等）将电源与触电者隔开，必要时进行人工呼吸。

（2）烫伤　被火、高温物体烫伤后，切勿用水冲洗，更不要将烫起的水泡挑破，可在烫伤处涂上烫伤药膏。必要时送医院治疗。

（3）割伤　先将伤口中的异物取出，不要用水冲洗伤口，涂上红药水或创可贴，必要时送医院救治。

（4）酸（或碱）伤　酸（或碱）溅入眼内，应立即用水冲洗，再用 2% 的 $Na_2B_2O_7$ 溶液（或 3% 的硼酸溶液）冲洗眼睛，然后用蒸馏水冲洗。酸（或碱）洒到皮肤上时，先应用大量水冲洗，再用饱和碳酸氢钠（或 2% 醋酸溶液）冲洗，最后再用水冲洗，涂敷氧化锌软膏（或硼酸软膏）。

（5）毒物误入口内　将 5～10mL 稀硫酸铜溶液加入一杯温水中，内服后再用手指伸入咽喉部，促使呕吐，然后立即送医院治疗。

（6）吸入刺激性或有毒气体　吸入如溴蒸气、氯化氢时，可吸入少量酒精和乙醚的混合蒸气解毒；因不慎吸入煤气、硫化氢气体时，应立刻到室外呼吸新鲜空气。

（7）伤势严重者，应立即送医院诊治。

4. 消防　实验室不慎起火后，不应惊慌失措，而应根据不同的着火情况，采用不同的灭火措施。由于物质燃烧需要一定的氧气（空气）和达到着火点（一定的温度），所以灭火的原则是降温或将燃烧的物质与空气隔绝。

化学实验中常用的灭火措施如下。

（1）小火用湿布、石棉布覆盖燃烧物即可灭火，大火应用泡沫灭火器灭火。对 Na、K、Mg、Al 等活泼金属等引起的着火，应用干燥的细沙覆盖灭火。有机溶剂着火，切勿用水灭火，而应用二氧化碳灭火器、沙子或干粉等灭火（表 1-1）。

表 1-1　常用灭火器种类及其适用范围

名　称	适　用　范　围
泡沫灭火器	用于一般失火及油类着火。此种灭火器是由 $Al_2(SO_4)_3$ 和 $NaHCO_3$ 溶液作用产生大量的 $Al(OH)_3$ 及 CO_2 泡沫，泡沫把燃烧物质覆盖与空气隔绝而灭火。因为泡沫能导电，所以不能用于扑灭电器设备着火
四氯化碳灭火器	用于电器设备及汽油、丙酮等着火。此种灭火器内装液态 CCl_4。CCl_4 沸点低，相对密度大，不会被引燃，所以把 CCl_4 喷射到燃烧物的表面，CCl_4 液体迅速气化，覆盖在燃烧物上而灭火
1211 灭火器	用于油类、有机溶剂、精密仪器、高压电气设备。此种灭火器内装 CF_2ClBr 液化气，灭火效果好
二氧化碳灭火器	用于电器设备失火及忌水的物质着火。内装液态 CO_2
干粉灭火器	用于油类、电器设备、可燃气体及遇水燃烧等物质的着火。内装 $NaHCO_3$ 等物质和适量的润滑剂和防潮剂。此种灭火器喷出的粉末能覆盖在燃烧物上，组成阻止燃烧的隔离层，同时它受热分解出 CO_2，能起中断燃烧的作用，因此灭火速度快

（2）在加热时着火，应立即停止加热，关闭煤气总阀，切断电源，将一切易燃易爆物品移至远处。

（3）电器设备着火，先切断电源，再用四氯化碳灭火器灭火，也可用干粉灭火器灭火。

（4）当衣服着火时，切勿慌张跑动，应赶快脱下衣服或用石棉布覆盖着火处，或在地上卧倒打滚，起到灭火的作用。

（5）必要时报火警。

书网融合……

习题

第二章　基础化学实验基本操作

一、常用玻璃仪器

基础化学实验常用的玻璃仪器见表 2-1。

表 2-1　常用玻璃仪器

仪器名称	规　格	用途及注意事项
烧杯　　锥形瓶（磨口）	以容积（mL）表示，一般有 50、100、200、400、1000、2000 等	加热时将烧杯放置在石棉网上，使受热均匀，所盛反应液体一般不能超过烧杯容积的 2/3
试管　　离心试管	分硬质试管、软质试管、普通试管和离心试管 普通试管以管外径(mm)×长度(mm)表示，一般有 12×150、15×100、30×200 等 离心试管以容积（mL）表示，一般有 5、10、15 等	普通试管用作少量试剂的反应器，便于操作和观察。离心试管还可用于定性中的沉淀分离 加热时不能骤冷，以防炸裂。反应液体一般不能超过试管容积的 1/2，加热时不能超过 1/3。离心管不能直火加热
量筒　　量杯	以容积（mL）表示，量筒有 10、20、50、100 等，量杯有 10，20 等	量取一定体积的液体用 不能直接加热，不可用作反应器
吸量管　移液管	以容积（mL）表示，有 1、2、5、10、25、50 等规格	精确量取一定体积的液体用 管口上无"吹出"字样者，使用时末端的溶液不允许吹出。不能加热
滴定管	滴定管分酸式和碱式，无色和棕色。以容积（mL）表示，有 25，50 等	滴定或量取准确体积的溶液时使用 酸式滴定管盛酸性溶液或氧化型溶液；碱式滴定管盛碱性溶液或还原型溶液 碱式滴定管不能盛放氧化剂，见光易分解的滴定液宜用棕色滴定管。不能加热和量取热的液体

续表

仪器名称	规　格	用途及注意事项
容量瓶	以容积（mL）表示，有 50、100、250、1000 等	用来配制准确浓度的溶液 不能受热，不得贮存溶液，不能在其中溶解固体，瓶塞与瓶是配套的，不能互换
漏斗　长颈漏斗　漏斗架	以口径（mm）大小表示，有 30、40、60 等	用于过滤操作 不能用火加热。加液体不能超过其容积的 2/3
吸滤瓶　布氏漏斗	布氏漏斗瓷质，以直径（cm）表示，有 6、8 等。吸滤瓶为玻璃制品，以容积（mL）表示，有 250，500 等	用于减压过滤 不能直接加热，滤纸要略小于漏斗的内径。使用时先开抽气泵，后过滤；过滤完毕，先拔掉抽滤瓶接管，后关抽气泵
蒸发皿	以口径（mm）或容积（mL）表示。材质有瓷质、石英或金属等，分有柄和无柄	蒸发液体用，还可用作反应器用 可耐高温、直接加热，但高温时不能骤冷。随液体不同可选用不同质地的蒸发皿
坩埚　泥三角	以容积（mL）表示。材质有瓷、石英、铁、镍、铂等 泥三角有大小之分，用铁丝弯成并套上瓷管	用于灼烧试剂 一般忌骤冷、骤热，依试剂性质选用不同材质的坩埚
干燥器	以外径（mm）表示大小，分普通干燥器和真空干燥器，内放干燥剂	保持物品干燥 为防止盖子滑动打碎，热的物品待稍冷后才能放入。盖的磨口处涂适量的凡士林，干燥剂要及时更换

续表

仪器名称	规　格	用途及注意事项
 研钵	以口径（mm）表示。材质有瓷、玻璃、玛瑙等	用于研磨固体物质 大块物质不能敲，只能压碎。不能用于加热，按固体的性质和硬度选用不同的研钵。放入量不宜超过容积的1/3
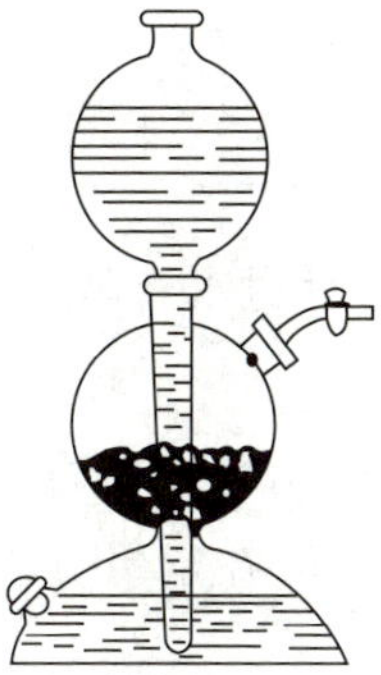 启普发生器	以容积（mL）表示	用于制备少量气体 不能加热，装入的固体反应物必须是较大的块状物，不适用颗粒细小的固体反应物
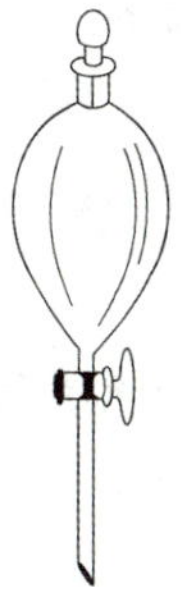 分液漏斗	以容积（mL）和形状（球形，梨形）表示	用于分离互不相溶的液体，或用作发生气体装置中的加液漏斗 不得加热，漏斗塞子、活塞不得互换
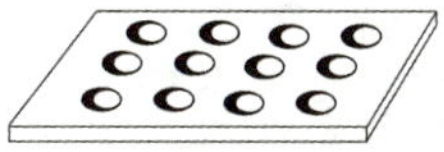 点滴板	材质有透明玻璃和瓷质，瓷质分白色、黑色	用于点滴反应，尤其是显色反应 白色沉淀用黑色板，有色沉淀或者溶液用白色板。不能加热

二、水及化学试剂的规格

1. 水的规格、制备及检验方法

（1）水的规格及合理选用　对化学及相关学科的科技工作者而言，依据任务和要求的不同，对水的纯度要求也不同。水是最常用的溶剂，水是许多物质的良好溶剂，大多无机化学反应都是在水溶液中进行的，以至今天的整个无机化学体系都是建立在水溶液体系之上的，都是"水化学"。所谓的物质的性质，如不特别说明，都是在水溶液中才具备的。

自然界中的水（天然水）含有较多杂质，一般在科学实验及工业生产过程中很少直接应用。经过初步处理后的自来水相对而言是比较纯净的，但仍含有较多可溶性的杂质，在实验室中常用作粗洗仪器用水、实验冷却水及无机制备前期用水等。

自来水再经过进一步处理后，依据不同的制备方法，可得到不同规格的纯水。我国已建立了实验室用水规格的国家标准——《分析实验室用水规格和试验方法》（GB/T 6682—2008）（以下简称《标准》）。《标准》中规定了实验室用水的技术指标、制备方法及检验方法。

实验室用水的规格见表 2 - 2。

表 2 - 2   实验室用 H_2O 的级别及主要指标

指 标 名 称	一 级	二 级	三 级
pH 范围(25℃)	-	-	5.0 ~ 7.5
电导率(κ, $\mu S/cm$, 25℃) ≤	0.1	1.0	5.0
吸光度(A)(254nm,1cm 光程) ≤	0.001	0.01	-
$\rho(SiO_2)$(mg/L) ≤	0.02	0.05	-
可氧化物的限度实验	-	符合	符合

《标准》只规定了一般技术指标，在具体的科学研究和工业生产过程中，有时对水有特殊的要求，还要检查其他有关项目，例如 Cl，Fe，Cu，Zn，Pb，Ca，Mg 等离子。

在实验过程中依据不同的任务和要求，应选择不同规格的纯水。普通的溶剂用水、普通仪器清洗用水及无机制备的前期用水仅需使用三级水。在仪器分析实验中常使用二级水。而在定量分析化学实验及有些精密仪器（如高效液相色谱仪）的实验中则需要一级水。

（2）纯水的制备   三级水可以采用蒸馏、反渗透或者去离子等方法制备，二级水可在三级水的基础上再经蒸馏制备，一级水可用二级水经过蒸馏、离子交换混合床和 0.2 μm 过滤膜的方法或者用石英装置进一步蒸馏而制得。

高纯水（如一级、二级水）应该用特殊塑料容器保存，不能用玻璃容器保存，以免玻璃中的杂质及钠盐慢慢溶于水而使水的纯度降低。

（a）蒸馏法   将水蒸发成水蒸气，再通过冷凝器将水蒸气冷凝下来，所得到的水就称蒸馏水。使用的蒸馏器由玻璃、铜、石英等材料制成。蒸馏法成本低，操作方便，但能量消耗大，只能除去水中不挥发性杂质，不能除去溶解在水中的气体。如要通过蒸馏法得到较高纯度的水，可在蒸馏水中加入少量高锰酸钾和氢氧化钡，再次进行蒸馏，这样可以除去水中极微量的有机杂质、无机杂质以及挥发性的酸性氧化物（如 CO_2）。这种水常称为二次蒸馏水。

（b）离子交换法   经过离子交换树脂处理后的水，叫离子交换水，因为溶于水的杂质离子被去掉，所以又称为去离子水，去离子水的纯度很高，常温下的电阻率可达 5×10^6 $\Omega \cdot cm$ 以上。离子交换树脂是一种人工合成的高分子化合物，其主要组成部分是交联成网状的高分子骨架，另一部分是连在其骨架上的许多可以被交换的活性基团。树脂的骨架特别稳定，它不受酸、碱、有机溶剂和一般氧化剂的作用。当它与水接触时，能吸附并交换溶解在水中的阳离子和阴离子。根据能交换的离子种类不同，离子交换树脂可分为阳离子交换树脂和阴离子交换树脂两大类。

阳离子交换树脂含有磺酸基—SO_3H，羧基—COOH 和酚羟基—OH 等，能和水溶液中的其他阳离子进行交换（称为 H 型）。磺酸是强酸，所以含磺酸基的树脂又称为强酸性阳离子交换树脂，用 R—SO_3H 表示，其中 R 代表树脂中网状骨架部分。R—OH 和 R—COOH 均为弱酸性阳离子交换树脂。强酸性阳离子交换树脂可在酸性、中性、碱性溶液中使用，交换速率快，与所有的阳离子均可进行交换。弱酸性阳离子交换树脂的交换能力受酸度影响较大，羧基在 pH > 4、酚羟基在 pH > 9.5 时才有离子交换能力，但它们的选择性较好。

阴离子交换树脂含有碱性的活性基团，如含有季胺基—$N(CH_3)_3^+$ 的强碱性阴离子交换树脂 R—$N(CH_3)_3^+OH^-$，含有叔胺基—$N(CH_3)_2$、仲胺基—$NH(CH_3)$、氨基—NH_2 的弱碱性阴离子交换树脂 R—$NH(CH_3)_2^+OH^-$、R—$NH_2(CH_3)^+OH^-$ 和 R—$NH_3^+OH^-$，它们所含的 OH^- 均可与水溶液中的其他阴离子进行交换（称为 OH 型），强碱性阴离子交换树脂在酸性、中性、碱性溶液中都能应用，强酸和弱酸的酸根离子均能被交换。弱碱性阴离子交换树脂在碱性溶液中就失去了交换能力。

制备去离子水，通常都使用强酸性阳离子交换树脂和强碱性阴离子交换树脂，预先将它们分别处理成 H 型和 OH 型。交换过程通常在离子交换柱中进行，水先经过阳离子树脂交换柱，水中的阳离子（Na^+、Ca^{2+}、Mg^{2+} 等）与树脂上的 H^+ 进行交换：

$$R-SO_3^-H^+ + Na^+ \rightleftharpoons RSO_3^-Na^+ + H^+$$

$$2R-SO_3^-H^+ + Ca^{2+} \rightleftharpoons (RSO_3^-)_2Ca^{2+} + 2H^+$$

$$2R-SO_3^-H^+ + Mg^{2+} \rightleftharpoons (RSO_3^-)_2Mg^{2+} + 2H^+$$

交换之后，树脂变成"钠型""钙型"或"镁型"，水呈现弱酸性。然后再将水通过阴离子树脂交换柱，水中的杂质阴离子（Cl^-、SO_4^{2-}、HCO_3^- 等）与树脂上的 OH^- 进行交换：

$$RN(CH_3)_3^+OH^- + Cl^- \rightleftharpoons RN(CH_3)_3^+Cl^- + OH^-$$

$$2RN(CH_3)_3^+OH^- + SO_4^{2-} \rightleftharpoons [RN(CH_3)_3^+]_2SO_4^{2-} + 2OH^-$$

交换之后，树脂变成"氯型"等，交换下来的 OH^- 和 H^+ 中和：

$$H^+ + OH^- \longrightarrow H_2O$$

从而将水中的可溶性离子全部去掉。

交换后水的纯度高低与所用树脂的量多少以及流经树脂时水的流速等因素有关。一般树脂量越多，水流越慢，得到的水的纯度就越高。

上述离子交换过程是可逆的。交换反应主要向哪个方向进行，与水中两类离子（如 H^+ 与 Na^+，OH^- 与 Cl^-）的浓度有关。当水中杂质离子较多，而树脂上的活性基团上的离子都是 H^+ 或 OH^- 时，则水中的杂质离子被交换占主导地位；但如果水中杂质离子减少而树脂上活性基团又大量被杂质离子占领时，水中的 H^+ 和 OH^- 反而会把杂质离子从树脂上交换下来。由于交换反应的这种可逆性，所以只用阳离子交换柱和阴离子交换柱串联起来处理后的水，仍然会含有少量的杂质离子。为提高水的纯度，可使水再通过一个由阴、阳离子交换树脂均匀混合的"混合柱"，其作用相当于串联了很多个阳离子交换柱与阴离子交换柱，而且在交换柱层的任何部位的水都是中性的，从而减少了逆反应的可能性。

树脂使用一定时间后，活性基团上的 H^+、OH^- 分别被水中的阳、阴离子所交换，从而失去了原先的交换能力，称之为"失效"。利用交换反应的可逆性使树脂再生，恢复其交换能力，这过程称之为"洗脱"或"再生"。

阳离子交换树脂的再生是加入适当浓度的酸（一般用 5% ~10% 的盐酸），其反应为：

$$RSO_3^-Na^+ + H^+ \rightleftharpoons RSO_3^-H^+ + Na^+$$

阴离子交换树脂的再生是加入适当浓度的碱（一般用 5% 的 NaOH），其反应为：

$$RN(CH_3)_3^+Cl^- + OH^- \rightleftharpoons RN(CH_3)_3^+OH^- + Cl^-$$

经再生后的树脂可以重复使用。混合离子交换树脂应该先用饱和食盐水溶液将其分离后，再分别进行再生。

（c）电渗析法　这是在离子交换基础上发展起来的一种方法。它是在外电场的作用下，利用阴、阳离子交换膜对溶液中离子的选择性透过而使杂质离子从水中分离出来的方法。该法除去杂质的效率较低，适用于要求不很高的分析工作。

2. 化学试剂的规格、选用及贮存

（1）化学试剂的分类及选用　化学试剂门类很多，世界各国对化学试剂的分类、分级的标准不尽一致，各国都有自己的国家标准及其他标准（行业标准、学会标准等）。我国化学试剂的标准有国家标准（GB）、化工部标准（HG）及企业标准（QB）三级。

化学试剂产品已有近万种，且随着科学技术和生产力的发展，新的试剂种类还将不断产生，到目前为止，还没有统一的分类标准。暂将化学试剂分为一般试剂、高纯试剂、标准试剂、专用试剂四大类。

（a）一般试剂　是实验室普遍使用的试剂，可分为四个等级及生化试剂等。一般试剂的分级、适用范围、标志及标签颜色列于表2-3。

表2-3　一般试剂的规格和适用范围

级　别	中文名称	英文符号	适 用 范 围	标签颜色
一　级	优级纯（保证试剂）	G. R.	精密分析实验	绿色
二　级	分析纯（分析试剂）	A. R.	一般分析实验	红色
三　级	化学纯	C. P.	一般化学实验	蓝色
四　级	实验试剂	L. R.	一般化学实验辅助试剂	棕色或其他颜色
生化试剂	生化试剂 生物染色剂	B. R.	生物化学及医用化学实验	咖啡色染色剂（玫瑰色）

指示剂也属于一般试剂。

实验试剂（四级），只能用于一般的化学实验及教学工作。

分析纯（二级）及化学纯（三级）试剂，适用于一般的分析研究及教学实验工作。

优级纯（一级）试剂，又称保证试剂，适用于精密的分析及研究工作。

生化试剂，用于各种生物化学实验。

（b）高纯试剂　特点是杂质含量低（比优级纯、基准试剂都低），主体含量通常与优级纯试剂相当，但规定检测的杂质项目比同种优级或基准试剂多1~2倍。高纯试剂主要用于微量分析中标准溶液的制备。

（c）标准试剂　是用于测量其他（待测）物质化学量的标准物质，其特点是主体含量高且准确可靠。标准试剂有时称为基准试剂（或基准物质），主要用于配制各种标准溶液。顾名思义，光谱纯试剂是光谱分析中的标准物质；色谱纯试剂用作色谱分析的标准物质。国产标准试剂的种类及用途列于表2-4。

表2-4　主要国产标准试剂的种类与用途

类　别	主　要　用　途
滴定分析第一基准试剂	工作基准试剂的定值
滴定分析工作基准试剂	滴定分析标准溶液的定值
杂质分析标准溶液	仪器及化学分析中作为微量杂质分析的标准
滴定分析标准溶液	滴定分析法测定物质的含量
一级pH基准试剂	pH基准试剂的定值和高精密度pH计的校准
pH基准试剂	pH计的校准（定位）
热值分析试剂	热值分析仪的标定
色谱分析标准	色谱法进行定性和定量分析的标准
临床分析标准溶液	临床化验
农药分析标准	农药分析
有机元素分析标准	有机物元素分析

（d）专用试剂　是指具有特殊用途的试剂。如气相色谱担体及固定液、液相色谱填料等。与高纯试剂相似之处是主体含量较高，且杂质含量很低。与高纯试剂的区别是在特定的用途中有干扰的杂质成分只需控制在不致产生明显干扰的限度以下。

（2）化学试剂的存放　化学试剂的贮存在实验室中是一项十分重要的工作。通常，化学试剂应贮存在干净、干燥和通风良好的地方，要远离火源，并注意防止灰分、水分和其他物质的污染，同时，依据试剂的性质应用不同的贮存方法。

固体试剂一般存放在易于取用的广口瓶内，液体试剂则存放在细口的试剂瓶中。一些用量小而使用频繁的试剂，如指示剂、定性分析试剂等可盛装在滴瓶中。见光易分解的试剂（如 $AgNO_3$、$KMnO_4$、饱和 Cl_2 水等）应装在棕色瓶中。对于 H_2O_2，虽然也是见光易分解的物质，但不能盛放在棕色的玻璃瓶中，因棕色玻璃中含有重金属氧化物成分，会催化 H_2O_2 的分解。因此通常将 H_2O_2 存放于不透明的塑料瓶中，放置于阴凉的暗处。试剂瓶的瓶盖一般都是磨口的，但盛强碱性试剂（如 NaOH、KOH）及 Na_2SiO_3 溶液的瓶塞应换成橡皮塞，以免长期放置互相粘连。易腐蚀玻璃的试剂（如氟化物等）应保存在塑料瓶中。

对于易燃、易爆、强腐蚀性、强氧化剂及剧毒品的存放应特别加以注意，一般需要分类单独存放，如强氧化剂要与易燃、可燃物分开隔离存放。低沸点的易燃液体要求在阴凉通风的地方存放，并与其他可燃物和易产生火花的器物隔离放置，更要远离明火。闪点在 $-4℃$ 以下的液体（如石油醚、苯、乙酸乙酯、丙酮、乙醚等）理想的存放温度为 $-4\sim4℃$；闪点在 25℃ 以下的（如甲苯、乙醇、丁酮、吡啶等）存放温度不得超过 30℃。

盛装试剂的试剂瓶都应贴上标签，并写明试剂的名称、纯度、浓度和配制日期，标签外面应涂蜡或用透明胶带等保护。

三、玻璃仪器的洗涤和干燥　微课1

1. 玻璃仪器的洗涤　化学实验过程中经常使用各种玻璃仪器。如果使用不洁净的仪器，仪器上的杂质和污物会对实验产生影响，使实验得不到正确的结果，甚至可导致实验失败。实验中所使用的仪器必须洁净，实验后要及时清洗仪器，否则，不清洁的仪器长期放置后，会导致以后的洗涤工作更加困难，因此，玻璃仪器的洗涤是化学实验中的一项重要内容。

玻璃仪器洁净的标准是用水洗冲后，仪器内壁能均匀地被水润湿而不沾附水珠，否则需要进一步清洗。

洗涤仪器的方法很多，一般应依据实验的要求、污物的性质及沾污的程度以及仪器的类型和形状来选择合适的洗涤方法。

一般来说，污物主要有可溶性污物和不溶性污物、灰尘、有机物及油污等。洗涤方法通常分为以下几种。

（1）一般洗涤　对于水溶性的污物，一般直接用水冲洗，冲洗不掉的物质，可以选用合适的毛刷洗，若毛刷很难刷到，可用碎纸捣成糊浆，放进容器，剧烈摇动，导致污物脱落，再用水冲洗干净。洗涤仪器时应该一个一个地洗，不要同时抓多个仪器一起洗，以免碰坏或摔坏仪器。

用自来水洗净的仪器，往往在仪器内还残留一些 Ca^{2+}、Mg^{2+}、Cl^- 等离子，要去除这些杂质离子，应用蒸馏水漂洗几次。用蒸馏水洗涤仪器应采用"少量多次"法，常用洗瓶进行挤压洗瓶使其喷出一股细蒸馏水流，均匀地喷射在仪器内壁上并不断转动仪器，再将水倒掉。如此重复几次即可。

已洗净的玻璃仪器应该是清洁透明的。凡已洗净的仪器，内壁不能用布或纸擦拭，否则布或纸上的纤维及污物会沾污仪器。

（2）洗液洗涤　对于有油污的仪器，可先用水冲洗掉可溶性污物，再用毛刷蘸取合成洗涤剂或肥皂液刷洗。用肥皂液或合成洗涤剂仍然刷洗不掉的污物，或因口小、管细不便用毛刷洗涤的仪器及不能用毛刷洗刷的特殊仪器（精密容量玻璃仪器），可用少量浓 HNO_3 或少量浓 H_2SO_4 或洗液洗涤。氧化型污物可选用还原性洗液洗涤；还原性污物则选用氧化性洗液洗涤。

实验室常用的洗液有铬酸洗液、$KMnO_4$ 碱性洗液、碱性乙醇洗液和乙醇 – 浓 HNO_3 洗液。这些洗液的配制方法见表 2 – 5。

表 2 –5　常用洗液的配制

洗 液 名 称	配 制 方 法	应用及注意事项
铬酸洗液	取 20g $K_2Cr_2O_7$（LR）于 50mL 烧杯中，加 40mL H_2O，加热溶解，冷后，缓缓加入 320mL 浓 H_2SO_4 即成（边加边搅），贮于磨口细口瓶中	用于洗涤油污及有机物，使用时防止被 H_2O 稀释。用后倒回原瓶，可反复使用，直至溶液变为绿色*
$KMnO_4$ 碱性洗液	取 4g $KMnO_4$（LR），溶于少量 H_2O 中，缓缓加入 100mL 10% NaOH 溶液	用于洗涤油污及有机物，洗后玻璃壁上附着的 MnO_2 沉淀，可用亚铁溶液或 Na_2SO_3 溶液洗去
乙醇–浓 HNO_3 洗液		用于沾有有机物或油污的结构较复杂的仪器，洗涤时先加少量乙醇于脏仪器中，再加入少量 HNO_3，即产生大量棕色 NO_2，将有机物氧化而破坏

注：*已还原为绿色的铬酸洗液，可加入固体 $KMnO_4$ 使其再生，这样，实际消耗的是 $KMnO_4$ 可减少铬对环境的污染。

若污物是无机物一般多选用铬酸洗液，若污物为有机物一般选用 $KMnO_4$ 溶液。洗涤仪器前，应尽可能倒尽仪器中的残留水分，然后向仪器中装入约 1/5 体积的洗液，将仪器倾斜并慢慢地转动，使内壁全部被洗液湿润，若能浸泡一段时间或用热的洗液洗涤，效果会更好。

洗液用后，应倒回原瓶，可反复多次使用。多次使用后，铬酸洗液会变成绿色（Cr^{3+}）；$KMnO_4$ 洗液会变成浅红或无色，底部有时会出现 MnO_2 沉淀，此时的洗液已不具备强氧化性，不能再继续使用。

洗液具有强的腐蚀性，使用时应注意安全，不要溅在皮肤、衣物上，不能用毛刷蘸取洗液刷洗仪器，如果不慎将洗液洒在衣物、皮肤时，应立即用水冲洗。废的洗液和洗液的首次冲洗液应倒入废液缸里，不能倒入水槽，以免腐蚀下水道。

少量的废洗液（含 Cr^{3+}）可加入废碱液或石灰使其生成 $Cr(OH)_3$ 沉淀，将此废渣埋于地下（指定地点），以防止铬的污染。

（3）特殊污垢的洗涤　有些仪器上常常沉积一些已知主要化学成分的污垢，这时我们就需要视污垢的性质选用合适的试剂，经与其化学作用而除去。例如，做银镜反应实验时在试管内壁沉积的银或铜，可用硝酸除去。一些常见污垢的处理方法见表 2－6。

表 2－6　常见污垢的处理方法

污 垢	处 理 方 法
碱土金属的碳酸盐、$Fe(OH)_3$、一些氧化剂如 MnO_2 等	用稀 HCl 处理，MnO_2 需要用 6mol/L 的 HCl
沉积的金属如银、铜	用 HNO_3 处理
沉积的难溶性银盐	用 $Na_2S_2O_3$ 洗涤，Ag_2S 则用热、浓 HNO_3 处理
沾附的硫黄	用煮沸的石灰水处理 $3Ca(OH)_2 + 12S \longrightarrow 2CaS_5 + CaS_2O_3 + 3H_2O$
高锰酸钾污垢	草酸溶液（沾附在手上也用此法）

污　　垢	处　理　方　法
残留的 Na_2SO_4、$NaHSO_4$ 固体	用沸水使其溶解后趁热倒掉
沾有碘迹	可用 KI 溶液浸泡；或用温热的稀 NaOH 或用 $Na_2S_2O_3$ 溶液处理
瓷研钵内的污迹	用少量食盐在研钵内研磨后倒掉，再用水洗
有机反应残留的胶状或焦油状有机物	视情况用低规格或回收的有机溶剂（如乙醇、丙酮、苯、乙醚等）浸泡；或用稀 NaOH、或用浓 HNO_3 煮沸处理
一般油污及有机物	用含 $KMnO_4$ 的 NaOH 溶液处理
被有机试剂染色的比色皿	可用体积比为 1∶2 的盐酸－乙醇处理

除了上述的清洗方法外，还有超声波清洗器清洗法。只要将需清洗的仪器放在配有合适洗涤剂的溶液中，接通电源，利用声波的能量，振动污垢，就可将仪器清洗干净。

2. 玻璃仪器的干燥　有些仪器洗涤干净后就可用来做实验，但有些化学实验，需要在无水条件下进行。例如，容量分析中的非水滴定，无结晶水的稀土配合物的合成等。此时，所需仪器必须是干燥的，根据不同的情况，可采用不同的方法将仪器干燥。

（1）晾干　对于不急用的仪器，洗净后将仪器倒立放置在适当的仪器架上，让其在空气中自然干燥。

（2）烘干　将洗净的仪器放入电热恒温干燥箱内加热烘干。

恒温干燥箱（简称烘箱）是实验室常用的干燥仪器（图 2-1），常用来干燥玻璃仪器或烘干无腐蚀性、热稳定性比较好的固体药品，但易燃品或刚用乙醇、丙酮淋洗过的仪器切勿放入烘箱内，以免爆炸。

烘箱使用方法如下。连接电源，并打开电源开关。调节控温仪，根据实验需求设置温度，加热指示灯亮，烘箱处于升温状态，并在显示屏上实时显示当前箱内温度。待温度达到预置温度并稳定后，放入样品，并根据需要选择是否启动风扇功能。使用完毕后，关闭干燥箱的加热功能和风扇功能，切断电源。

玻璃仪器干燥时，先洗净并将水尽量倒干，放置时应平放或使仪器口朝上，带塞的瓶子应打开瓶塞，有时将仪器放在托盘里。一般在 105℃ 加热 15 分钟左右即可干燥。最好让烘箱降至常温后再取出仪器。如果热时就要取出仪器，应注意用干布垫手，防止烫伤。热玻璃仪器不能碰水，以防炸裂。热仪器自然冷却时，器壁上常会凝上水珠，这可以用吹风机吹冷风助冷而避免。烘干的药品一般取出后应放在干燥器里保存，以免在空气中又吸收水分。

（3）吹干　用热或冷的空气流将玻璃仪器吹干，所用仪器是电吹风机或"玻璃仪器气流干燥器"。用吹风机吹干时，一般先用热风吹玻璃仪器的内壁，待干后再吹冷风使其冷却。如果先用易挥发的溶剂如乙醇、乙醚、丙酮等淋洗一下仪器，将淋洗液倒净，然后用吹风机用冷风－热风－冷风的顺序吹，则会干得更快。另一种方法是将洗净的仪器直接放在气流烘干器里进行干燥。

（4）烤干　用煤气灯小心烤干。一些常用的烧杯、蒸发皿等可置于石棉网上用小火烤干，烤干前应先擦干仪器外壁的水珠。试管烤干时应使试管口向下倾斜，以免水珠倒流炸裂试管（图 2-2）。烤干时应先从试管底部开始，慢慢移向管口，不见水珠后再将管口朝上。把水汽赶尽。

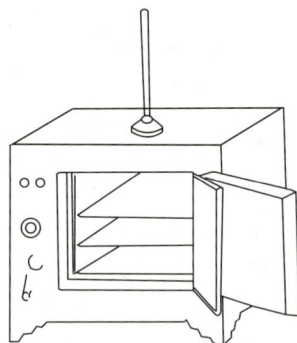

图2-1　电热恒温干燥箱

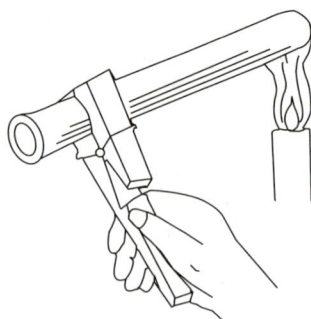

图2-2　烤干试管

　　还应注意的是，一般带有刻度的计量仪器，如移液管、容量瓶、滴定管等不能用加热的方法干燥，以免热胀冷缩影响这些仪器的精密度。玻璃磨口仪器和带有活塞的仪器洗净后放置时，应该在磨口处和活塞处（如酸式滴定管、分液漏斗等）垫上小纸片，以防止长期放置后黏上不易打开。

四、干燥器的使用

　　易吸水潮解的固体或灼热后的坩埚等应放在干燥器内，以防吸收空气中的水分。干燥器是一种有磨口盖子的厚质玻璃器皿，磨口上涂有一层薄薄的凡士林，以防水汽进入，并能很好地密合。干燥器的底部装有干燥剂（变色硅胶、无水氯化钙等），中间放置一块干净的带孔瓷板，用来盛放被干燥物品。打开干燥器时，应左手按住干燥器，右手按住盖的圆顶，向左前方（或向右）推开盖子，如图2-3（a）所示。温度很高的物体（例如灼烧至恒重的坩埚等）放入干燥器时，不能将盖子完全盖严，应该留一条很小的缝隙，待冷后再盖严，否则易被内部热空气冲开盖子，或者由于冷却后的负压使盖子难以打开。搬动干燥器时，应用两手的拇指同时按住盖子，以防盖子因滑落而打碎，如图2-3（b）所示。

（a）　　　　　　　（b）

图2-3　干燥器的使用

五、加热

　　1. 加热装置　在实验室的加热操作中，常使用酒精灯、酒精喷灯、煤气灯或电炉等直接加热，有时也采用水浴、油浴、砂浴或空气浴等间接加热。

　　（1）酒精灯　由灯罩、灯芯和灯壶三部分组成，如图2-4所示。使用时先要加酒精，即应在灯熄灭情况下，牵出灯芯，借助漏斗将酒精注入，最多加入量为灯壶容积的2/3。必须用火柴点燃，绝不能用另一个燃着的酒精灯去点燃，以免洒落酒精引起火灾（图2-5）。熄灭时，用灯罩盖上即可，不要用嘴吹。片刻后，还应将灯罩再打开一次，以免冷却后盖内负压使以后打开困难。

酒精灯的加热温度通常为 400～500℃，适用于不需太高加热温度的实验。

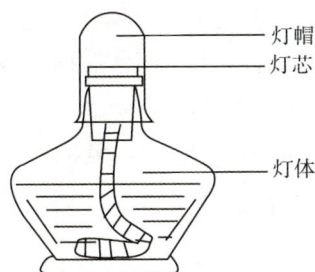

图 2 - 4　酒精灯

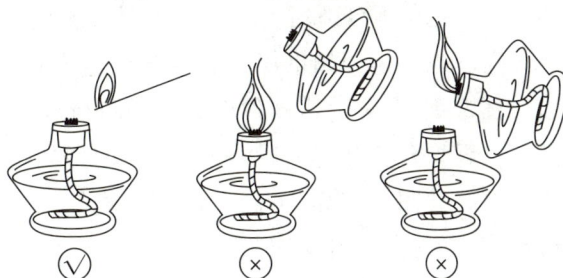

图 2 - 5　点燃方法

（2）酒精喷灯　有挂式和座式两种，构造见图 2 - 6 和图 2 - 7，它们的使用方法相似。应先在酒精灯壶或储罐内加入酒精，注意在使用过程中不能续加，以免着火。在预热盘中加满酒精并点燃（挂式喷灯应将储罐下面的开关打开，从灯管口冒出酒精后再关上；在点燃喷灯前先打开），等酒精燃烧完将灯管灼热后，打开空气调节器并用火柴将灯点燃。酒精喷灯是靠气化的酒精燃烧，所以温度较高，可达 700～900℃。用完后关闭空气调节器，或用石板盖住灯口即可将灯熄灭。挂式喷灯不用时，应将储罐下面的开关关闭。

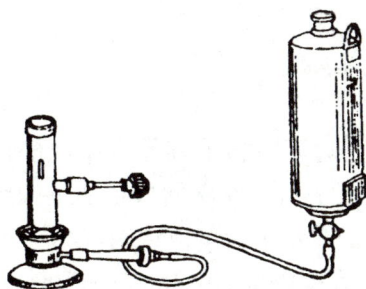

图 2 - 6　挂式酒精喷灯

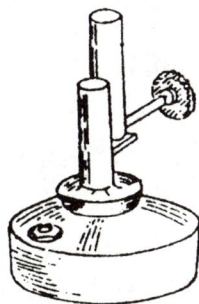

图 2 - 7　座式酒精喷灯

座式喷灯最多使用半小时，挂式喷灯也不可将罐里的酒精一次用完。若需连续使用，应将喷灯熄灭，冷却，添加酒精后再次点燃。

使用时注意灯管必须灼热后再点燃，否则易造成液体酒精喷出引起火灾。

（3）煤气灯　在有煤气（天然气）的地方，煤气灯是化学实验室中最常用的加热装置。它的样式虽多，但构造原理基本相同，主要由灯管和灯座组成，如图 2 - 8 所示。灯管下部有螺旋与灯座相连，并开有作为空气入口的圆孔。旋转灯管，可关闭或打开空气入口，以调节空气进入量。灯座侧面为煤气入口，用橡皮管与煤气管道相连。灯座侧面（或下面）有螺旋形针阀，可调节煤气的进入量。

使用时应先关闭煤气灯的空气入口，将燃着的火柴移近灯口时再打开煤气管道开关，将煤气灯点燃（切勿先开气后点火）。然后调节煤气和空气的进入量，使二者的比例合适，得到分层的正常火焰，如图 2 - 9 所示。火焰大小可用管道上的开关控制。关闭煤气管道上的开关，即可熄灭煤气灯（切勿吹灭）。

煤气灯的正常火焰分三层（图 2 - 9）：外层 1，煤气完全燃烧，称为氧化焰，呈淡紫色；中层 3，煤气不完全燃烧，分解为含碳的化合物，这部分火焰具有还原性，称为还原焰，呈淡蓝色；内层 4，煤气和空气进行混合并未燃烧，称为焰芯。正常火焰的最高温度在还原焰顶部上端与氧化焰之间的 2

处，温度可达 800～900℃。

当空气和煤气的比例不合适时，会产生不正常火焰。如果火焰呈黄色或产生黑烟，说明煤气燃烧不完全，应调大空气进入量；如果煤气和空气的进入量过大，火焰会脱离灯管在管口上方临空燃烧，称为临空火焰［图 2－10（a）］，这种火焰容易自行熄灭；若煤气进入量很小（或煤气突然降压）而空气比例很高时，煤气会在灯管内燃烧，在灯口上方能看到一束细长的火焰并能听到特殊的嘶嘶声，这种火焰叫侵入火焰［图 2－10（b）］，片刻即能把灯管烧热，不小心易烫伤手指。遇到后两种情况时，应关闭煤气阀，重新调节后再点燃。

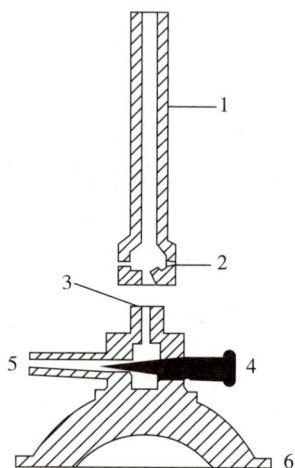

图 2－8　煤气灯构造

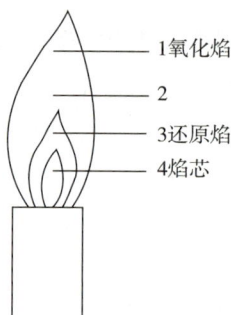

图 2－9　正常火焰

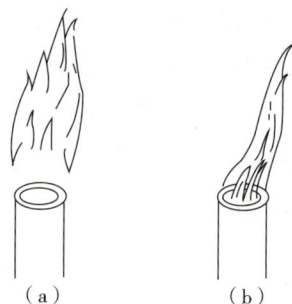

图 2－10　不正常火焰

（4）水浴　水浴常在水浴锅中进行［图 2－11（a）］，有时为了方便常用规格较大的烧杯等代替［图 2－11（b）］。水浴锅一般为铜制外壳，内壁涂锡。盖子由一套不同口径的铜圈组成，可以按加热器皿的外径任意选用，使用时，锅下加热，受热器皿悬置在水中，可保持液温到95℃左右的恒温。使用水浴应注意如下事项：① 水浴锅内存水量应保持在总体积的2/3左右；② 受热玻璃器皿不能触及锅壁或锅底；③ 水浴锅不能作油浴或砂浴用。　微课2

（5）油浴　油浴锅一般由生铁铸成，有时也可用大烧杯代替。油浴适用于100～250℃加热，反应物的温度一般低于油浴液的20℃左右，常用的油浴如下。

① 甘油：可以加热到140～150℃，温度过高分解。

② 植物油：如菜油、蓖麻油和花生油，可以加热到220℃。常加入1％的对苯二酚等抗氧化剂，便于久用。温度过高时会分解，达到闪点可能燃烧，所以，使用时要十分小心。

③ 石蜡：能加热到200℃左右，冷到室温则成为固体，保存方便。

④ 液体石蜡：可加热到200℃左右，温度稍高并不分解，但较易燃烧。

使用油浴应特别小心防止着火；当油受热冒烟时，应立即停止加热；油量应适量，不可过多，以免油受热膨胀而溢出；浴锅外不能沾油，如若外面有油，应立即擦去；如遇油浴着火，应立即拆除热源，并用石棉网等盖灭火焰，切勿用水浇。

⑤ 硅油：硅油在250℃时仍较稳定，透明度好，只是价格昂贵。

（6）砂浴　砂浴通常采用生铁铸成的砂浴盘。盘中盛砂子，使用前先将砂子加热熔烧，以去掉有机物。加热温度在80℃以上者可以使用，特别适用于加热温度在220℃以上者，砂浴的缺点是传热慢，温度上升慢，且不易控制。因此，砂层要薄些。特别注意，受热器不能触及浴盘底部。

（7）空气浴　沸点在80℃以上的液体原则上均可采用空气浴加热。最简单的空气浴可用下法制作：

取空的铁罐一只（用过的罐头盒即可），罐口边缘剪光后，在罐的底层打数行小孔。另将圆形石棉片（直径略小于罐的直径2～3mm）放入罐中，使其盖在小孔上，罐的四周用石棉布包裹。另取直径略大于罐口的石棉板（厚2～4mm）一块，在其中挖一个洞（洞的直径略大于被加热容器的颈部直径），然后对切为二，加热时用以盖住罐口。使用时将此装置放在铁三脚架或铁支台的铁环上，用灯焰加热即可。注意蒸馏瓶或其他受热器在罐中切勿触及罐底，其正确的位置如图2－12所示。

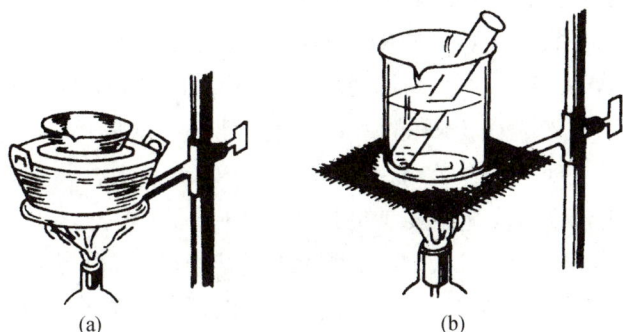

（a）　　　　　　　（b）

图2－11　水浴加热　　　　　　　　　　图2－12　空气浴

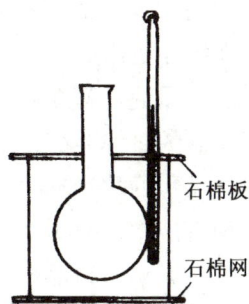

（8）电热套加热　电热套是一种较好的热源，它是由玻璃纤维包裹着电热丝织成的碗状半圆形的加热器，有控温装置可调节温度。由于它不是明火加热，因此，可以加热和蒸馏易燃有机物，也可加热沸点较高的化合物，适应加热温度范围较广。

2. 液体的加热

（1）在试管中加热液体　在试管中加热液体时，液体量不应超过试管容积的1/3，还必须用试管夹夹持试管，管口稍向上倾斜（图2－13），注意管口不要对着别人和自己，以免被沸腾的溶液喷出烫伤。加热时，应先加热液体的中上部，再加热底部，并上下移动，使各部分液体均匀受热。

（2）加热烧杯、烧瓶中的液体　加热时必须在仪器下面垫上石棉网（图2－14），使仪器受热均匀。加热烧瓶时还应该用铁夹将其固定。加热的液体量不应超过烧杯容积的1/2和烧瓶容积的1/3。烧杯加热时还要适当加以搅拌以免爆沸，烧瓶加热时也要视情况放入1～2粒沸石。

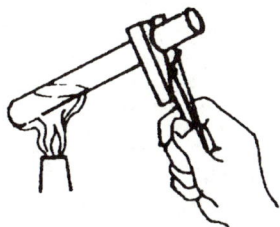

图2－13　加热试管中的液体　　　　　图2－14　加热烧杯中的液体

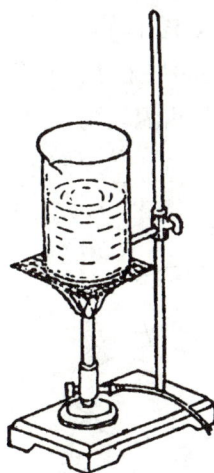

（3）蒸发、浓缩与结晶　蒸发、浓缩与结晶是物质制备实验中常用的操作之一，通过此步操作可将产品从溶液中提取出来。

例如，大部分无机化合物的溶解度随温度升高而增大，因此从溶液中使晶体析出的主要方法是蒸发和冷却。蒸发掉过多的溶剂，使溶液浓缩达到饱和或过饱和后析出；也可以采用冷却方法，直接降低物质的溶解度而使其析出。通常情况下，这两者是结合起来使用的。

蒸发浓缩通常在蒸发皿中进行，蒸发皿具有大的蒸发表面，有利于液体的蒸发。里面所盛液体量不应超过其容量的 2/3。如果无机物对热稳定，可以用煤气灯直接加热（应先均匀预热），一般情况下采用水浴加热，以使蒸发过程比较温和平稳。注意不要使瓷蒸发皿骤冷，以免炸裂。

需要蒸发到什么程度，应视物质的溶解度而定。若物质的溶解度随温度变化不大，为了获得较多的晶体，应在结晶析出后继续蒸发（如熬盐）。若物质在高温时溶解度很大而在低温时变小，又分为两种情况：若物质的溶解度大时，应蒸发至溶液表面出现晶膜（液面上漂浮一层固体），冷却即可析出晶体；若物质的溶解度小，则不必蒸发至出现晶膜，就可冷却结晶。某些结晶水合物在不同温度下析出时所带结晶水的数目不同，制备此类化合物时应注意要满足其结晶条件。

在过饱和溶液中，加入一小粒晶体（称为"晶种"）、摩擦器壁或搅拌溶液，常可加速晶体析出。析出晶体的颗粒大小与结晶条件有关。如果溶液浓度高、快速冷却并加以搅拌，则会析出细小晶体，静置溶液并缓慢冷却则有利于大晶体生成。从纯度上看，快速生长的细小晶体纯度较高，因为晶体上不易裹带母液和其他杂质；而大块晶体的纯度较低。但细小晶体会形成稠厚的糊状物，挟带母液不易洗净，也会影响纯度的提高。

当第一次结晶的纯度不合要求时，可将晶体加入适量的溶剂溶解，再次进行蒸发、结晶。这样第二次得到的晶体的纯度就会更高。这种操作叫作重结晶。根据纯度要求可以进行多次结晶。利用重结晶提纯物质，只适用于那些溶解度随温度上升而增大的物质，对于那些溶解度受温度影响小的物质则不适用。

3. 固体的加热

（1）在试管中加热固体　在试管中加热固体，应该用铁架台和铁夹固定试管或用试管夹夹持试管，管口略向下倾斜（图 2-15），以防止凝结在管口处的水珠倒流到灼热的管底使试管破裂。

（2）固体的灼热　高温灼烧或熔融固体使用的仪器是坩埚。根据所装物料的性质及需加热的温度选用不同材质的坩埚（如瓷坩埚、氧化铝坩埚、金

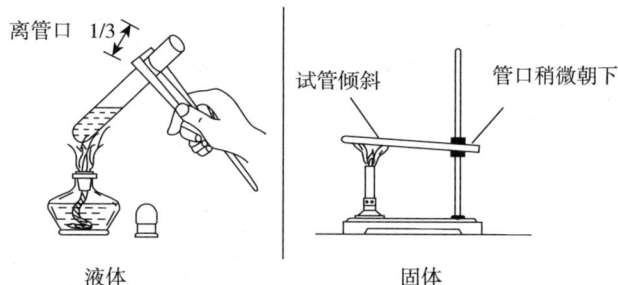

图 2-15　加热试管中的固体

属坩埚等）。加热时，将坩埚置于泥三角上，用氧化焰灼烧（图 2-16），不要使用还原焰以免坩埚外部结上炭黑。先用小火将坩埚均匀加热，然后加大火焰灼烧坩埚底部。根据实验要求控制灼烧温度和时间。夹取高温下的坩埚时，必须使用干净的坩埚钳。用前先在火焰上预热一下，再去夹取。坩埚钳使用后，应使尖端朝上（图 2-17）放在桌子上，以保证坩埚钳尖端洁净。

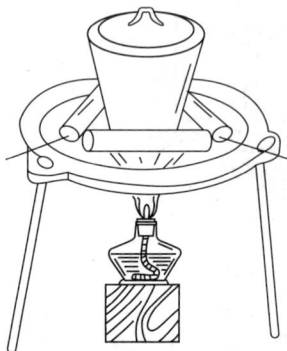

图 2-16　灼烧坩埚

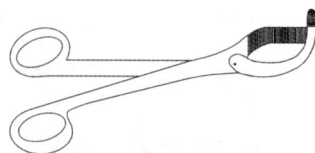

图 2-17　坩埚钳

用煤气灯灼烧温度一般可达 700~800℃，若需在更高温度下灼烧可使用马弗炉。用马弗炉可精确地控制灼烧温度和时间。

六、化学试剂的取用 微课3 微课4

取用药品前，应看清标签。取用时，注意勿使瓶塞污染，如果瓶塞的顶是扁平的，取出后可倒置桌上；如果不是扁平的，可用食指和中指（或中指和无名指）将瓶塞夹住（或放在清洁的表面皿上），绝不可将瓶塞横置于桌上。

固体药品需用清洁、干燥的药匙（有塑料、玻璃或牛角的）取用，不得用手直接拿取。

药匙的两端为大小两个匙（取用的固体要放入小试管时，可用小匙）。

液体药品一般用量筒量取，或用滴管吸取。用滴管将液体滴入试管中时，应用左手垂直地拿持试管，右手持滴管橡皮头将滴管放在试管口的正中上方，然后挤捏橡皮头，使液体恰好滴入试管中（图 2-18）。绝不可将滴管伸入试管中，否则，滴管口易碰上试管壁，并可能沾上其他液体，再将此滴管放回药品瓶中则会沾污药品。若所用的是滴瓶上的滴管，使用后应立即

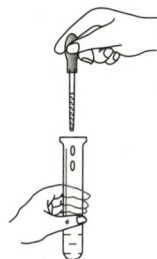

图2-18　用滴管加液体

插回原来的滴瓶中。不得把沾有液体药品的滴管横放或倒置，以免液体流入滴管的橡皮头而污染。

用量筒量取液体时，应左手持量筒，右手持药品瓶（药品标签应在手心处），瓶口紧靠量筒口边缘，慢慢注入液体（图 2-19）到所指刻度。读取刻度时，将量筒置于水平桌面上视线应与凹液面在同一水平面上（图 2-20）。如果不慎倾出了过多的液体，只好把它弃去或给他人使用，不得倒回原瓶。

图2-19　用量筒量取液体

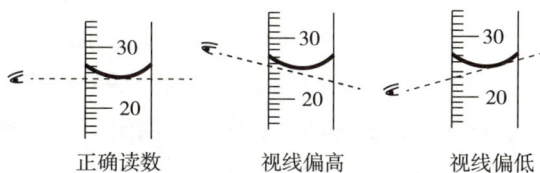

正确读数　　视线偏高　　视线偏低

图2-20　观看量筒内液体的容积

药品取用后，必须立即将瓶塞盖好。实验室中药品瓶安放，一般有一定的次序和位置，不要任意更动。若需移动药品瓶，使用后应立即放回原处。

取用浓酸、浓碱等腐蚀性药品时，务必注意安全。如果酸、碱等洒在桌上，应立即用湿布擦去，如果沾到眼睛或皮肤上要立即用大量清水冲洗。

七、容量瓶、滴定管、移液管的操作方法

常用的容量仪器除量筒外，尚有容量瓶、滴定管和移液管等。量筒量取液体，只能限于对体积要求不需十分精确的情况，而容量瓶、移液管和滴定管则有较高的精确度，容积在 100mL 以下的这些量器的精确限度一般可到 0.01mL。

1. 容量瓶　容量瓶主要是用来精确地配制一定体积、一定浓度的溶液的量器。例如，用来配制一定浓度的溶液。在容量瓶的颈部有一刻度线，在一定温度时，瓶内到达刻度线的液体的体积是一定的。

使用时，先将容量瓶洗净，再将一定量的固体溶质放入烧杯中用少量蒸馏水溶解，然后，将烧杯中的溶液沿玻璃棒小心地注入容量瓶中（图2-21）。再从洗瓶中挤出少量水淋洗烧杯及玻璃棒2~3次，并将每次淋洗的水注入容量瓶中。最后，加水到标线处。操作中需注意，当液面将接近标线时，应使用滴管小心地逐滴将水加到标线处（观察时视线、液面与标线均应在同一水平面上）。塞紧瓶塞，将容量瓶倒转数次（此时必须用手指压紧瓶塞，以免脱落），并在倒转时加以摇荡（图2-22），以保证瓶内溶液浓度上下各部分均匀。注意：瓶塞是磨口的，各自配套的，不能张冠李戴，一般可用橡皮圈系在瓶颈上。

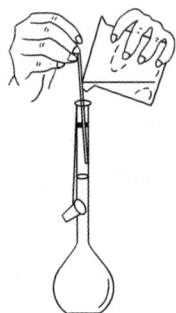

图2-21 定量转移操作   图2-22 容量瓶的拿法

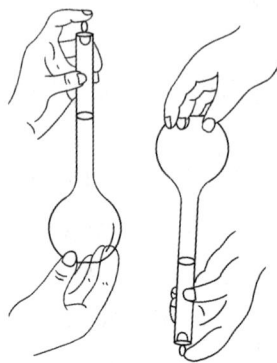

2. 滴定管   滴定管主要是滴定时用来精确量度液体的量器，刻度由上而下，与量筒刻度相反。常用滴定管的容量限度为50mL和25mL，刻度为0.1mL，而读数可估计到0.01mL。滴定管的阀门有两种（图2-23），一种玻璃活塞（a），另一种是装在橡皮管中的玻璃小球（b）。对前者，旋转玻璃活塞，可使液体沿活塞当中的小孔流出，但切勿将活塞横向移动，以致活塞松开或脱出，使液体从活塞旁边漏失。对后者，用大拇指与食指稍微捏挤玻璃小球旁侧的橡皮管，使之形成一条隙缝，液体即沿此隙缝流出。若液体对玻璃有侵蚀作用，如碱液，只能使用带橡皮管的滴定管（碱式滴定管）。若液体能侵蚀橡皮，如$KMnO_4$、I_2、$AgNO_3$溶液等，则必须使用带玻璃塞的滴定管（酸式滴定管）。

使用酸式滴定管时，玻璃塞需涂上润滑脂（一般可用凡士林代替）薄层。其涂法（图2-23）如下：先将活塞取下，将活塞筒及活塞洗净并用滤纸碎片将水吸干，然后分别在活塞筒的小口一端内壁及活塞的大头一端表面各涂一层很薄的润滑脂（活塞筒及活塞的中间小孔处不得沾有润滑脂）。再将活塞小心塞好，旋转活塞数次，使润滑脂均匀地分布在磨口面上。最后装上一些水检查，活塞应转动灵活，出水流畅，两侧不漏水。

滴定管在装入滴定溶液前，除了需用洗涤液、水及蒸馏水依次洗涤清洁外，还需用少量滴定溶液（每次约5mL）洗涤3次，以免滴定溶液的浓度被管内残留的水所稀释。洗涤滴定管时，应先将管平持（上端略向上）并不断转动，使洗涤的水或溶液与内壁的所有部分能充分接触。然后右手将滴定管持直，左手开放阀门，使洗涤的水或溶液通过阀门下面的一段玻璃管流出，起到洗涤作用。在洗涤带有玻璃活塞滴定管时，还需注意用手托住活塞部分（或用橡皮圈圈牢活塞），以防止活塞脱落而打碎。

滴定管装好溶液后，必须把滴定管阀门下端的气泡赶出，以免使用时带来读数误差。赶去气泡的方法如

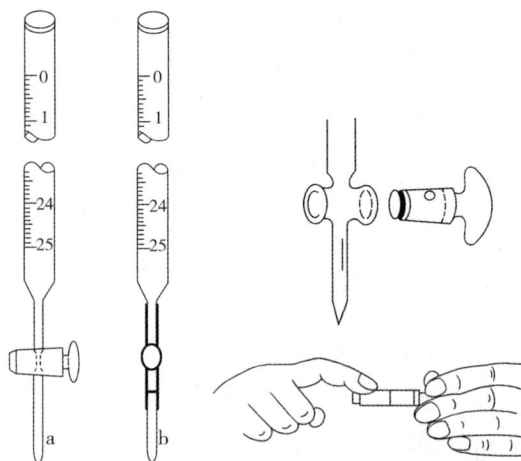

图2-23 滴定管的阀门安装法和类型

下：将滴定管阀门开到最大，利用溶液的气流把气泡赶去。对于碱式滴定管，也可把橡皮管向上弯折，然后稍微捏挤玻璃球旁侧的橡皮管，气泡即易被管中溶液压出（图 2 – 24）。

滴定管应保持垂直。滴定前后均需记录读数，终读数与初读数之差就是所消耗溶液的体积，习惯称为"用量"。初读数应调节在刻度刚为"0"或"0"下面的刻度上。读数时最好在滴定管的后面衬一张白纸片，视线必须与液面在同一水平面上，观察溶液弯月面底部所在的位置，仔细读到小数点后两位数字。视线不平或者没有估计到小数点后第二位数字，都会影响测定的精确程度。例如，图 2 – 25 所示的读数应记作 24.43，不能误读为 24.34 或 24.53，也不能简化为 24.4。

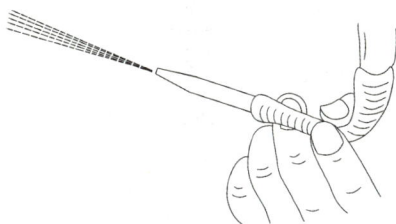

图 2 – 24 逐去气泡

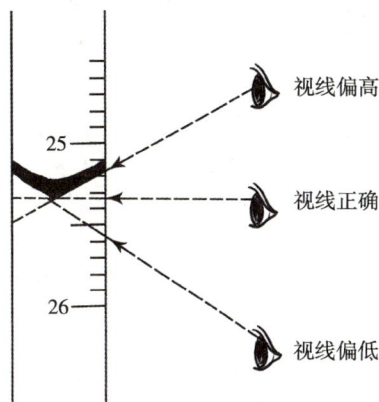

图 2 – 25 刻度的读数

滴定开始前，先把悬挂在滴定管尖端的液滴除去。滴定时用左手控制阀门，右手持锥形瓶（瓶口应接近滴定管尖端，不要过高或过低），并不断旋摇底部，使溶液均匀混合，充分反应（图 2 – 26 ～ 图 2 – 28）。滴定速度一般以控制溶液能成滴流下为宜。

临近滴定终点时（这时每滴入一滴溶液，指示剂的颜色复原较慢），滴定速度要放得很慢，最后要一滴一滴或半滴半滴地滴入，防止过量，并且要用洗瓶挤少量水淋洗瓶壁，以免有残留的液滴未起反应。

为了便于判断终点时指示剂颜色的变化，可把锥形瓶放在白色瓷板或白纸上观察。最后，必须待滴定管内液面完全稳定，方可读数（在滴定刚完毕时，可能有少量沾在滴定管壁上的溶液尚未下流）。

图 2 – 26 酸式滴定管的操作   图 2 – 27 两手操作姿势   图 2 – 28 在烧杯中的滴定操作

3. 移液管 用移液管移取液体时，把移液管的尖端部分插入待吸取的液体中，用洗耳球把液体慢慢吸入管中，待溶液上升到标线以上约 2cm 处时，立即用示指（不要用大拇指）堵住管口。将移液管持直并移出液面（图 2 – 29），微微移动示指或用大拇指和中指轻轻转动移液管，使管内液体的弯月面慢慢下降到标线处（注意：视线、液面、标线均应在同一水平面上），即压紧管口。若管尖挂有液滴，可使管尖与盛溶液器壁接触使液滴落下。再把移液管移入另一承接容器（如锥形瓶）中。并使管尖与

容器内壁接触。将移液管直立，承接容器倾斜30°左右，松开示指，让液体沿容器壁自由流出（图2 - 30），待管内液体不再流出，稍停（约15秒钟）后，把移液管拿开。若移液管上没有刻有"吹"字，此时遗留在管内的液滴不能吹出，因移液管的容量只计算自由流出液体的体积，刻制标线时已把液滴留在管内的因素考虑在内了。

图2 -29   用洗耳球吸取溶液   图2 -30   移液管的使用

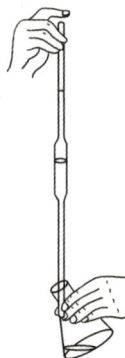

移液管在使用前的洗涤方法与滴定管相仿，除分别用洗涤液、水及蒸馏水洗涤外，还需用少量要移取的液体洗涤，可先慢慢地吸入少量洗涤的水或液体至移液管中，用示指按住管口，然后将移液管平持，放松示指，转动移液管，使洗涤的水或液体与管口以下的内壁充分接触。再将移液管持直，让洗涤水或液体流出，如此反复洗涤数次。

用移液管吸取有毒或有恶臭的液体时必须用配有吸气橡皮球或其他装置的移液管。

此外，为了精确地移取小量的不同体积（如1.00mL、2.00mL、5.00mL 等）的液体，也常用标有精细刻度的吸量管。吸量管的使用方法与移液管相仿。

八、试纸和滤纸 微课5

1. 用试纸检验溶液的酸碱性   常用pH 试纸检验溶液的酸碱性。将小块试纸放在干燥清洁的点滴板上，再用玻璃棒蘸取待测的溶液，滴在试纸上，观察试纸的颜色变化（不能将试纸投入溶液中检验），将试纸呈现的颜色与标准色板颜色对比，可以知道溶液的pH。

pH 试纸分为两类：一类是广泛 pH 试纸，其变色范围为pH 1 ~ 14，用来粗略地检验溶液的pH；另一类是精密 pH 试纸，用于比较精确地检验溶液的pH，精密试纸的种类很多，根据不同的需求选用。广泛 pH 试纸的变化为1 个pH 单位，而精密 pH 试纸变化小于 1 个pH 单位。

2. 用试纸检验气体   常用pH 试纸或石蕊试纸检验反应所产生气体的酸碱性。用蒸馏水润湿试纸并沾附在干净玻璃棒的尖端，将试纸放在试管口的上方（不能接触试管），观察试纸颜色的变化。用 KI 淀粉试纸来检验 Cl_2，此试纸是用 KI 淀粉溶液浸泡在碎滤纸上，晾干后使用。当 Cl_2 遇到试纸，将 I^- 氧化为 I_2，I_2 立即与试纸上的淀粉作用，使试纸变蓝。用 $Pb(OAc)_2$ 试纸来检验 H_2S 气体，此试纸是用 $Pb(OAc)_2$ 溶液浸泡后晾干使用。生成的 H_2S 气体遇到试纸后，生成黑色 PbS 沉淀而使试纸呈黑褐色。用 $KMnO_4$ 试纸来检验 SO_2 气体。

3. 滤纸   化学实验室中常用的有定量分析滤纸和定性分析滤纸两种，按过滤速度和分离性能的不同，又分为快速、中速和慢速三种。在实验过程中，应当根据沉淀的性质和数量，合理地选用滤纸。

我国国家标准《化学分析滤纸》（GB/T 1914 -93）对定量滤纸和定性滤纸产品的分类、型号和技术指标以及实验方法等都有规定。滤纸产品按质量分为 A 等、B 等、C 等。我们将 A 等产品的主要技术指标列于表 2 – 7。

表 2 – 7　定量和定性分析滤纸 A 等产品的主要技术指标及规格

指　标　名　称		快　速	中　速	慢　速
过滤速度/s		≤35	≤70	≤140
型号	定性滤纸	101	102	103
	定量滤纸	201	202	203
分离性能（沉淀物）		氢氧化铁	碳酸锌	硫酸钡（热）
湿耐破度/mmH$_2$O		≥130	≥150	≥200
灰分	定性滤纸	≤0.13%		
	定量滤纸	≤0.009%		
铁含量（定量滤纸）		≤0.003%		
定量/g · m^{-2}		80.0 ± 4.0		
圆形纸直径/cm		5.5、7、9、11、12.5、15、18、23、27		
方形纸尺寸/cm		60 × 60、30 × 30		

　　定量滤纸又称为无灰滤纸。以直径 12.5 cm 定量滤纸为例，每张滤纸的质量约为 1 g，在灼烧后其灰分的质量不超过 0.1 mg（小于或等于常量分析天平的感量），在重量分析法中可以忽略不计。滤纸外形有圆形和方形两种。常用的圆形滤纸有 ø7 cm、ø9 cm、ø11 cm 等规格，滤纸盒上贴有滤速标签。方形滤纸都是定性滤纸，有 30 cm × 60 cm、30 cm × 30 cm 等规格。

九、溶解、蒸发与结晶

　　1. 溶解与熔融　将固体物质转化为液体，通常采用溶解与熔融两种方法。

　　（1）溶解　溶解就是把固体物质溶于水、酸、碱等溶剂中制备成溶液。溶解固体时，应依据固体物质的性质选择适当的溶剂，并用加热、搅拌等方法促进溶解。

　　（2）熔融　熔融是将固体物质与某种固体熔剂混合，在高温下加热，熔化为液态。酸熔法是用酸性熔剂分解碱性物质；碱熔法是用碱性熔剂分解酸性物质。熔融一般在高温下进行，根据熔剂的性质和温度选择合适的坩埚（如铁坩埚、镍坩埚、白金坩埚、刚玉坩埚等），将固体物质与熔剂在坩埚中混匀后，送入高温炉中灼烧熔融，冷却后用水或酸浸取溶解。

　　2. 蒸发与浓缩　为了能从溶液中析出该物质的晶体或增大浓度，需对溶液进行蒸发、浓缩。水溶液的蒸发一般用蒸发皿。

　　在无机制备中，蒸发、浓缩一般在水浴上进行。若溶液很稀，物质对热的稳定性又较好时，可先放在低温电炉上或在石棉网上用煤气灯直接加热蒸发，然后再放在水浴上加热蒸发。蒸发皿内所盛放的液体不应超过其容量的 2/3。当水分不断蒸发，溶液就不断浓缩，蒸发到一定程度后冷却，就可析出晶体。

　　有机溶剂的蒸发应在通风橱中进行并回收溶剂，视溶剂的沸点和易燃性，注意选用合适的温度。最常用的是水浴，切不可用煤气灯直接加热有机溶剂。蒸发时须用沸石，以防止暴沸。

　　3. 结晶与重结晶　当溶液蒸发到一定浓度后冷却，即有晶体析出。结晶物质溶液的浓度达到饱和程度。物质在溶液中的饱和程度与物质的溶解度和温度有关。晶体的大小与溶质的溶解度、溶液浓度、冷却速度等因素有关。如液体的加热一节中所述，如果希望得到较大颗粒状的晶体，则不宜蒸发至太浓，此时溶液的饱和程度较低，结晶的晶核少，晶体易长大。反之，溶液饱和程度较高，结晶的晶核多，晶体快速形成，得到的是细小晶体。从纯度来看，缓慢生长的大晶体纯度较低，而快速生成的细小晶体，纯度较高。因为大晶体的间隙易包裹母液或杂质，因而影响纯度；但晶体太小且大小不均匀时，易形成糊状物，夹带母液较多，不易洗净，也影响纯度。因此晶体颗粒要求大小适中且应均匀，才有利于得到纯度较高的晶体。

如果第一次结晶所得物质的纯度不符合要求时，可进行重结晶。其方法是在加热的情况下使被纯化的物质溶于尽可能少的溶剂中，形成饱和溶液，并趁热过滤，除去不溶性杂质，然后使滤液冷却，被纯化物质即结晶析出，而杂质则留在母液中，过滤便得到较纯净的物质。若一次重结晶还达不到要求，可以再次重结晶。重结晶是提纯固体物质常用的重要方法之一，它适用于溶解度随温度有显著变化的化合物的提纯。

十、固液分离

化学实验中经常会遇到沉淀和溶液分离或晶体与母液分离等情况，分离方法主要有倾泻法、过滤法、离心分离等方法。

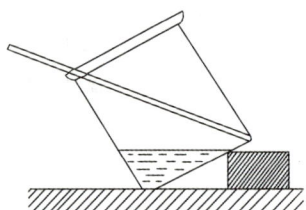

图 2-31   倾泻法

1. 倾泻法   如沉淀的相对密度较大或晶体颗粒较大时，沉淀很快沉降到容器底部时，可用倾泻法进行固-液分离。操作方法是待沉淀完全沉降后，小心地将沉淀上层清液慢慢地倾入另一容器中，倾倒时用一洁净的玻棒引流（图 2-31）。如果沉淀需要洗涤，则另加适量洗涤剂（如蒸馏水）搅拌均匀，静置沉降后再倾泻，反复几次，直至符合要求为止。

2. 过滤法   过滤法是最常用的一种固-液分离方法。过滤法利用沉淀和溶液在过滤器上穿透能力的不同，使沉淀留在滤器上而溶液透过滤器进入接受器中，使沉淀和溶液分离。因沉淀的性状、大小的不同，可选用不同型号的滤纸或砂芯漏斗等过滤器，采用常压、减压、热过滤等过滤方法。

（1）常压过滤   常压过滤是用滤纸紧贴在 60°角的圆锥玻璃漏斗上作为过滤器。当沉淀物作为胶状或微细晶体时，常压过程效果较好。

（a）准备过滤器   首先选一张半径比漏斗圆锥高度稍低的圆形滤纸（若为方形滤纸则要剪圆），然后把滤纸对折两次，将滤纸展开为 60°角的圆锥，从三层滤纸一边下面两层撕去一小角（如图 2-32），平整地放入干燥、洁净的漏斗中，使滤纸的圆锥面与漏斗相吻合。叠好的滤纸放入漏斗后用去离子水润湿，再以干净的玻棒轻压滤纸，使之紧贴漏斗壁，其间不应有空气泡，以保证溶液能快速通过滤纸。一般滤纸边应低于漏斗边 5mm。

（b）过滤   准备好的漏斗放在漏斗架上，下面用烧杯或其他盛器承接滤液，漏斗颈末端紧贴在承接器的内壁，以加快滤液的流速。将玻棒指向滤纸三层的一边，用玻棒引流让上层清液慢慢倾入过滤器（图 2-33）。倾入液体的高度要注意比滤纸边缘低 $0.5\sim1cm$。待漏斗中的液体流尽时再逐次将液体倾入漏斗。溶液倾倒完后，用洗瓶挤少量水淋洗盛放沉淀的容器及玻棒，并将洗涤水全部滤入承接器中（图 2-34）。若需洗涤沉淀，则用洗瓶挤出洗涤液在滤纸的三层部分离边缘稍下的地方，自上而下洗涤。并借此将沉淀集中在滤纸圆锥体的下部（图 2-35），如此洗涤多次。

强酸、强碱或强氧化剂会破坏滤纸结构，它们的固-液分离要以石棉纤维或玻璃丝代替滤纸，或改用玻璃砂芯漏斗。

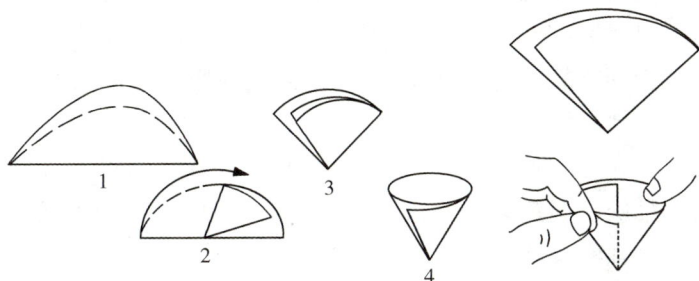

图2-32　滤纸的折叠方法及安装

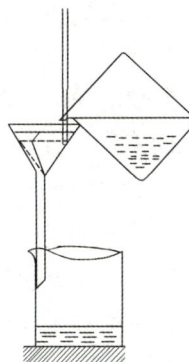

图2-33　常压过滤

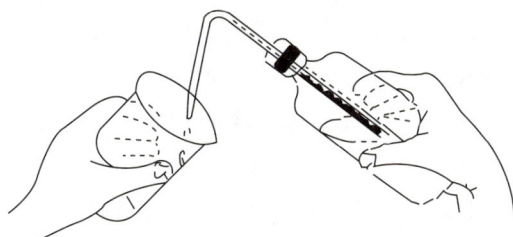

图2-34　淋洗器皿

图2-35　沉淀在漏斗中的洗涤

（2）减压过滤（抽滤或吸滤）　减压过滤是采用水泵或真空泵抽气使滤器两边产生压差而快速过滤并抽干沉淀上溶液的过滤方法。它不适宜过滤细小颗粒的晶体沉淀和胶状沉淀，前者会堵塞滤纸孔而难于过滤，后者会透过滤纸且堵塞滤纸孔。　📱 微课6

减压过滤装置如图2-36，它由吸滤瓶、布氏漏斗、安全瓶和玻璃抽气管组成。玻璃抽气管是一个简单的减压水泵，其内有一窄口，当水急速流经窄口时，把装置内的空气带出而形成一定的真空度，使吸滤瓶内的压力减小，瓶内与布式漏斗液面间产生压差而使过滤速度大大加快。

吸滤操作步骤如下。

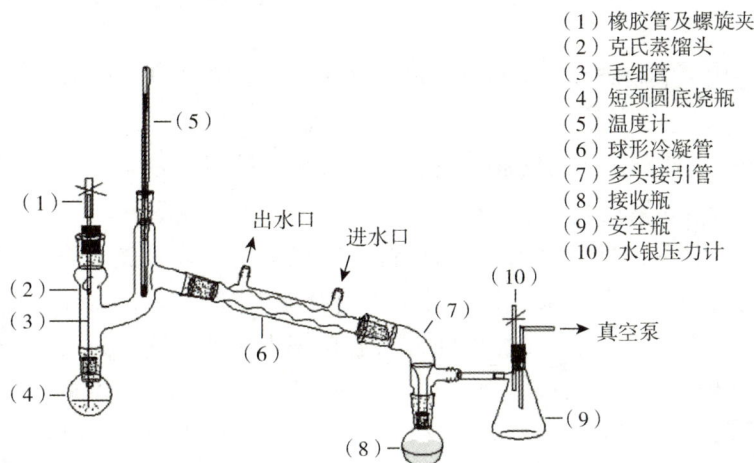

（1）橡胶管及螺旋夹
（2）克氏蒸馏头
（3）毛细管
（4）短颈圆底烧瓶
（5）温度计
（6）球形冷凝管
（7）多头接引管
（8）接收瓶
（9）安全瓶
（10）水银压力计

图2-36　减压过滤装置

（a）剪贴滤纸　将滤纸剪成比布氏漏斗内径略小但又能盖住瓷板上所有小孔的圆，平铺在瓷板上，以少量去离子水将滤纸润湿，微开水阀，轻轻抽吸，使滤纸紧贴在瓷板上。

（b）过滤　先将澄清的溶液以玻棒引流倒入布氏漏斗中，溶液量不要超过漏斗容积的2/3，开大水阀，待溶液滤完后再将沉淀转入漏斗，抽滤至干，沉淀平铺在瓷板的滤纸上。注意：吸滤瓶内液面不能达到支管的水平位置，否则滤液会被水泵抽出，因此滤液过多时，中途应拔掉吸滤瓶上的橡皮管，取下漏斗，把吸滤瓶的支管口向上，从瓶口倒出滤液，再装好继续过滤。

（c）关闭水阀　在抽滤过程中不能突然关闭水阀，停止抽滤时应先拔掉吸滤瓶支管上的橡皮管，使吸滤瓶与安全瓶脱离，然后再关闭水阀，否则水会倒吸。

沉淀的洗涤和抽干：在布氏漏斗内洗涤沉淀，首先应停止抽滤，然后加入少量洗涤液，让它缓缓通过沉淀，再接上吸滤瓶的橡皮管，微开水阀抽吸，最后开大水阀抽干，此时可用一个干净的平顶瓶塞挤压沉淀，帮助抽干。如此反复数次，直至达到要求为止。

（d）沉淀的取出　沉淀抽干以后，先将吸滤瓶与安全瓶拆开，再关闭水阀，然后取下漏斗，将漏斗的颈口向上，轻轻敲打漏斗边缘，或在漏斗颈口用力一吹，即可使沉淀脱离漏斗，倾入预先准备好的滤纸或容器上。欲得干燥沉淀可用干燥滤纸将水分吸干或放入恒温烘箱内烘干。

（3）热过滤　如果溶质的溶解度明显地随温度的降低而降低，但又不希望它在过滤过程中析出晶体时，可采用热过滤。其做法是把玻璃漏斗放在铜质的热漏斗内（图2-37），热漏斗内装热水以维持溶液的温度，趁热过滤。也可以在过滤前把玻璃漏斗放在水浴上用蒸气加热后快速过滤。

如过滤的溶液有强酸性或强氧化性，为了避免溶液和滤纸作用，应采用玻璃砂芯漏斗（图2-38）。由于碱易与玻璃作用，所以玻璃砂漏斗不宜过滤强碱性溶液。过滤时，不能引入杂质，不能用瓶盖挤压沉淀，其他操作要求基本如上述步骤。

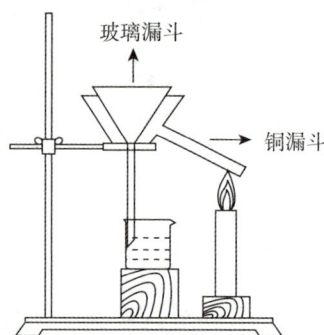

图2-37　热过滤　　　　图2-38　玻璃砂芯漏斗

3. 离心分离　是用离心机将少量沉淀和溶液分离的简便快速方法。实验中常用的电动离心机如图2-39所示。

离心分离时，选用大小相同、所盛混合物的量大致相等的离心试管，对称地放在离心机套筒内，如果只有一支要离心分离的试管，则可用另一支大小相同、盛有同量水的离心试管与之相配，以保持离心机平衡，然后盖上盖子。启动离心机时先调到变速器的最低档，转动后再逐渐加速，2~5分钟后，断开电源，让其自然停止，切不可加以外力强迫它停止转动。离心分离操作完毕后，轻轻取出试管，不要摇动，将一支干净的滴管排气后伸入离心管的液面下，慢慢吸取清液。在吸取过程中吸管口始终不离开液面而又不接触沉淀。欲得较纯净的沉淀，还需用洗涤洗液加入沉淀中，用玻棒搅拌均匀后再离心分离，反复数次直至达到要求。

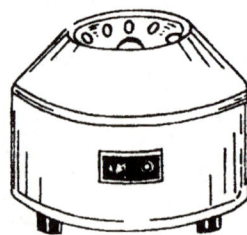

图2-39　电动离心机

十一、天平的使用 微课7

实验中由于对质量准确度的要求不同，需使用不同类型的天平进行称量。常用的天平有台天平

（也叫台秤）、化学天平和分析天平等。一般来说，台天平的感量（称量的精确程度）是 0.1g，化学天平（扭力天平）的感量是 0.001g，而分析天平的感量则为 0.0001g。

台天平，如图 2-40 所示。使用台天平前需先把游码放在刻度尺的零处，检查天平的摆动是否平衡。如果平衡，则指针摆动时所指示的标尺上的左右格数应相等，当指针静止时应指在标尺的中线。如果不平衡，可以调节平衡螺母，使之平衡。

称量时，将要称的物品放在左台上（左盘内），然后在右台上（右盘内）添加砝码。砝码通常从大的加起，如果偏重，就换放小的砝码，5（或 10）g 以下的砝码用游码代替，直到天平平衡为止。台天平的砝码和游码可以用干净的手指直接拿取和移动。

称固体药品时，应在两台上（或两盘内）各放一张质量相仿的蜡光纸，然后用药匙将药品放在左台（左盘）的纸上（称 NaOH、KOH 等易潮解或有腐蚀性的固体时，应衬以表面皿）。称液体药品时，要用已称过重量的容器盛放药品，称法同前。

1. 电子天平的结构

电子天平结构如图 2-41 所示。

2. 电子天平的四种状态

（1）工作状态　屏幕显示"Busy"，此时仪器不接受任何指令。

（2）准备状态　不切断电源，仅关闭电子天平，显示器显示"Standby"字样，此时开启后无需重新预热，随时可以称样。

（3）调零状态　见后面的 CAL 键说明。

（4）静止状态　切断电源，仪器处于静止状态。

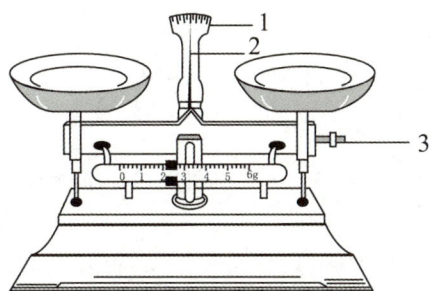

图 2-40　台天平

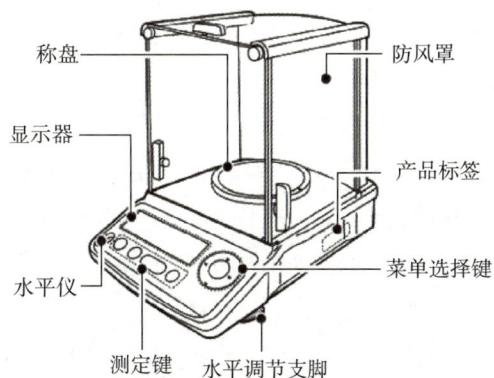

图 2-41　电子天平的结构示意图

3. 电子天平的功能键

（1）On/Off 键　称重后，暂时关闭天平时，按"On/Off"键，屏幕显示"Standby"，天平处于准备状态；再次称重时，按"On/Off"键，则随时可以进行称量，不必重新预热。

（2）Print 键　若天平连接有打印机，称样后按此键可打印称量结果。

（3）CAL 键　内部校正键，按"CAL"键，屏幕显示"C"，几秒后显示"CC"，同时听到声频信号，声频停止后，屏幕显示"0.0000g"。若按"CAL"键，屏幕显示"CE"，则应按下"TARE"键后再按下"CAL"键至屏幕显示"0.0000g"。

（4）TARE 键　去皮键，若用称量纸或称量瓶称取试样时，可将称量纸或称量瓶置于天平秤盘上，按此键去除皮重，显示器显示"0.0000g"。

4. 电子天平的使用方法

（1）水平调节　电子天平开机前，应观察电子天平水平仪内的水泡是否位于圆环的中央，假如不在中央位置，需要通过电子天平的水平调节支脚调节。

（2）预热　电子天平在初次接通电源或长时间断电后开机时，至少需要30分钟的预热时间或根据使用说明书来操作。因此，实验室万分之一电子天平在通常情况下，不要经常切断电源。

（3）称量　①按下"ON/OFF"键，接通电子天平的显示器等待电子天平自检，当显示器显示"0.0000g"时，自检过程结束，电子天平就可进行称量。②放置称量纸或称量瓶，按下"Tare"键去皮，待显示器显示"0.0000g"时，在称量纸上加所要称量的样品，待显示器读数稳定并显示重量单位"g"时，即可读出所称样品重量。③称量完毕，按下"ON/OFF"键，使天平处于准备状态，若当天不再使用该天平，可切断天平电源。

5. 电子天平室使用守则　天平室要求安静、防震、避光、整齐、清洁、干燥。

（1）不得在天平室内大声喧哗，严禁吸烟、饮食。

（2）在天平室内走动要轻，避免天平部件因震动受损。

（3）天平应避免阳光及强光直照，室内温度保持在20℃左右，相对湿度应为50%～70%。

（4）天平台面、地面要保持清洁，不宜用水擦地面。

（5）进入天平室后应严格遵守天平操作规程，实验完毕，复原天平，填写天平使用登记本，经教师检查允许后方可离开。

十二、酸度计的使用

酸度计（又称pH计）是一种通过测量电势差的方法来测定溶液pH的仪器，除可以测量溶液的pH外，还可以测量氧化还原电对的电极电势值（mV）及配合电磁搅拌进行电位滴定等。实验室常用的酸度计有25型、pHS－2型、pHS－2C型和pHS－3型等，各种型号的仪器结构虽有不同，但基本原理相同。

1. 仪器工作原理　酸度计是用电位法测定溶液pH的测量仪器，也叫pH计。其工作原理主要是利用一对电极在不同pH溶液中能产生不同的电动势，并在仪器上指示出测量结果，如图2－42。

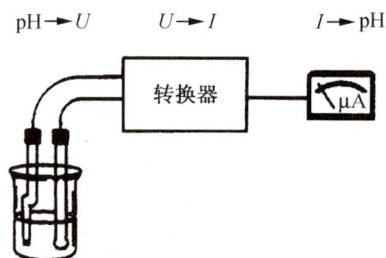

图2－42　酸度计工作原理示意图

这对电极中的一个是指示电极，其电极电势随被测溶液的pH变化，一般采用玻璃电极（图2－43）。另一个是参比电极，要求其电极电势值恒定，与被测溶液的pH无关，一般采用甘汞电极（图2－44）。

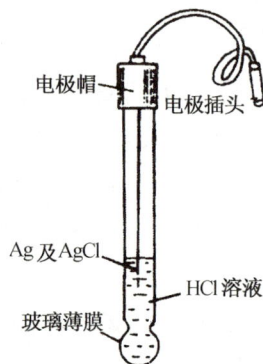

图2－43　玻璃电极示意

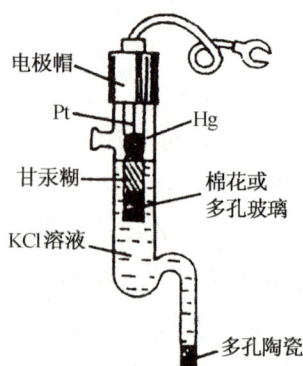

图2－44　甘汞电极示意图

玻璃电极的下端是由一种特殊导电玻璃吹制成的空心小球泡。球内装有 0.1mol/L HCl（或一定的 pH 的缓冲溶液）和 Ag-AgCl（覆盖 AgCl 的 Ag 丝）。将它插入一待测溶液中，便组成了原电池的一端，可表示为：

$$Ag，AgCl(s)|HCl(0.1mol/L)|玻璃|待测溶液$$

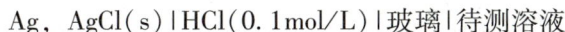

导电玻璃薄膜把两种溶液分开，被测溶液的 H^+ 离子与电极球泡表面水化层进行离子交换，球泡内层也产生电极电势。由于球内氢离子浓度是固定的，而外层氢离子浓度在变化，因此内外层的电势差也在变化，所以该电极的电势随待测溶液 pH 的不同而改变。在 25℃时，

$$E_G = E_G^\ominus + 0.05917 \lg\{c(H^+)/c\}$$
$$= E_G^\ominus - 0.05917\text{pH}$$

式中，E_G、$E_G^\ominus$ 分别表示上述玻璃电极的电势和标准电势。对给定的玻璃电极，$E_G^\ominus$ 为一常数；c（H^+）为被测溶液的氢离子浓度。

甘汞电极的组成：

$$Hg|Hg_2Cl_2(s)|KCl|$$

电极反应：

$$Hg_2Cl_2 + 2e \rightleftharpoons 2Hg + 2Cl^-$$

该电极的稳定性好，其电极电势不随被测溶液 pH 的变化而变化，在一定温度下为一定值。

将玻璃电极和甘汞电极插入待测溶液中组成原电池，与伏特表连接，可以测定该电池的电动势 E。在 25℃时，

$$E_池 = E_正 - E_负 = E_{甘汞} - E_G = E_{甘汞} - (E_G^\ominus - 0.05917\text{pH})$$
$$\text{pH} = \frac{E - E_{甘汞} + E_G^\ominus}{0.05917}$$

式中，$E_{甘汞}$ 表示甘汞电极电势，常用的是饱和甘汞电极，在 25℃时其电势为 0.2415 V；$E_G^\ominus$ 的数值可用一已知 pH 的标准缓冲溶液代替待测溶液组成原电池。为免除计算的麻烦，酸度计把所测电动势读数直接用 pH 表示出来。

温度对 pH 测定值的影响，可根据能斯特（W. Nernst）方程式给予校正。在 pH 计中已配有温度补偿器补偿（校正）。

pHS-25 型酸度计外观如图 2-45 所示。

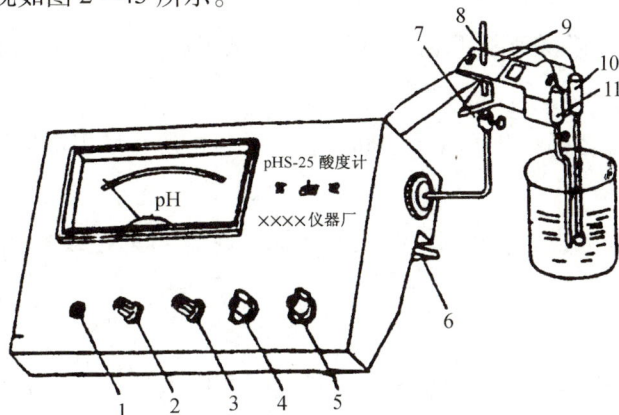

图 2-45　PHS-25 型酸度计

1. 电源指示灯；2. 温度补偿器；3. 定位调节器；4. 功能选择器；5. 量程选择器；6. 仪器支架；

7. 电极杆固定圈；8. 电极杆；9. 电极夹；10. pH 玻璃电极；11. 甘汞参比电极

2. 使用操作方法

（1）仪器安装 首先按图 2-45 所示的方式装上电极杆和电极夹，并按需要的位置固定，然后装上电极，支好仪器后背支架。将量程开关置于中间位置后接通电源。

（2）酸度计的检查 通过下列操作方法，可初步判断仪器是否正常。

（a）将"功能选择器"开关置于"+mV"或"-mV"。此时电极插座不能插入电极。

（b）"量程选择器"开关置于中间位置，开仪器电源开关，此时电源指示灯应亮。表针位置在未开机时的位置。

（c）将"量程选择器"开关置"0~7"档，指示电表的示值应为 0mV（±10mV）位置。

（d）将"功能选择器"置"pH"档，调节"定位"，使电表示值应能小于 6pH。

（e）将"量程选择器"开关置"7~14"，调节"定位"，使电表示值应能大于 8pH。

当仪器经过以上方法检查，都能符合要求后，则可认为仪器的工作基本正常。

（3）仪器的校正 pH 玻璃电极在使用前必须在蒸馏水中浸泡 24 小时以上。参比电极在使用前必须拔去橡皮塞和橡皮套。pH 玻璃电极和甘汞参比电极在使用时，必须注意内电极与球泡之间及内电极与陶瓷芯之间是否有气泡停留，如有，则必须排除。

仪器在测未知溶液 pH 之前需要校正。仪器的校正可按如下步骤进行：

（a）用去离子水清洗电极，电极用滤纸吸干后，即可把电极放入一已知 pH 的标准缓冲溶液中，调节"温度"调节器，使所指定的温度为待测溶液的温度。

（b）置"功能选择器"开关于所测 pH 标准缓冲溶液的范围这一档（如对 pH=4 或 pH=6.86 的溶液则置"0~7"档）。

（c）调节"定位"旋钮，使电表指示该缓冲溶液的准确 pH。

注意：经上述步骤定位的仪器，"定位"旋钮不应再有任何变动。

（d）移去标准缓冲溶液，用去离子水小心淋洗电极，并用滤纸吸干电极。

（4）pH 的测量

（a）把电极插在未知溶液之内，稍稍摇动烧杯，使之缩短电极响应时间。

（b）调节"温度"定位器至指示溶液的温度。

（c）置"功能选择器"开关于"pH"。

（d）置"量程选择器"开关于被测溶液的可能 pH 范围。此时仪器所指示的 pH 即为未知溶液的 pH。

（e）测定完毕，将量程选择器开关置于中间位置，关闭电源，取出电极洗净保存。

（5）测量电极电势

（a）电极接线：把甘汞电极与 pHS-25 型酸度计标"+"号的接线柱相连，待测电极与标"-"号的接线柱相连。

（b）接电源：打开开关，预热 5 分钟。

（c）根据电极电势的极性置"功能选择器"开关，若原电池的正、负极与仪器上标明的"+""-"号相同，开关指向"+mV"；若原电池的正、负极与仪器上标明的"+""-"号不一致，就扳向"-mV"。

（d）置"量程选择器"开关于被测电池的可能电动势范围，当开关指向"0～7"处，如果指针向左偏出刻度，表示电动势大于700mV，就要把开关扳向"7～14"，此时仪器所指示的数值即被测电池的电动势。

书网融合……

微课1　　微课2　　微课3　　微课4

微课5　　微课6　　微课7　　习题

第二部分　实验内容

第三章　基础性实验

实验一　冰点降低法测定葡萄糖的摩尔质量

PPT

目的要求 •••

1. 掌握冰点降低法测定摩尔质量的实验方法。
2. 通过实验进一步理解拉乌尔定律。

实验原理 •••

冰点是溶液（或溶剂）的蒸气压等于其纯溶剂的固相的蒸气压时的温度，溶液的冰点降低是由于溶液的蒸气压小于同温下纯溶液的蒸气压而造成的。

溶液的冰点决定于其浓度和溶剂性质。溶液浓度越大，冰点下降值越大。

对非电解质溶液：$\Delta T_f = K_f \cdot m$

对电解质溶液：$\Delta T_f = i \cdot K_f \cdot m$

式中，ΔT_f 为溶液的冰点下降值；i 为范特荷甫系数；m 为质量摩尔浓度，指溶液中溶质的物质的量除以溶剂的质量（mol/kg），也可以写为 mol（溶质）/kg（溶剂）。如有 ag 溶质（即 a/Mmol）溶于 Ag 溶剂中，则溶液的质量摩尔浓度 m 应为 $m = \dfrac{a \times 1000}{A \times M}$

式中，M 为溶质的摩尔质量，将 m 代入拉乌尔公式得：

$$\Delta T_f = K_f \cdot \frac{a \times 1000}{A \times M}$$

$$或 \ \Delta T_f = i \cdot K_f \cdot \frac{a \times 1000}{A \times M}$$

式中，K_f 是摩尔冰点降低常数，它仅决定于溶剂的性质，而与溶质的性质无关，所以不同的溶剂有不同的 K_f 值。

仪器和药品 •••

1. 仪器

（1）1/10 刻度温度计　读至小数点后第二位（可用放大镜帮助读数）。

（2）细搅棒　上下搅动测定液，使温度均匀。

（3）测定管　内盛欲测之液体或溶液。

（4）空气套管　精测冰点时使用。

（5）粗搅棒　用以搅动冰水及盐的混合物。

（6）厚玻璃烧杯　内盛冰水及盐以降温用。

（7）橡皮塞。

2. 药品 葡萄糖、粗盐、冰。

实验内容 •••

1. 测葡萄糖溶液的冰点

（1）准备工作。在厚玻璃大烧杯中，装入碎冰和少量水（两者约占烧杯中总体积3/4），然后加入适量粗盐作降温冷却用。

（2）在电子天平上先准确称量油光纸的重量，再用加重法称取葡萄糖 2.3～2.5g（精密称量至小数点后三位），小心用油光纸包好，防止散失。

（3）测葡萄糖溶液的冰点。将称好的葡萄糖小心倒入干燥的测定管中，并注意使纸上葡萄糖完全进入管内，然后用 25mL 移液管准确吸取 25mL 蒸馏水，沿管壁加入测定管中，轻轻振荡（注意切勿溅出）待葡萄糖完全溶解后，装上塞子（包括温度计与细搅棒），然后直接插入冰浴中（图 3-1），此时开始用粗搅棒搅动冰水液，同时以细搅棒轻轻地搅拌溶液，但不要碰及管壁与温度计，以免摩擦生热影响实验结果。在降温过程中，会产生过冷现象（即到冰点时并不结冰），当温度继续下降至某一温度后又迅速上升至某一点而达恒定，此时的温度即为冰点，可通过放大镜读出准确读数。

冰点的测定须重复两次，两次测定结果的差值，不得超过 0.02℃（否则测第三次）。溶液的冰点取两次结果的平均值。

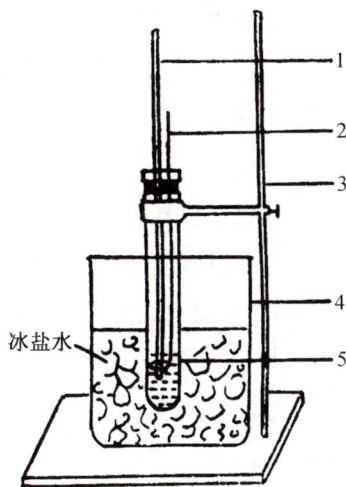

图 3-1 冰点测定装置
1. 温度计；2. 搅拌棒；
3. 铁站架；4. 烧杯；5. 测量管

冰盐水

2. 测纯溶剂（水）的冰点 弃去测定管内溶液，先以自来水洗净测定管，再用少量蒸馏水洗涤，然后加入 25mL 蒸馏水同上法测定水的冰点（取测定结果的平均值）。

测定次数	冰点（℃）		溶质质量 g	溶剂质量 g	ΔT_f
	蒸馏水	葡萄糖溶液			
1					
2					
3					

结果计算 $$M = K_f \frac{a \times 1000}{A \times \Delta T_f}$$

3. 数据记录及结果处理

（1）按实验所得结果计算葡萄糖的分子量。

（2）由实验所得摩尔质量与理论值比较，计算相对误差。

（3）数据记录及结果处理。

实 验 指 导

预习要求 •••

1. 什么叫冰点？纯溶剂的冰点与溶液的冰点有什么不同？

2. 溶液的冰点为什么低于纯溶剂的冰点？稀溶液冰点下降值遵守什么规律？

3. 本实验为什么要用冰盐水作冷冻环境？只用冰行不行？

4. 本实验方法中，为什么要测纯溶剂的冰点？

5. 能否用冰点降低法测定挥发性物质的摩尔质量？

基本操作 •••

1. 学习制备冰盐水冷冻环境。

2. 正确使用电子天平。

3. 正确使用移液管。

4. 1/10 刻度温度计的读数。

5. 测定液体的冰点。

注意事项 •••

1. 测定管需干燥。

2. 葡萄糖需定量转移至测定管内。

3. 定量的溶剂放入测定管时，不要溅出管外。

4. 1/10 刻度温度计读数，需用放大镜读取。

5. 温度计前端的球玻璃极薄（可提高测温灵敏度），切勿将温度计代替搅拌棒用。

6. 测纯水冰点时，有时温度计会与冰冻结在一起，此时应让冰融化后再取出温度计。

7. 先测溶液冰点，测定管必须洗净后再测纯水冰点。

报告格式 •••

1. 目的

2. 原理

3. 实验数据处理、结果与讨论

实验后思考 •••

1. 溶液的浓度越大，冰点下降值越大，实验误差也小，所以实验时溶液的浓度越大越好，是否正确？为什么？

2. 如果待测葡萄糖中夹杂一些不溶性杂质，对测得的摩尔质量有何影响？

3. 如果将定量的蒸馏水放入测定管时，损失了一些，对测得的摩尔质量又有何影响？

4. 将实验所得的结果与理论值比较，分析产生误差的原因。

Experiment 1　The Usage of Depression of Freezing Point to Determine the Glucose′s Molecular Weight

◆ Objectives

1. Learn the cryoscopy technique to determine the molecular weight.

2. Understand Raoult′s Law.

◆ Principles

The freezing point is the temperature at which the vapor pressure of a solution(or solvent) equals the vapor pressure of its pure solid phase solvent. When the vapor pressure of the solution is less than the solvent at the same temperature, the freezing point is lowering. In general, solution has a lower freezing point than does the pure solvent.

The freezing point is determined by the concentration of the solution and the properties of the solvent. And the lowering of the freezing point is proportional to the number of dissolved particles (molecules or ions).

The amount of the depression is given by:

$\Delta T_f = K_f \cdot m$ (nonelectrolyte solution)

or $\Delta T_f = i \cdot K_f \cdot m$ (electrolyte solution)

Here, ΔT_f denotes the amount of the depression; i denotes van't Hoff mole number; m denotes molality which means the moles of the solute divide by the quantity of the solvent (mol/kg or mol(solute)/1000g(solvent)). For example, if "a" gram of solute (a/Mmol) is soluble in "A" gram of solvent, then the molality $m = \dfrac{a \times 1000}{A \times M}$ (M is the molecular weight of the solute). Putting the formula of m above into Raoult's expressions, there exists:

$$\Delta T_f = K_f \cdot \frac{a \times 1000}{A \times M}$$

$$\text{or} \quad \Delta T_f = i \cdot K_f \cdot \frac{a \times 1000}{A \times M}$$

Here, K_f is mole cryoscopic constant, which is only determined by the property of the solvent, but does not depend on the property of the solute. So different solvent has different value of K_f.

◆ Equipment and Chemicals

1. Equipment

(1) 1/10 graduated thermometer: Read to the second decimal digit.

(2) Thin stirring rod.

(3) Test tube: Fill in the liquid to be determined.

(4) Air coated tube: Determine the freezing point accurately.

(5) Thick stirring rod: Stir the compound of ice water with salts.

(6) Thick wall beaker: Fill in some ice water to decrease the temperature.

(7) Rubber plug.

2. Chemicals glucose, coarse salt, ice

◆ Procedure

1. Determine the freezing point of the glucose solution

(1) Preparation: Fill some pieces of ice and a small quantity of water into a thick walled beaker. (The volume of these two substances is about 3/4 of the whole beaker). Then add proper amount of crude salt to decrease the temperature.

(2) First weigh the oiled paper on the electronic balance, then weigh 2.3 ~ 2.5g glucose with weight-adding method (the weight is made with a precision of up to three decimal places). Pack the glucose carefully and avoid dispersing outside.

(3) Determine the freezing point.

Put the glucose into a completely dry determining tube, then add 25mL distilled water from the inner wall of the tube with transfer pipette, shaking slightly (Be careful not to splash the solution out). When the glucose has been dissolved completely, pack the plug including the thermometer and the thin stirring rod, and insert them into the ice bath. Now please stir the ice water with thick stirring rod and stir the glucose solution with thin stirring rod slightly. Take care not to touch the wall of the tube and the thermometer during the stirring process. If not, the heat produced from friction will affect the result. In the temperature decreasing process, super-cooling phenomenon happens (Don't freeze at freezing point). But with the temperature going on decreasing to a certain degree, it increases quickly until reaches a certain point and then keeps at this point stably. The point at this certain temperature is called freezing point. Read and record accurately through magnifying glass.

Repeat the above process. The difference of the two readings should be no more than 0.02℃. The average of the two values is the freezing point.

2. Determine the freezing point of the pure solvent (water)

Discard the solution in the determining tube. Wash the tube with tap water and then rinse it with distilled water. After adding 25mL distilled water, determine the freezing point of the water using the same method mentioned in operation (1).

3. Data record and result dealing

(1) Calculate the molecular weight of the glucose according to the result.

(2) Compare the result with the theoretical value. Calculate the relative error.

(3) Data record and result dealing.

Experiment number	freezing point(℃)		solute (g)	solvent (g)	ΔT_f
	distilled water	glucose solution			
1					
2					
3					

Result: $M = K_f \dfrac{a \times 1000}{A \times \Delta T_f}$

◆ **Instructions**

1. Requirements

(1) Understand the concept of the freezing point. Tell the difference between the freezing point of pure solvent and the solution.

(2) Why the freezing point of solution is lower than the pure solvent? Which law does the amount of depression of a dilute solution obey?

(3) Why could the ice salt solution be used to lower the temperature? If only ice water is used, what's the result?

(4) Why the freezing point of the pure solvent should be determined in this experiment?

(5) Can the cryoscopy technique be used to determine the molecular weight of a volatile substance?

2. Operation

(1) Learn to prepare the ice salt solution.

(2) Learn to use electronic balance correctly.

(3) Learn to use the transfer pipette.

(4) Learn to record the temperature value from 1/10 graduated thermometer.

（5）Determine the freezing point of a liquid.

3. Notes

（1）The determining tube should be dry.

（2）The measured glucose should be transferred into a determining tube quantificationally.

（3）Notice not to splash the quantitative solvent out of the determining tube.

（4）Use the magnifying glass to read the temperature from the 1/10 graduated thermometer.

（5）The glass of the bulb at the front tip of the thermometer is too thin（to promote the sensitivity when measuring temperature）. So be sure not to replace the stirring rod with thermometer.

（6）When determining the freezing point of the pure water, the thermometer may be frozen with ice. Notice to make the ice melt before taking it out.

（7）After determining the freezing point of the solution and before determining the freezing point of the pure water, the determining tube should be washed carefully.

4. Report Format

（1）Objectives

（2）Principles

（3）Data record and calculation

5. Questions

（1）Larger is the amount of depression and smaller is the error, if the concentration is higher. So the higher concentration is better for this experiment. Do you think it is correct?

（2）How will be on the result of the molecular weight of glucose when it was mixed with some insoluble impurity?

（3）What influence will be on the result of the molecular weight of losing some distilled water when you add quantitative distilled water to a determining tube?

（4）Compare the experiment result with the theoretical value. Analyze the reason for the error.

实验二　化学反应速率和化学平衡

PPT

目的要求 •••

1. 了解浓度、温度和催化剂对反应速率的影响。

2. 测定过二硫酸铵[$(NH_4)_2S_2O_8$]与碘化钾（KI）反应的反应速率，并计算反应级数、反应速率常数及反应的活化能。

实验原理 •••

在均相反应中，反应速率决定于反应物的本性、浓度、温度和催化剂。反应速率的快慢可以单位时间内反应物浓度的减少或生成物浓度的增加来表示，本实验利用不同浓度的$(NH_4)_2S_2O_8$氧化 KI 生成 KI_3，KI_3与淀粉生成蓝色加合物作为反应完成的标志，出现蓝色的时间越短，表明反应速率越快，出现蓝色的时间越长说明反应速率越慢。

在水溶液中，$(NH_4)_2S_2O_8$与 KI 发生以下反应：

$$S_2O_8^{2-} + 2I^- \longrightarrow 2SO_4^{2-} + I_2 \tag{1}$$

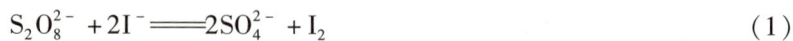

这个反应的平均反应速率可用下式表示：

$$V = -\frac{\Delta[S_2O_8^{2-}]}{\Delta t} = k[S_2O_8^{2-}]^m[I^-]^n$$

式中，V 为平均反应速率；$\Delta[S_2O_8^{2-}]$ 为 Δt 时间内 $S_2O_8^{2-}$ 的浓度变化；$[S_2O_8^{2-}]$ 和 $[I^-]$ 分别为 $S_2O_8^{2-}$ 与 I^- 的起始浓度；k 为反应速度常数；m 和 n 则为反应级数。

为了测定 Δt 时间内 $S_2O_8^{2-}$ 的浓度变化，在将 $(NH_4)_2S_2O_8$ 溶液和 KI 溶液混合的同时，加入一定体积的已知浓度的 $Na_2S_2O_3$ 溶液和淀粉溶液。这样在反应（1）进行的同时，还发生以下反应：

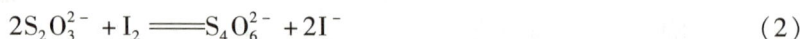

$$2S_2O_3^{2-} + I_2 \longrightarrow S_4O_6^{2-} + 2I^- \tag{2}$$

反应（2）的速率比反应（1）快得多，所以由反应（1）生成的 I_2 立即与 $S_2O_3^{2-}$ 作用生成了无色的 $S_4O_6^{2-}$ 和 I^-。但是一旦 $Na_2S_2O_3$ 耗尽，反应（1）生成的微量 I_2 就立即与淀粉作用，使溶液显蓝色。

从反应式（1）和（2）可以看出，$S_2O_8^{2-}$ 减少 1 mol 时，$S_2O_3^{2-}$ 则减少 2 mol，

即 $$\Delta[S_2O_8^{2-}] = \frac{\Delta[S_2O_3^{2-}]}{2}$$

记录从反应开始到溶液出现蓝色所需要的时间 Δt。由于在 Δt 时间内 $S_2O_3^{2-}$ 全部耗尽，所以 $\Delta[S_2O_3^{2-}]$ 实际上就是反应开始时 $Na_2S_2O_3$ 的浓度，进而可以计算反应速率 $-\dfrac{\Delta[S_2O_8^{2-}]}{\Delta t}$。

对反应速率表示式 $V = k[S_2O_8^{2-}]^m[I^-]^n$ 两边取对数，得

$$\lg V = m\lg[S_2O_8^{2-}] + n\lg[I^-] + \lg k$$

当 $[I^-]$ 不变时，以 $\lg V$ 对 $\lg[S_2O_8^{2-}]$ 作图，可得一直线，斜率即为 m。同理，当 $[S_2O_8^{2-}]$ 不变时，以 $\lg V$ 对 $\lg[I^-]$ 作图，可求得 n。

求出 m 和 n，可由 $k = \dfrac{V}{[S_2O_8^{2-}]^m[I^-]^n}$ 求得反应速率常数 k。

反应速率常数 k 与反应温度 T 一般有以下关系：

$$\lg k = A - \frac{E_a}{2.303RT}$$

式中，E_a 为反应的活化能；R 为摩尔气体常数；T 为热力学温度。测出不同温度时的 k 值，以 $\lg k$ 对 $\dfrac{1}{T}$ 作图，可得一直线，由直线斜率 $\left(-\dfrac{E_a}{2.303RT}\right)$ 可求得反应的活化能 E_a。

仪器和药品 •••

1. 仪器　计时器 1 只，100℃ 温度计 1 支，水浴 2 只；烧杯，150mL 3 只，100mL 2 只；量筒，100mL 3 只，10mL 2 只。

2. 药品　$(NH_4)_2S_2O_8$ 0.20mol/L；KI 0.2mol/L；$Na_2S_2O_3$ 0.01mol/L；KNO_3 0.2mol/L；$(NH_4)_2SO_4$ 0.2mol/L；0.2% 淀粉溶液。

实验内容 •••

1. 浓度对化学反应速率的影响，求反应级数　在室温下，用 3 只量筒分别量取 20mL 0.2mol/L KI 溶液，8mL 0.01mol/L $Na_2S_2O_3$ 溶液和 4mL 0.2% 淀粉溶液，都加到 250mL 烧杯中，混合均匀。再用另 1 只量筒量取 20mL 0.2mol/L $(NH_4)_2S_2O_8$ 溶液，快速加到烧杯中。同时开动秒表，并不断搅拌。当溶

液刚出现蓝色时，立即停秒表。记下时间及室温。

用同样的方法按照表 1 中的用量进行另外四次实验。为了使每次实验中的溶液的离子强度和总体积保持不变，不足的量分别用 0.2mol/L KNO_3 溶液和 0.2mol/L $(NH_4)_2SO_4$ 溶液补足。

算出各实验中的反应速率 V，并填入表中。

用表中实验 Ⅰ、Ⅱ、Ⅲ 的数据以 $\lg V$ 对 $\lg[S_2O_8^{2-}]$ 作图，求出 m；用实验 Ⅰ、Ⅳ、Ⅴ 的数据以 $\lg V$ 对 $\lg[I^-]$ 作图，求出 n。

求出 m 和 n 后，再算出各实验的反应速率常数 k，把计算结果填入表 1 中。

表 1　浓度对化学反应速率的影响

实 验 序 号		Ⅰ	Ⅱ	Ⅲ	Ⅳ	Ⅴ
反应温度						
试剂用量（mL）	0.2mol/L $(NH_4)_2S_2O_8$ 溶液	20	10	5	20	20
	0.2mol/L KI 溶液	20	20	20	10	5
	0.01mol/L $Na_2S_2O_3$ 溶液	8	8	8	8	8
	0.2% 淀粉溶液	4	4	4	4	4
	0.2mol/L KNO_3 溶液	/	/	/	10	15
	0.2mol/L $(NH_4)_2SO_4$ 溶液	/	10	15	/	/
$V_{总} = 52\text{mL}$						
起始浓度 mol/L	$(NH_4)_2S_2O_8$ 溶液					
	KI 溶液					
	$Na_2S_2O_3$ 溶液					
反应时间，Δt, s						
$S_2O_8^{2-}$ 的浓度变化，$\Delta[S_2O_8^{2-}]$, mol/L						
反应的平均速率，$V = -\dfrac{\Delta[S_2O_8^{2-}]}{\Delta t}$						
反应速度常数，$k = \dfrac{V}{[S_2O_8^{2-}]^m[I^-]^n}$						

2. 温度对化学反应速度的影响，求活化能　在 250mL 烧杯中加入 10mL KI 溶液，4mL 淀粉，8mL $Na_2S_2O_3$ 和 10mL KNO_3 溶液。

在另一小烧杯中加入 20mL $(NH_4)_2S_2O_8$ 溶液，并把它们同时放在冰水浴中冷却。等烧杯中的溶液都冷到 0℃ 时，把 $(NH_4)_2S_2O_8$ 加到 KI 等混合溶液中，同时开动秒表，并不断搅拌。当溶液刚出现蓝色时，立即停秒表，记下反应时间。

在约 10℃、20℃、30℃ 的条件下，重复以上实验。这样就可以得到 4 个温度（0℃、10℃、20℃、30℃）下的反应时间。算出 4 个温度下的反应速度及反应速率常数，把数据及计算结果填入表 2 中。

表 2　温度对化学反应速率的影响

实 验 序 号	Ⅰ	Ⅱ	Ⅲ	Ⅳ
反应温度℃				
反应时间 s				
反应速率 V				
反应速度常数 k				
$\lg k$				
$\dfrac{1}{T}$				

注：T 为热力学温度。

用表中各次实验的 $\lg k$ 对 $\dfrac{1}{T}$ 作图，求出反应（1）的活化能。

3. 温度对化学平衡的影响   取 1 只带有两个玻璃球的平衡仪（平衡双球），如图 3-2 所示，玻璃球中二氧化氮和四氧化二氮处于平衡状态，其反应如下：

$$2NO_2 \rightleftharpoons N_2O_4 + 57 \text{ kJ}$$

NO_2 为深棕色气体，N_2O_4 为无色气体，这两种气体混合物则视二者的相对含量而具有淡棕至深棕的颜色。

将 1 只玻璃球浸入热水中，另 1 只玻璃球浸入冷水中。观察二只玻璃球中气体颜色的变化。试以观察到的现象指出各玻璃球中气体平衡向哪一方移动？并用吕·查德里原理说明之。

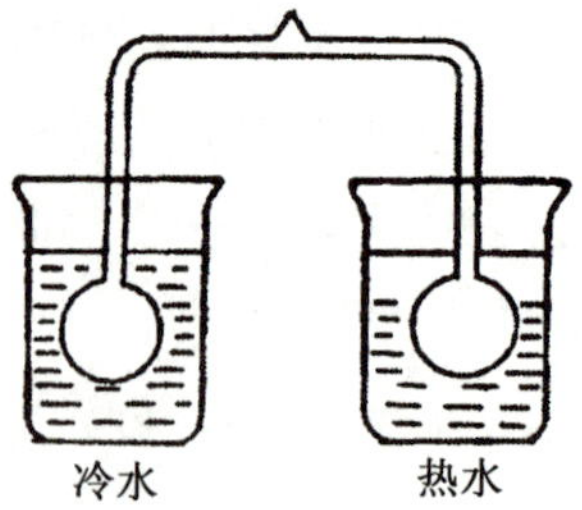

图 3-2   化学平衡的移动

平衡球	气体颜色变化	化学平衡移动的方向
在热水中		
在冷水中		

实 验 指 导

预习要求 ●●●

1. 何谓化学反应速率，影响化学反应速率的因素有哪些？本实验中如何试验浓度、温度对反应速率的影响？

2. 试说明质量作用定律和吕·查德里原理。

3. 根据反应方程式，是否能确定反应级数？举例说明。

4. 本实验是测定 $S_2O_8^{2-} + 2I^- \longrightarrow 2SO_4^{2-} + I_2$ 的反应速率常数，为什么反应物混合时，同时要加一定量的 $Na_2S_2O_3$？

5. 表 1 中哪些项目可以预先计算好填入，哪些需要实验后经计算才能填入？

6. 为什么用 KNO_3 和 $(NH_4)_2SO_4$ 溶液补足溶液的体积？能否用水补充？

基本操作 ●●●

1. 掌握秒表的使用。

2. 学会使用冰浴、热水浴来控制所需要的温度。

注意事项 ●●●

1. 因本实验是利用 $[S_2O_3^{2-}]$ 浓度来衡量反应产生的 $[I_2]$ 浓度，从而计算消耗的 $[S_2O_8^{2-}]$ 浓度。所以准确添加 $Na_2S_2O_3$ 的量是本实验成败的关键。

2. 经计算得出的 5 个 k 值其最大值和最小值之间的差值不得超过 0.5。

3. $\lg k$ 对 $\dfrac{1}{T}$ 作图时，比例、布局必须合适。

报告格式 ●●●

1. 目的。

2. 原理（用最简短的文字表达）。

3. 实验内容 按实验讲义内容填写 3 份表格，并计算反应速率 V，反应速率常数 k，$\lg k$，$\dfrac{1}{T}$。另

用表 2 中各次实验 $\lg k$，对 $\dfrac{1}{T}$ 作图，求出反应的活化能。

实验后思考 •••

试判断下列说法是否正确。

1. 速度方程式中各物质浓度的幂次方等于反应方程式中各物质分子式前的系数时，该反应即为基元反应。

2. 一个反应达到平衡的标志，是正逆反应的速度常数相等。

3. 反应速率常数越大，反应速率也越大，其值相等。

4. 活化能越大，反应速率越小。

5. 反应级数和反应分子数的概念是相等的。

Experiment 2 Chemical Reaction Rate and Chemical Equilibrium

◆ Objectives

1. Understand the effect of concentration, temperature and catalyst on the reaction rate.

2. Determine the rate at which ammonium persulfate $(NH_4)_2S_2O_8$ reacts with potassium iodide (KI), and calculate the reaction order, rate constant and activation energy.

◆ Principles

In the homogeneous phase reactions, the reaction rate is determined by the nature of the reactant, the temperature, the concentration and catalyst. The rate law can be formulated by the change in concentration of one of the reactants or products which occurs in the small time interval after the reagents are mixed.

In this experiment, when $(NH_4)_2S_2O_8$ oxidizes KI to KI_3, KI_3 then reacts with starch solution and a blue complex is formed, which can be regarded as the end of the reaction. Shorter period of the appearance of blue means, quicker the reaction is. Otherwise, slower the reaction rate is.

In aqueous solution, the reaction between $(NH_4)_2S_2O_8$ and KI is:

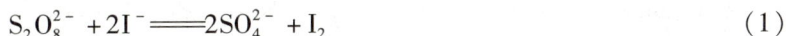

$$S_2O_8^{2-} + 2I^- \Longrightarrow 2SO_4^{2-} + I_2 \tag{1}$$

Where the average reaction rate law has the form

$$V = -\frac{\Delta[S_2O_8^{2-}]}{\Delta t} = k[S_2O_8^{2-}]^m[I^-]^n$$

V is the average reaction rate; $\Delta[S_2O_8^{2-}]$ is the change in concentration of $S_2O_8^{2-}$ during Δt time interval; $[S_2O_8^{2-}]$ and $[I^-]$ are the initial concentration of $S_2O_8^{2-}$ and I^- respectively; k is reaction rate constant; m and n are reactant orders.

In order to determine the value of $\Delta[S_2O_8^{2-}]$, mix the solution of $(NH_4)_2S_2O_8$ with KI, add starch solution and definite amount of $Na_2S_2O_3$ with definite concentration value. So when reaction (1) is going on, the other reaction as the following also exists：

$$2S_2O_3^{2-} + I_2 \Longrightarrow S_4O_6^{2-} + 2I^- \tag{2}$$

The rate of reaction (2) is much more quicker than reaction (1), so the product of I_2 in reaction (1) will react with $S_2O_3^{2-}$ immediately, forming $S_4O_6^{2-}$ (colorless) and I^- as soon as the reactant $S_2O_3^{2-}$ in reaction (2) is exhausted, the product of I_2 produced in reaction (2) will react with starch solution at once, then blue color appears.

From the reaction (1) and reaction (2), we can see that:

$$\Delta[S_2O_8^{2-}] = \frac{\Delta[S_2O_3^{2-}]}{2}$$

Record the time interval Δt from the beginning of the reaction to the appearance of blue color. Because $S_2O_3^{2-}$ has been completely exhausted in Δt time, $\Delta[S_2O_3^{2-}]$ equals to the initial concentration of $Na_2S_2O_3$. So the reaction rate has the form of $-\dfrac{\Delta[S_2O_8^{2-}]}{\Delta t}$. Taking the logarithm of reaction rate formulation $V = k[S_2O_8^{2-}]^m[I^-]^n$ gives:

$$\lg V = m\lg[S_2O_8^{2-}] + n\lg[I^-] + \lg k$$

When $[I^-]$ is constant, make the graph (assign $\lg V$ to y-coordinate, $\lg[S_2O_8^{2-}]$ to x-coordinate and $n\lg[I^-] + \lg k$ to intercept) and we can get the value of the slope m; when $[S_2O_8^{2-}]$ is constant, make the graph (assign $\lg V$ to y-coordinate, $\lg[I^-]$ to x-coordinate and $m\lg[S_2O_8^{2-}] + \lg k$ to intercept) and then we can get the value of the slope n.

Then reaction rate constant k can be obtained as following form:

$$k = \frac{V}{[S_2O_8^{2-}]^m[I^-]^n}$$

With reaction rate constant k and reaction temperature T, we can predict the activation energy of chemical reaction, then the following form is obtained:

$$\lg k = A - \frac{E_a}{2.303RT}$$

E_a represents the activation energy of the reaction, R is universal gas constant, T is absolute temperature (or the temperature on the Kelvin scale). A plot of $\lg k$ as a function of $1/T$ should be a straight line with the slope of $-\dfrac{E_a}{2.303RT}$. So the value of E_a is obtained.

◆ Equipment and Chemicals

1. Equipment stopwatch, thermometer, water bath, five beaker (three with volume of 150mL, two with volume of 100mL), measuring cylinder.

2. Chemicals $(NH_4)_2S_2O_8$ 0.20mol/L; KI 0.2mol/L; $Na_2S_2O_3$ 0.01mol/L; KNO_3 0.2mol/L; $(NH_4)_2SO_4$ 0.2mol/L; 0.2% starch solution.

◆ Procedure

1. Concentration effect on the reaction rate and calculation of the reaction order At the room temperature, measure 20mL 0.2mol/L KI solution, 8mL 0.01mol/L $Na_2S_2O_3$ solution and 4mL 0.2% starch solution with three measuring cylinders. Then put solutions into a beaker (250mL) and mix homogeneously. Add 20mL 0.20mol/L $(NH_4)_2S_2O_8$ solution with measuring cylinder to the beaker quickly, turning on stopwatch at the same time, and stirring continuously. As soon as blue colour appears, please turn off the stopwatch, record the passing time interval Δt and room temperature.

Do the other four experiments according to the given amount in table 1 in the same way. In order to keep the ionic strength and the total volume constant in each portion of the mixture, we add 0.20mol/L $(NH_4)_2SO_4$ solution respectively to supplement the inadequate part.

Calculate the value of the reaction rate V, and fill in Table 1. Make the graph ($\lg V$ to $\lg[S_2O_8^{2-}]$) with the data in experiment I, IV, V, then calculate the value of m; make the graph($\lg V$ to $\lg[I^-]$) with the data in experiment i, iv, v, then calculate the value of n. With the value of m, n, we can get value of reaction rate constant k in each experiment.

Table 1 Concentration effect on the chemical reaction rate

Experiment number		I	II	III	IV	V
Experiment temperature						
Reagents (mL)	0.20mol/L $(NH_4)_2S_2O_8$	20	10	5	20	20
	0.20mol/L KI	20	20	20	10	5
	0.01mol/L $Na_2S_2O_3$	8	8	8	8	8
	0.2% starch solution	4	4	4	4	4
	0.20mol/L KNO_3	/	/	/	10	15
	0.20mol/L $(NH_4)_2SO_4$	/	10	15	/	/
$V_{total}=52mL$						
Initial concentration (mol/L)	$(NH_4)_2S_2O_8$					
	KI					
	$Na_2S_2O_3$					
$\Delta t, s$						
$\Delta[S_2O_8^{2-}]$, mol/L						
$V=-\dfrac{\Delta[S_2O_8^{2-}]}{\Delta t}$						
$k=\dfrac{V}{[S_2O_8^{2-}]^m[I^-]^n}$						

2. Temperature effect on the reaction rate and calculation of the activation energy Add 10mL KI solution, 4mL starch solution, 8mL $Na_2S_2O_3$ solution and 10mL KNO_3 solution to a beaker (250mL).

Add 20mL $(NH_4)_2S_2O_8$ solution into another small beaker, then put the two beakers into a water bath at 0℃. When the solution in the beakers is cooled to 0℃, mix the solution in the two beakers and turn on the stopwatch, stirring continuously. Record the passing time interval Δt when the solution just turns blue.

At the temperature of about 10℃, 20℃, 30℃, repeat the above experiment. Finally we can get four values of Δt at four different temperatures. Calculate the reaction rate and reaction rate constant at four temperature, and fill in Table 2.

Table 2   Temperature effect on the chemical reaction rate

Experiment number	I	II	III	IV
Reaction temperature (℃)				
Reaction time interval (s)				
Reaction rate (V)				
Reaction rate constant k				
lg$k$				
$1/T$				

Draw the graph (lg$k$ to $1/T$) with the data obtained in the above experiments and obtain the value of E_a in reaction 1.

3. Temperature effect on the chemical equilibrium   Take a balance apparatus with two glass globes shown in figure 3 – 2. In it NO_2 and N_2O_4 are initially at equilibrium：

$$2NO_2 \Longrightarrow N_2O_4 + 57kJ$$

NO_2 is a dark brown gas while N_2O_4 is a colorless gas. The color of the mixture of the two gases is determined by the relative content of the each one.

Put a glass globe into hot water, another into cold water. Observe the color change of the glass. Please justify how the system removes. Explain the phenomenon using Lechatelier's principle.

Balance globes	Color change of gas	Direction of equilibrium remove
Hot water		
Cold water		

◆ **Instructions**

1. Requirements

（1）What is chemical reaction rate? Which factors will affect the reaction rate? How to validate the effects of concentration, temperature and the reaction rate?

（2）Please explain the law of mass action and Lechatelier's principle.

（3）Illustrate if you can get the reaction order according to the reaction equations.

（4）The purpose of this experiment is to test the reaction rate constant of reaction, $S_2O_8^{2-} + 2I^- \Longrightarrow 2SO_4^{2-} + I_2$. Please give a reason why we should add a definite amount of $Na_2S_2O_3$ when mixing the reagents.

（5）Which items in table 1 can be calculated before hand, which ones need to be calculated after the experiment?

（6）Why should we add KNO_3 solution and $(NH_4)_2SO_4$ solution to supplement the inadequate volume? Can we use water?

2. Operation

（1）Grasp how to use stopwatch.

（2）Learn how to control the temperature with ice water bath and hot water bath?

3. Notes

（1）In this experiment we use $[S_2O_3^{2-}]$ to determine the concentration of the product I_2, then calculate the consuming concentration of $S_2O_8^{2-}$. So adding $Na_2S_2O_3$ accurately is the key step of the experiment.

（2）The purpose of the experiment is to determine constant, so the value of the time and the concentration should be read correspondent to the precision of the apparatus.

（3）The difference between the maximum value of k and the minimum value of k among the five value of k should be no more than 0. 5.

（4）Make the graph（$\lg k$ to $1/T$）on the graph paper, and the proportion and configuration should be apposite.

4. Report format

（1）Objectives

（2）Principles（with concise works）

（3）Content：Fill in three tables, and calculate the reaction rate V, reaction rate constant k, $\lg k$, $1/T$. Make the graph（$\lg k$ to $1/T$）with the data in Table 2 and calculate the activation energy.

Words

concentration	浓度	activation energy	活化能
coordinate	坐标	phase	相
equilibrium	平衡	precision	精度
coefficient	系数	homogeneous	均匀的
reaction rate	反应速度	proportion	比例
constant	常数	oxidize	氧化
ionic strength	离子强度	apposite	合适的
stopwatch	秒表	starch	淀粉
water bash	水浴	power	幂
reaction order	反应级数	logarithm	对数

实验三　电解质溶液

PPT

目的要求 ●●●

1. 了解弱酸、弱碱的电离平衡及平衡移动的基本原理。
2. 练习缓冲溶液的配制并试验其性质。
3. 了解盐类水解平衡及其平衡移动的基本原理。
4. 了解难溶电解质的多相电离平衡及其平衡移动的基本原理。

实验原理 ●●●

1. 弱电解质电离平衡及其平衡移动　如果在弱电解质的电离平衡体系（如醋酸）中加入与弱电解质有共同离子的强电解质（如醋酸钠）时，由于该离子(Ac^-)浓度的增加，而使电离平衡向生成弱电解质（醋酸）的方向移动。因此，同离子效应使得弱电解质的电离度降低（使醋酸的酸性减弱）。

2. 缓冲溶液　在醋酸溶液中加入醋酸钠或在氨水中加入氯化铵以后，能使溶液中的氢离子或氢氧根离子浓度在加入其他酸或碱时仅有微小的改变，这种由弱酸或弱碱和它们的盐所组成的溶液叫做缓冲溶液。

3. 盐类水解反应　盐类离子与水反应生成弱电解质和 H^+ 或 OH^- 的反应称为盐的水解反应。加入水解产物可抑制水解反应，稀释盐溶液或提高温度可促进水解反应。

4. 难溶电解质多相电离平衡及其平衡移动　难溶电解质与它的饱和溶液是一个多相电离平衡体系，在一定温度下，其中有关离子浓度幂次方的乘积等于该物质的溶度积。

如果设法在此平衡体系中，加入某一种物质与平衡体系中的某一离子相结合，从而使某一离子的浓度降低，离子浓度幂次方的乘积小于溶度积，因而沉淀溶解。例如 $CaCO_3$ 溶于酸是由于 H^+ 与 CO_3^{2-} 结合生成弱电解质 H_2CO_3，而 H_2CO_3 又分解为水和二氧化碳。

仪器和药品

1. 仪器　酸度计，移液管，滴定管。

2. 药品

（1）固体 NH_4Cl

（2）酸　HAc 1mol/L，0.1mol/L；HCl 6mol/L，0.1mol/L；柠檬酸 0.1mol/L；$H_2C_2O_4$ 0.1mol/L。

（3）碱　$NH_3 \cdot H_2O$ 2mol/L，1mol/L，0.1mol/L，6mol/L；NaOH 0.1mol/L。

（4）盐　$MgCl_2$ 0.1mol/L；Na_3PO_4 0.01mol/L；NH_4Cl 饱和溶液 0.1mol/L；$CaCl_2$ 0.1mol/L；NaAc 0.5mol/L；$Cu(NO_3)_2$ 0.1mol/L；Na_2HPO_4 0.2mol/L；$(NH_4)_2C_2O_4$ 0.1mol/L；$SbCl_3$ 0.1mol/L。

3. 其他　酚酞指示剂，甲基橙指示剂，百里酚蓝指示剂。

实验内容

1. 弱酸、弱碱的电离平衡及其平衡移动（同离子效应）

（1）取两支小试管，各加 1mL 蒸馏水，2 滴 2mol/L $NH_3 \cdot H_2O$ 溶液及 1 滴酚酞指示剂。摇荡均匀后溶液呈何色？在一支试管中加 NH_4Cl 固体少许，摇荡后与另一支试管比较，有何变化？何故？

（2）在试管中加入 2mL 1mol/L HAc 溶液，甲基橙指示剂 1 滴，观察溶液的颜色。然后再加少量固体 NaAc，观察指示剂颜色的变化，说明了什么？

（3）取两支小试管，各加 5 滴 0.1mol/L $MgCl_2$ 溶液，在其中一支试管中再加 5 滴饱和 NH_4Cl 溶液，然后分别在这两支试管中加 5 滴 2mol/L $NH_3 \cdot H_2O$ 溶液，观察两试管发生的现象有何不同，何故？

2. 缓冲溶液

（1）在试管中加 3mL 0.1mol/L HAc 和 3mL 0.1mol/L NaAc 溶液配成 HAc-NaAc 缓冲溶液，加百里酚蓝指示剂 5 滴，混合后，注意观察溶液的颜色。然后把溶液分盛三支试管中。在 a 试管中加 5 滴 0.1mol/L HCl，b 试管中加 5 滴 0.1mol/L NaOH，c 试管中加 5 滴 H_2O，观察溶液颜色的变化。然后再在 a 试管中加入过量 0.1mol/L HCl，b 试管中加入过量 0.1mol/L NaOH，注意观察溶液颜色的变化，以 c 管颜色作对照比较，并做出结论。

百里酚蓝指示剂的变色范围：

pH	小于2.8	2.8~9.6	大于9.6
颜色	红	黄	蓝

（2）缓冲溶液的配制与观察指示剂的变色范围　按下表内溶液的用量，配制 pH = 2.2~5.0 缓冲溶液（两人合配一份）。

pH	0.1mol/L 柠檬酸溶液，mL	0.2mol/L 磷酸氢二钠溶液，mL
2.2	19.60	0.40
3.0	15.89	4.11
4.0	12.29	7.71
4.4	11.29	8.71
5.0	9.70	10.30

取试管 5 支，各加入上面配好的缓冲溶液各 4mL，按 pH 由小到大的顺序把试管排好，往每个试管中滴加甲基橙指示剂 1 滴，把溶液充分摇匀后，观察试管中溶液的颜色并确定甲基橙的变色范围。

（甲基橙变色范围 pH 3.0~4.4，红→橙黄）

（3）缓冲溶液的缓冲性能　用移液管吸取 25mL 0.1mol/L 氨水溶液及 25mL 0.1mol/L NH_4Cl 溶液，配成缓冲溶液，用酸度计测定 pH（与事先计算的 pH 相比较）。

在上面配好的缓冲溶液中，加入 0.5mL 0.1mol/L HCl（约 10 滴）后，用酸度计测定其 pH，再加 1mL 0.1mol/L NaOH（约 20 滴），再测定其 pH（与计算值相比较）。

缓 冲 溶 液	pH 计算值	pH 测定值
25mL 0.1mol/L $NH_3 \cdot H_2O$ 25mL 0.1mol/L NH_4Cl		
加入 0.5mL 0.1mol/L HCl		
加入 1mL 0.1mol/L NaOH		

3. 盐类的水解

（1）取小试管 4 支，分别注入 5 滴 1mol/L Na_2CO_3、$FeCl_3$、NaCl、NH_4Ac 溶液，用 pH 试纸测定其酸碱性。指出哪些盐类发生水解反应，写出离子反应式。

（2）将少量 $SbCl_3$ 固体〔或取 0.1mol/L $Bi(NO_3)_3$ 溶液 2 滴〕加到盛有约 1mL 水的试管中，有何现象发生？用 pH 试纸试其酸碱性，滴加 6mol/L HCl 使溶液刚好澄清，再加水稀释又有何现象？用平衡移动原理解释这一系列现象。

4. 沉淀的生成和溶解　取两支小试管各加入 1mL 0.1mol/L $CaCl_2$ 溶液，在一试管内加 0.1mol/L $(NH_4)_2C_2O_4$ 溶液 5 滴，在另一试管内加 0.1mol/L $H_2C_2O_4$ 溶液 5 滴，比较其结果。在加有 $(NH_4)_2C_2O_4$ 的试管内，再加 6mol/L HCl 溶液 5 滴，有何现象产生？再在此试管内加稍过量的 6mol/L $NH_3 \cdot H_2O$，有何现象？为了断定此沉淀是 $Ca(OH)_2$ 还是 CaC_2O_4 可另取一个试管，加同量的 $CaCl_2$ 及 $NH_3 \cdot H_2O$ 溶液，有何现象？与上面的试管比较，你能否断定上面的沉淀是什么？

实 验 指 导

预习要求 •••

1. 什么叫同离子效应？本实验通过哪几个实验验证同离子效应的存在？

2. 在实验室中若想得到较高浓度的 S^{2-} 离子溶液，H_2S 和 Na_2S 水溶液你选择哪一种？如果只有 H_2S 水溶液，如何使 S^{2-} 离子浓度增大？

3. 缓冲溶液具有哪些性能？

4. 本实验将各为 3mL 0.1mol/L 的 HAc 和 NaAc 溶液相混合配制成 HAc–NaAc 缓冲溶液，你能根据 HAc 的电离平衡常数估计出该缓冲溶液的 pH 吗？

5. 怎样抑制或促进水解反应？

6. 根据平衡移动的原理，在沉淀的生成和溶解实验中，你能否预测在什么条件下（酸性还是碱性）有利于 CaC_2O_4 沉淀？

基本操作 •••

1. 进一步练习使用 pH 计。

2. 熟练使用滴定管和移液管。

3. 学会使用点滴板。

注意事项 •••

1. 在使用点滴板之前，须将其清洗干净，并且不要用手直接拿取 pH 试纸，以防污染。

2. "缓冲溶液的缓冲性能" 一项中，也可每人各配 1 份缓冲溶液，分别加 0.5mL 0.1mol/L 的 HCl 和 NaOH 后，再分别测定其 pH。

3. Bi^{3+} 离子的水解实验，必须用固体或浓度较高的 $Bi(NO_3)_3$ 溶液做试剂才可使实验现象明显。

报告格式 •••

1. 目的

2. 原理

3. 列表说明实验内容、现象、原理和结论

实验后思考 •••

1. 欲配制 pH 为 3 的缓冲溶液，已知有下列物质，问选择哪种弱电解质及其盐？

（1）HCOOH $K_a = 1.77 \times 10^{-4}$

（2）HAc $K_a = 1.76 \times 10^{-5}$

（3）$NH_3 \cdot H_2O$ $K_b = 1.79 \times 10^{-5}$

2. 试讨论相同浓度的两种盐溶液，哪一种 pH 较大？

（1）NaAc，NaCN

（2）$NaHCO_3$，Na_2CO_3

3. 利用平衡移动原理，试判断下列哪种难溶盐可以加入某种强酸（如 HNO_3）来溶解。

$MgCO_3$、Ag_3PO_4、$BaSO_4$、$BaSO_3$、$AgCl$、MgC_2O_4

Experiment 3   Electrolyte Solution

◆ Objectives

1. Learn the ionization equilibrium of weak acids, weak bases and principle of the shift of the equilibrium.

2. Practice how to prepare buffer solution and test its properties.

3. Learn the hydrolysis equilibrium of salts and the principle of the shift of the equilibrium.

4. Learn the multiphase ionization equilibrium of some slightly soluble electrolytes and the principle of the shift of the equilibrium.

◆ Principles

1. The ionization equilibrium of weak electrolytes and its shifting   A weak electrolyte dissociates to a lesser extent when one of its ions is present in solution. For example, if sodium acetate, a strong electrolyte, is added to the acetic acid solution, the equilibrium will shift in the direction of forming acetic acid. It is called the common ion effect which makes the degree of ionization decreasing and the acidity of acetic acid decreasing.

2. Buffer solution   Common ion effect has application in preparation of buffer solution. If add NaAc to HAc solution or add NH_4Cl to $NH_3 \cdot H_2O$, the concentration of $[H^+]$ and $[OH^-]$ change a little even if some other acids or bases are added. This kind of solution consisting of a weak acid or a weak base and its salt is named buffer solution.

3. Hydrolysis reaction of salts   The ions of salts react with water to form weak electrolytes and H^+ ion or OH^- ion. Such reaction is called hydrolysis reaction. Adding a product of hydrolysis can restrain the reaction, but diluting the salt solution or improving the temperature can strengthen the reaction.

4. The multiphase equilibrium of slightly soluble electrolyte and the shift of the equilibrium   The slightly soluble electrolyte and its saturated solution make a multiphase equilibrium system. At certain temperature, the product of the powers of concentration of some relative ions is equal to the solubility product of this substance.

If try to add a certain substance which can combine with an ion in this equilibrium system, then the concentration of the ion will decrease. So the product of the powers of concentration is less than the solubility product, with the result of the precipitate dissolving. For example, $CaCO_3$ is soluble in acid because H^+ ion combines with CO_3^{2-} ion to form a weak electrolyte H_2CO_3, which can decompose to water and CO_2.

◆ Equipment and Chemicals

1. Equipment   acidometer, transfer pipette, burette

2. Chemicals

    (1) Solid NH_4Cl

    (2) Acid   HAc 1mol/L, 0.1mol/L; HCl 6mol/L, 0.1mol/L

             citric acid 0.1mol/L; $H_2C_2O_4$ 0.1mol/L

    (3) Base $NH_3 \cdot H_2O$ 2mol/L, 0.1mol/L, 6mol/L; NaOH 0.1mol/L;

    (4) Salt $MgCl_2$ 0.1mol/L, Na_3PO_4 0.1mol/L;

             NH_4Cl saturated solution 0.1mol/L;

             $CaCl_2$ 0.1mol/L; NaAc 0.5mol/L;

             $Cu(NO_3)_2$ 0.1mol/L; Na_2HPO_4 0.2mol/L;

             $(NH_4)_2C_2O_4$ 0.1mol/L; $SbCl_3$ 0.1mol/L;

3. Other   Phenolphthalein indicator, methyl orange indicator, thymol blue indicator.

◆ Procedures

1. The ionization equilibrium of weak bases and the shift of the equilibrium (common ion effect)

(1) Take two test tubes, then add 1mL distilled water, 2 drops of 2mol/L $NH_3 \cdot H_2O$ and 1 drop of phenolphthalein indicator to each tube. Shake them and observe the color of the solutions. Then add a little amount of solid NH_4Cl to one tube. Please compare it with the other tube without NH_4Cl. Explain the reason of change.

(2) Add 2mL 2mol/L HAc solution to a test tube. Observe the color of the solution when 1 drop of methyl orange indicator is added. Then add a little amount of solid NaAc to it. What color appears? Please draw a con-

clusion from that change.

（3）Take two tubes containing 5 drops of $MgCl_2$ solution each. Add 5 drops of saturated NH_4Cl solution to one tube. Then add 5 drops of 2mol/L $NH_3 \cdot H_2O$ to each tube. Observe the different phenomena in these two tubes.

2. Buffer solution

（1）Add 3mL 0.1mol/L HAc and 3mL 0.1mol/L NaAc to a test tube to prepare a buffer solution HAc − NaAc. Mix 5 drops of thymol blue indicator with it. Observe the color of the solution. Then divide the solution into three tubes（a,b and c）. Add 5 drops of 0.1mol/L HCl to tube a,add 5 drops of 0.1mol/L NaAc to tube b, add 5 drops of H_2O to tube c. Observe the color change. Then add excessive amount of 0.1mol/L HCl to tube a, add excessive amount of 0.1mol/L NaOH to tube b. Pay attention to the color change. Then compare with the color in tube c. Please draw a conclusion from above.

The range of the color change of thymol blue indicator is as follows：

pH	color
less than 2.8	red
8 ~9.6	yellow
more than 9.6	blue

（2）Prepare the buffer solution and observe the range of the color change of the indicator. Prepare the buffer solutions（pH =2.2 ~5.0）according to the data shown in the table below：

pH	0.1mol/L citric acid solution,ml	0.2mol/L Na_2HPO_4 solution,ml
2.2	19.60	0.40
3.0	15.89	4.11
4.0	12.29	7.71
4.4	11.29	8.71
5.0	9.70	10.30

Add 4mL of the prepared buffer solutions mentioned above to five test tubes respectively. Line up these tubes according to the sequence with the value of pH increasing. Then add one drop of methyl orange indicator to each tube. Observe the color change after shaking and tell the range of the color change of methyl orange（The range of color change of methyl orange is pH:3.0 ~4.4,red→orange）.

（3）The property of the buffer solution

Take 25mL 0.1mol/L $NH_3 \cdot H_2O$ and 25mL 0.1mol/L NH_4Cl with transfer pipette to prepare a buffer solution. Determine the value of pH with acidometer（compare the determined value with the calculated value）.

Add 0.5mL 0.1mol/L HCl（about 10 drops）to the buffer solution above,and determine the value of pH with acidometer. Then add 1mL 0.1mol/L NaOH（about 20 drops）and determine the value of pH.

Buffer solution	Determined pH	Calculated pH
25mL 0.1mol/L $NH_3 \cdot H_2O$		
25mL 0.1mol/L NH_4Cl		
Add 0.5mL 0.1mol/L HCl		
Then add 1mL 0.1mol/L NaOH		

3. Hydrolysis of salts

（1）Add 5 drops of 1mol/L Na_2CO_3, $FeCl_3$, $NaCl$, NH_4Ac to each of four test tubes respectively. Determine the value of pH of the solution with pH test paper. Tell which salts are hydrolytic and write out the reaction equation.

（2）Add a little amount of solid $SbCl_3$（or 2 drops of 0.1mol/L $Bi(NO_3)_3$ solution）to a test tube containing 1mL water. What happens? Determine its value of pH with pH test paper. Then add 6mol/L HCl until the solution just turns transparent. Dilute the solution with water. What happens then? Explain the series of phenomena with the principle of shift of the equilibrium.

4. Forming and dissolving the precipitates

Add 1mL 0.1mol/L $CaCl_2$ to two test tubes respectively. Then add 5 drops of 0.1mol/L $(NH_4)_2C_2O_4$ to one tube, and 5 drops of 0.1mol/L $H_2C_2O_4$ solution to another tube. Compare the changes in two tubes. After that, if 5 drops of HCl solution (6mol/L) is added to the first tube, what happens? Then a little excessive amount of 6mol/L $NH_3 \cdot H_2O$ is added, what happens again? In order to judge the precipitate is $Ca(OH)_2$ or CaC_2O_4, take another tube and add equal amount of $CaCl_2$ and $NH_3 \cdot H_2O$. According to the phenomena in different test tubes, can you tell what the precipitate is?

◆ Instructions

1. Requirements

（1）What is common ion effect? Which experiment can test the common ion effect?

（2）To prepare a concentrated S^{2-} solution, which aqueous solution will you choose, H_2S or Na_2S? If only H_2S aqueous solution is offered, how to increase the concentration of S^{2-}?

（3）What is buffer solution? What properties does the buffer solution have?

（4）In this experiment, buffer solution HAc−NaAc is made up of 3mL 0.1mol/L HAc and NaAc. Can you evaluate the value of pH of the buffer solution according to the ionization equilibrium constant of HAc?

（5）What is the hydrolysis reaction of salt? How to restrain or strengthen this kind of reaction?

（6）According to the principle of the shift of the equilibrium, can you forecast the best condition for the forming of the precipitate CaC_2O_4 in the experiment above?

2. Operation

（1）Practice to use the acidometer

（2）Practice the operation of burette and transfer pipette

（3）Learn how to use the dropping board.

3. Notes

（1）To prevent from pollution, the dropping board should be cleaned with water before used and pH test paper can't be taken with hands directly.

（2）Every student can prepare a buffer solution in the experiment of "The properties of the buffer solution". Then add 0.5mL 0.1mol/L HCl and NaOH respectively. Finally, determine the value of pH of each sample.

（3）In the experiment of hydrolysis reaction of Bi^{3+} ion, solid $Bi(NO_3)_3$ or concentrated $Bi(NO_3)_3$ solution should be used in order to observe the phenomena obviously.

4. Report Format

（1）Objectives

（2）Principles

（3）Procedures

5. Questions

（1）To prepare a buffer solution（pH =3），which is the best choice in the following weak electrolytes and their salts?

（a）HCOOH K_a =1.77 × 10^{-4}

（b）HAc K_a =1.76 × 10^{-5}

（c）NH$_3$·H$_2$O K_b =1.79 × 10^{-5}

（2）The solutions in the same couple are in the same concentration，which has a higher value of pH?

（a）NaAc，NaCN

（b）NaHCO$_3$，Na$_2$CO$_3$

（3）Try to judge which slightly soluble salt can be soluble in a certain strong acid（such as HNO$_3$）according to the principle of shift of the equilibrium.

MgCO$_3$，Ag$_3$PO$_4$，BaSO$_4$，BaSO$_3$，AgCl，MgC$_2$O$_4$

实验四　沉淀平衡

PPT

目的要求 ●●●

1. 了解沉淀平衡及沉淀平衡的移动。

2. 根据溶度积规则判断

（1）沉淀的生成和溶解。

（2）沉淀的转化和分步沉淀。

3. 测定溶度积。

实验原理 ●●●

1. 在难溶盐的饱和溶液中，未溶解的固体与溶解后形成的离子间存在着平衡，若以 AB 代表难溶盐，A$^+$、B$^-$代表溶解后的离子，它们之间存在下列平衡：

$$AB(s) \rightleftharpoons A^+(aq) + B^-(aq)$$

利用沉淀的生成可以将有关离子从溶液中除去，但不可能完全除去。

在沉淀平衡中，若增加 A$^+$或 B$^-$的浓度，平衡向生成沉淀的方向移动，有沉淀析出，这种现象叫同离子效应。

据溶度积能判断沉淀的生成与溶解，当[A$^+$][B$^-$] > K_{sp}，则有沉淀析出；[A$^+$][B$^-$] = K_{sp}，溶液达到饱和，但仍无沉淀析出；[A$^+$][B$^-$] < K_{sp}，溶液未饱和，没有沉淀析出。

实验中有关难溶电解质的 K_{sp} 如下：

难溶电解质	Pb(SCN)$_2$	PbCl$_2$	PbI$_2$	PbCrO$_4$	CuS	PbS	Ag$_2$CrO$_4$
K_{sp}	2 × 10^{-5}	1.6 × 10^{-6}	7.1 × 10^{-9}	1.8 × 10^{-14}	1.3 × 10^{-36}	8.8 × 10^{-29}	1.1 × 10^{-12}

2. 硫代乙酰胺的分子结构式为 $CH_3\overset{\displaystyle S}{\overset{\|}{C}}$—NH$_2$，水解后产生 H$_2$S，与 PbCl$_2$ 沉淀的反应如下：

$$CH_3CSNH_2 + 2H_2O \longrightarrow CH_3COONH_4 + H_2S$$

$$PbCl_2(s) \Longleftrightarrow 2Cl^- + Pb^{2+}$$

$$H_2S \Longleftrightarrow 2H^+ + S^{2-}$$

$$Pb^{2+} + S^{2-} \longrightarrow PbS(S)（黑色）$$

3. 如果在溶液中有两种或两种以上的离子都可以与同一种沉淀剂反应生成难溶盐，沉淀的先后次序是根据所需沉淀剂离子浓度的大小而定。所需沉淀剂离子浓度小的先沉淀出来，所需沉淀剂离子浓度大的后沉淀出来，这种先后沉淀的现象，称为分步沉淀。

使一种难溶电解质转为另一种难溶电解质的过程称为沉淀的转化，一般说来，溶解度较大的难溶电解质容易转化为溶解度较小的难溶电解质。

仪器和药品

1. 仪器 离心机、烧杯。

2. 药品

（1）酸 HNO_3 6mol/L。

（2）碱 NaOH 0.2mol/L；$NH_3 \cdot H_2O$ 2mol/L。

（3）盐 $Pb(NO_3)_2$ 0.1mol/L，0.001mol/L；KI 0.1mol/L，0.001mol/L；NH_4Cl 1mol/L；

NH_4CNS 0.5mol/L；$FeCl_3$ 0.1mol/L；K_2CrO_4 0.1mol/L；

Na_2S 0.1mol/L；Na_2CO_3 0.1mol/L；NaCl 0.1mol/L；

$MgCl_2$ 0.2mol/L；$(NH_4)_2C_2O_4$ 饱和溶液；$AgNO_3$ 0.1mol/L；

$CuSO_4$ 0.1mol/L；$CaCl_2$ 0.1mol/L；$BaCl_2$ 0.3mol/L；

硫代乙酰胺溶液。

3. 其他 pH 试纸 pH 5.5～9.0。

实验内容

1. 沉淀平衡与同离子效应

（1）沉淀平衡 取 0.1mol/L $Pb(NO_3)_2$ 溶液 10 滴，加 0.5mol/L 硫氰酸铵溶液至沉淀完全，振荡试管（由于 $Pb(SCN)_2$ 容易形成过饱和溶液，可用玻棒摩擦试管内壁，或剧烈摇动试管）。离心分离，在离心液中加 0.1mol/L K_2CrO_4 溶液，振荡试管，有什么现象？试说明在沉淀移去后离心液中是否有 Pb^{2+} 存在？

（2）同离子效应 在试管中加 1mL 饱和 PbI_2 溶液，然后加 5 滴 0.1mol/L KI 溶液，振荡片刻，观察有何现象产生？为什么？

2. 溶度积规则应用

（1）沉淀的生成

①在试管中加 1mL 0.1mol/L $Pb(NO_3)_2$ 溶液，然后加 1mL 0.1mol/L KI 溶液，观察有无沉淀生成。试以溶度积规则解释之。

②在试管中加 1mL 0.001mol/L $Pb(NO_3)_2$ 溶液，然后加 1mL 0.001mol/L KI 溶液，观察有无沉淀生成。试以溶度积规则解释之。

③在离心管中加 2 滴 0.1mol/L Na_2S 溶液和 5 滴 0.1mol/L K_2CrO_4 溶液。加 5mL 蒸馏水稀释，再加 5 滴 0.1mol/L $Pb(NO_3)_2$ 溶液，观察首先生成的沉淀是黑色还是黄色？离心分离，再向离心液中滴加 0.1mol/L $Pb(NO_3)_2$ 溶液，会出现什么颜色的沉淀？根据有关溶度积数据加以说明。

（2）沉淀的溶解

①取 0.3mol/L $BaCl_2$ 溶液 5 滴，加饱和草酸铵溶液 3 滴，此时有白色沉淀生成，离心分离，弃去溶液，在沉淀上滴加 6mol/L HCl 溶液，有何现象？写出反应式。

②取 0.1mol/L $AgNO_3$ 溶液 10 滴，加 0.1mol/L NaCl 溶液 10 滴，离心分离，弃去溶液，在沉淀上滴加 2mol/L 氨水溶液，有何现象？写出反应式。

③取 0.1mol/L $FeCl_3$ 溶液 5 滴，加 0.2mol/L NaOH 溶液 5 滴，生成 $Fe(OH)_3$ 沉淀；另取 0.1mol/L $CaCl_2$ 溶液 5 滴，加 0.1mol/L Na_2CO_3 溶液 5 滴，得 $CaCO_3$ 沉淀。分别在沉淀上滴加 6mol/L HCl，观察它们的现象。写出反应式。

④在试管中加 10 滴 0.2mol/L $MgCl_2$ 溶液，再滴加 2mol/L 的氨水，观察有何现象？然后再滴加 1mol/L NH_4Cl 溶液，又有何现象发生？写出反应式。

⑤在置有 5 滴 0.1mol/L $CuSO_4$ 的试管中加 0.1mol/L 的 Na_2S 溶液 5 滴，观察有何现象？然后向该试管中滴加 10 滴 6mol/L HNO_3 溶液，并微热之，观察有何现象？写出反应式。

（3）沉淀的转化

①取 0.1mol/L $Pb(NO_3)_2$ 溶液 5 滴，加 3 滴 0.1mol/L NaCl 溶液，有白色沉淀生成，再加 5 滴硫代乙酰胺溶液，水浴加热，有何现象，为什么？

②在置有 10 滴 0.1mol/L $AgNO_3$ 溶液的试管中，加 10 滴 0.1mol/L K_2CrO_4 溶液，然后滴加 0.1mol/L NaCl 溶液 10 滴，观察有何现象产生。写出反应式。

3. 氢氧化镁溶度积的估算 取 50mL 烧杯 1 只，加 0.2mol/L $MgCl_2$ 溶液 25mL，烧杯底部衬一黑纸。在 $MgCl_2$ 溶液中逐滴滴入 0.2mol/L 氢氧化钠溶液，并不断搅拌，直到开始有沉淀产生（在强光下观察），氢氧化钠溶液不能过量，为什么？放置，用 pH 试纸测定溶液的 pH，计算 $[OH^-]$ 和 K_{sp}。

实 验 指 导

预习要求

1. 难溶电解质与弱电解质在性质上有哪些相同与不同之点？要区别电离度和溶解度的概念。

2. 沉淀平衡与弱电解质的电离平衡有哪些相同的地方？

3. 沉淀平衡中的同离子效应与电离平衡中的同离子效应是否相同？

4. 沉淀生成的条件是什么？

5. 什么叫分步沉淀？怎样根据溶度积的计算来判断本实验中沉淀先后次序？

6. 沉淀的溶解有哪几种方法？

7. 书写沉淀反应式时应注意什么？

8. 下列物质中，哪些难溶于水？

$Pb(Ac)_2$，$Pb(SCN)_2$，$Pb(NO_3)_2$，$PbCrO_4$，PbI_2，$BaCl_2$，BaC_2O_4，$AgCl$，$[Ag(NH_3)_2]Cl$，Ag_2CrO_4，$FeCl_3$，$Fe(OH)_3$，$CaCO_3$，Na_2CO_3，$Mg(OH)_2$，$CuSO_4$，CuS。

9. 能否通过比较 $PbCl_2$，PbI_2，$PbSO_4$，$PbCrO_4$，PbS 的 K_{sp} 大小来说明有关沉淀转化的原因，为什么？

基本操作

1. 学习离心分离操作。

2. 巩固水浴加热。

注意事项 •••

1. 离心机的使用及注意事项

（1）记住自己放入的位置。

（2）离心管应对称放置，以防止由于重量不均衡引起振动而造成轴的磨损。只有一份溶液需离心时，应再取一支空白离心管，加入与试样体积相同的蒸馏水，与试样离心管一起对称放入离心机进行离心。

（3）停止离心操作时，不可用手去按住离心机的轴，应让其自然停止转动。

2. 溶液和沉淀分离的操作。取一毛细吸管，先捏紧其橡皮头，然后插入试管中，插入的深度以尖端接近沉淀而不接触沉淀为限。然后慢慢放松橡皮头，吸出溶液移去，这样反复数次尽可能把溶液移去，留下沉淀。

3. 水浴加热是在一定温度范围内进行较长时间加热的一种方法。此外，还有油浴、沙浴、蒸气浴等。水浴加热可用铜制水锅，也可用烧杯代替。在烧杯中放入一定量的水，在电炉（或其他热源）上加热后即为水浴。将需进行水浴加热的样品试管放入其中即可。

4. 离心管是用来进行离心分离的试管，标有刻度且管底玻璃较薄，整个试管的厚度不匀。所以离心管不能用直火加热，只能在水浴中加热。

报告格式 •••

1. 目的

2. 内容

例：

$$Pb^{2+} + SCN^- \longrightarrow Pb(SCN)_2 \downarrow \xrightarrow[\text{分离}]{\text{离心}} \begin{cases} 溶液 + CrO_4^{2-} \longrightarrow PbCrO_4 \downarrow 黄 \\ \\ 沉淀\ Pb(SCN)_4 \downarrow \end{cases}$$

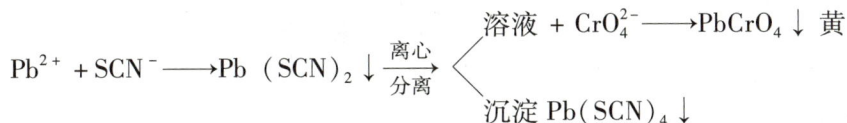

结论：说明离心液中仍含有 Pb^{2+} 离子。

实验后思考 •••

1. 能否应用吕·查德里原理来确定改变沉淀反应的条件所导致的结果？能否运用这一原理作定量的推测？

2. 反应条件（如温度和浓度）怎样影响沉淀反应的自发性和完成程度？

3. 用同时含有 Fe^{2+} 和 Cu^{2+} 的 $PbCl_2$ 溶液如何制取 $Pb(Ac)_2$？

Experiment 4　Precipitation Equilibrium

◆ Objectives

1. Understand the precipitation equilibrium and the shift of the equilibrium.

2. On the rule of the solubility product, please judge：

（1）The formation and dissolution of the precipitates.

（2）Transform of the precipitates and fractional precipitation.

3. Determine the value of the solubility product.

◆ Principles

1. In the saturated solution of some slightly soluble salts, there is an equilibrium between the solid phase that doesn't dissolve and its ions that are formed after part of the solid dissolving. Here, AB denotes slightly soluble salt, A^+ and B^- denote ions that are formed after part of AB dissolving. There exists an equilibrium as following:

$$AB(s) \Longrightarrow A^+ + B^- \text{(aqueous)}$$

So with the formation of the precipitates, most of the relative ions can be separated from the solution, but it is impossible to remove all the ions.

In the above equilibrium, if the concentration of $[A^+]$ or $[B^-]$ is increased, the equilibrium shifts to the formation of precipitate AB. This is called "common ion effect".

Solubility product can be regarded as the standard rule to judge about the formation or the dissolution of precipitates. When $[A^+][B^-] > K_{sp}$, the precipitate is forming; when $[A^+][B^-] = K_{sp}$, the solution just turns saturated, so the precipitate is not forming yet; When $[A^+][B^-] < K_{sp}$, the solution is unsaturated, so no precipitate appears.

K_{sp} of some relative slightly soluble electrolytes are listed is the table below:

slightly soluble electrolyte	$Pb(SCN)_2$	$PbCl_2$	PbI_2	$PbCrO_4$	CuS	PbS	Ag_2CrO_4
K_{sp}	2×10^{-5}	1.6×10^{-6}	7.1×10^{-9}	1.8×10^{-14}	1.3×10^{-36}	8.8×10^{-29}	1.1×10^{-12}

2. The molecule structure formula of thioacetamide is $CH_3\overset{\overset{S}{\|}}{C}-NH_2$. It hydrolyzes to form H_2S, so can react with precipitate $PbCl_2$:

$$CH_3CSNH_2 + 2H_2O \longrightarrow CH_3COONH_4 + H_2S$$
$$PbCl_2(s) \Longrightarrow 2Cl^- + Pb^{2+}$$
$$H_2S \Longrightarrow 2H^+ + S^{2-}$$
$$Pb^{2+} + S^{2-} \longrightarrow PbS(s, black)$$

3. If there are two kinds or more than two kinds of ions which can react with a precipitating reagent to form slightly soluble salts, the sequence of the formation of precipitates is dependent on the needed concentration of the precipitating reagent ions. The lowest concentration of the ion needed means the corresponding precipitate will form first, then another precipitate will form with the greater concentration of the ion needed. We call it fractional precipitation.

The process of transforming a slightly soluble electrolyte to another is called precipitate transform. As a common fact, the slightly soluble electrolyte with higher value in solubility can easily be transformed to those with lower value in solubility.

◆ Equipment and Chemicals

1. Equipment Centrifugal machine, beaker.

2. Chemicals

(1) Acid HNO_3 6mol/L

（2）Base　NaOH 0. 2mol/L;NH$_3$ · H$_2$O 2mol/L;

（3）Salts　Pb(NO$_3$)$_2$ 0. 1mol/L,0. 001mol/L;KI 0. 1mol/L,0. 001mol/L;

　　　　　NH$_4$Cl 1mol/L;NH$_4$CNS 0. 5mol/L;FeCl$_3$ 0. 1mol/L

　　　　　K$_2$CrO$_4$ 0. 1mol/L; Na$_2$S 0. 1mol/L;Na$_2$CO$_3$ 0. 1mol/L;NaCl 0. 1mol/L;

　　　　　MgCl$_2$ 0. 2mol/L;(NH$_4$)$_2$C$_2$O$_4$ saturated solution;

　　　　　AgNO$_3$ 0. 1mol/L;CuSO$_4$ 0. 1mol/L;CaCl$_2$ 0. 1mol/L

　　　　　BaCl$_2$ 0. 3mol/L;thioacetamide solution

3. Others　pH test paper,pH 5. 5 ~9. 0

◆ Procedures

1. Precipitation equilibrium and common ion effect

（1）Precipitation equilibrium　Add 0. 5mol/L NH$_4$CNS solution to 10 drops of 0. 1mol/L Pb(NO$_3$)$_2$ solution until precipitate is forming completely. Shaking the tube (Because Pb(SCN)$_2$ easily occurs over-saturated). We can rub the inner wall of the test tube with a glass rod, also we can shake the tube tempestuously. Separate the precipitate with centrifugal machine. Then add 0. 1mol/L K$_2$CrO$_4$ to the above centrifugal solution. What happens? Try to illustrate if there exists Pb^{2+} in the centrifugal solution after separating the precipitate.

（2）Common ion effect　Add 1mL saturated PbI$_2$ solution to a test tube,then add 5 drops of 0. 1mol/L KI solution. Shake the tube for a moment. What happens?

Explain the reason.

2. The application of the solubility product

（1）Formation of the precipitates

（a）Add 1mL 0. 1mol/L Pb(NO$_3$)$_2$ solution to a test tube,then add 1mL 0. 1mol/L KI solution. Observe if there are precipitates forming? Explain the reason.

（b）Add 1mL 0. 001mol/L Pb(NO$_3$)$_2$ solution to a test tube,then add 1mL 0. 001mol/L KI solution. Observe if there are precipitates forming? Explain the reason.

（c）Add 2 drops of 0. 1mol/L Na$_2$S solution and 5 drops of 0. 1mol/L K$_2$CrO$_4$ to a centrifugal tube. Dilute with 5mL distilled water. Then add 5 drops of 0. 1mol/L Pb(NO$_3$)$_2$ solution. Observe the color of the precipitate first appears (black or yellow?) Separate with centrifugal machine. Then drop 0. 1mol/L Pb(NO$_3$)$_2$ solution to the centrifugal solution above which was separated from the precipitate. What happens? (Observe the color of the precipitate). Illustrate the phenomena on the value of the solubility product.

（2）Dissolution of the precipitates

（a）Mix 3 drops of saturated oxalate amine solution with 5 drops of 0. 3mol/L BaCl$_2$ solution. White precipitate is forming. Separate with centrifugal machine and then discard the solution. Add 6mol/L HCl solution to the precipitate. What happens? Write out the reaction equation.

（b）Mix 10 drops of 0. 1mol/L NaCl solution with 10 drops of 0. 1mol/L AgNO$_3$ solution. Separate the precipitate with centrifugal machine. Discard the solution. Then add 2mol/L NH$_3$ · H$_2$O to the precipitate. What happens? Write out the reaction equation.

（c）Mix 5 drops of 0. 2mol/L NaOH solution with 5 drops of 0. 1mol/L FeCl$_3$ solution to form precipitate Fe(OH)$_3$;In another tube,mix 5 drops of 0. 1mol/L Na$_2$CO$_3$ with 5 drops of 0. 1mol/L CaCl$_2$ solution to form precipitate $CaCO_3$. Add 6mol/L HCl drop by drop to the each precipitate mentioned above. What happens? Write out the reaction equation.

(d) Add 2mol/L $NH_3 \cdot H_2O$ drop by drop to a test tube containing 10 drops of 0. 2mol/L $MgCl_2$ solution. Then with the addition of 1mol/L NH_4Cl solution, what happens? Write out the reaction equation.

(e) Add 5 drops of 0. 1mol/L Na_2S solution to a test tube containing 5 drops of 0. 1mol/L $CuSO_4$ solution. What happens? Then add 10 drops of 6mol/L HNO_3 solution and heat the solution slightly. What happens then? Write out the reaction equation.

(3) Precipitate transform

(a) Add 3 drops of 0. 1mol/L NaCl solution to 5 drops of 0. 1mol/L $Pb(NO_3)_2$ solution. White precipitate is forming. Then add 5 drops of thioacetamide solution and heat it over water bath. What happens? Explain the reason.

(b) Add 10 drops of 0. 1mol/L K_2CrO_4 solution to 10 drops of 0. 1mol/L $AgNO_3$ solution. Then with the addition of 0. 1mol/L NaCl solution, what happens? Write out the reaction equation.

3. Predict the value of the solubility product of $Mg(OH)_2$ Add 25mL of 0. 2mol/L $MgCl_2$ to a 50mL beaker which is underlaid a piece of black paper at the bottom. Then add 0. 2mol/L NaOH solution to $MgCl_2$ solution drop by drop. Stir constantly until precipitate is forming. (Please observe on the blazing sunshine directly). Notice that NaOH solution should not be excessive. (Why?) Determine the value of pH of the solution with pH test paper. Calculate $[OH^-]$ and K_{sp}.

◆ Instructions

1. Requirements

(1) Tell the different and the common properties between the slightly soluble electrolyte and the weak electrolyte. Notice to distinguish the concepts of degree of ionization from solubility.

(2) Tell the common points between the precipitation equilibrium and the ionization equilibrium.

(3) Is the common ion effect in the precipitation equilibrium as same as that in the ionization equilibrium?

(4) What is the condition of the formation of the precipitate?

(5) What is fractional precipitation? How to judge about the sequence of the formation of the precipitates in a certain experiment based on the value of the solubility product?

(6) Please tell several methods of dissolving the precipitates.

(7) What should be noticed when writing the precipitation reaction equation?

(8) Among the following substances, which ones will be precipitated from the water?

$Pb(Ac)_2$　$Pb(SCN)_2$　$Pb(NO_3)_2$　$PbCrO_4$　PbI_2　$BaII_2$　BaC_2O_4　AgCl

$[Ag(NH_3)_2]Cl$　Ag_2CrO_4　$FeCl_3$　$Fe(OH)_3$　$CaCO_3$　Na_2CO_3　$Mg(OH)_2$　$CuSO_4$　CuS

(9) Can we explain the precipitate transform by comparing the K_{sp} of $PbCl_2, PbI_2, PbSO_4, PbCrO_4, PbS$?

2. Operation

(1) Learn to use centrifugal machine.

(2) Practice to heat over the water bath.

3. Notes

(1) How to use the centrifugal machine.

(a) Remember the centrifugal tube place.

（b）Centrifugal tube should be placed symmetrically to prevent the friction caused by weight unbalance. When a solution is to be centrifugated, another empty tube should be added the same volume of distilled water and be placed symmetrically to balance the centrifugal machine.

（c）When finishing the separation, please don't stop the axis of the centrifugal machine by your hand but make it stop naturally.

（2）How to separate the precipitate from the solution

Pinch the rubber part of a capillary tube, then insert it into a test tube. Be sure that the tip of the capillary tube is near the surface of the precipitate but not touch it. Then loosen the rubber part, transfer the absorbed solution from the precipitate. Repeat several times to transfer the solution as possible as you can.

（3）Heating over water bath is a kind of method when the sample should be heated for a long time at a certain range of temperature. There are also air bath sand bath and vapor bath. Water bath can be in a copper boiler or in a beaker（Heat some water in a beaker, then put the sample that need to be heated into the beaker）.

（4）Centrifugal tube is marked with graduation. Its bottom part is too thin. Because the thickness of the whole tube is not well - proportioned, it can't be heated directly on fire, but only in the water bath.

4. Report Format

（1）Objectives

（2）Procedures

$$\text{e. g. } Pb^{2+} + SCN^- \longrightarrow Pb(SCN)_2 \downarrow \xrightarrow{\text{separate}} \begin{cases} \text{solution} & + CrO_4^{2-} \longrightarrow PbCrO_4 \downarrow \text{ (yellow)} \\ \text{precipitate } Pb(SCN)_4 \downarrow \end{cases}$$

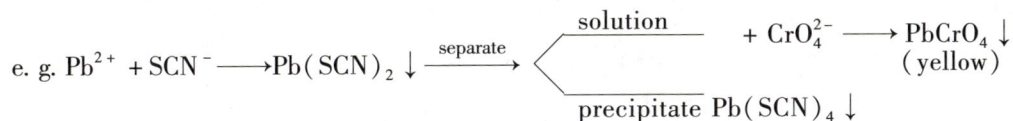

Conclusion: There exists Pb^{2+} ion in the solution after separation.

5. Questions

（1）Can we confirm the results if changing the precipitation reaction condition with Le Chatelier's principle? Can this principle be used to make some quantitative speculations?

（2）How do the reaction conditions（such as temperature and concentration）affect the spontaneity and degree of completion of the precipitation reaction?

（3）How to get $Pb(Ac)_2$ from $PbCl_2$ solution which consists of Fe^{2+} and Cu^{2+}?

实验五　氧化还原

PPT

目的要求 •••

1. 理解氧化还原反应的实质，了解常用的氧化剂和还原剂。

2. 通过实验，了解某些金属电极在电极电势表中的位置，从而加深对电极电势物理意义的认识。

3. 应用标准电极电势比较氧化剂和还原剂的相对强弱。

4. 了解浓度、酸度和温度对氧化还原反应的影响。

5. 了解原电池的装置及利用原电池产生电流进行电解。

实验原理 •••

氧化还原反应是物质得失电子的过程，反映在元素的氧化数发生变化。反应中得到电子的物质称为氧化剂，反应后氧化数降低被还原；反应中失去电子的物质称为还原剂，反应后氧化数升高被氧化。氧化还原是同时进行的，其中得失电子数相等。

电极电势是判断氧化剂和还原剂相对强弱的标准，并可用以确定氧化还原反应进行的方向。电极电势表，是各种物质在水溶液中进行氧化还原反应规律性的总结，溶液的浓度、酸度、温度均影响电极电势的数值。一般说，在表中上方的还原态是较强的还原剂，可使其下方的氧化态还原，表下方的氧化态是较强的氧化剂，可使其上方的还原态氧化。

将化学能转化为电能的装置称为原电池。

利用原电池产生的电流，可电解 Na_2SO_4 溶液。

仪器和药品 •••

1. 仪器 铜棒—导线—铜棒，锌棒—导线—铜棒

盐桥（充有饱和 KCl 和琼脂溶液的 U 形管）

2. 药品

（1）固体 Pb 粒；Zn 粒

（2）酸 HNO_3 2mol/L，浓；HCl 1mol/L；H_2SO_4 3mol/L，1mol/L；HAc 6mol/L；H_2S 溶液。

（3）碱 NaOH 40%，6mol/L。

（4）盐 $KMnO_4$ 0.01mol/L；$FeSO_4$ 0.5mol/L；$NaSO_3$ 0.1mol/L；$Na_2C_2O_4$ 0.1mol/L；

$NaBr$ 1mol/L；$K_2Cr_2O_7$ 0.1mol/L；NaI 1mol/L；$Fe_2(SO_4)_3$ 0.025mol/L；

$Pb(NO_3)_2$ 1mol/L；$CuSO_4$ 1mol/L；Na_2SO_4 0.5mol/L；

$MnSO_4$ 0.2mol/L；$AgNO_3$ 0.1mol/L。

3. 其他 红色石蕊试纸；H_2O_2 3%；$CHCl_3$。

实验内容 •••

1. 氧化剂和还原剂

（1）取试管 2 支，各加 5 滴 0.01mol/L $KMnO_4$ 溶液和 3 滴 3mol/L H_2SO_4，然后在第一支试管中加 1～2 滴 3% 的 H_2O_2 溶液，第二支试管中加 2～3 滴 0.5mol/L $FeSO_4$ 溶液，观察现象。指出反应中的氧化剂和还原剂。

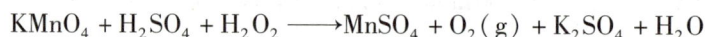

$$KMnO_4 + H_2SO_4 + H_2O_2 \longrightarrow MnSO_4 + O_2(g) + K_2SO_4 + H_2O$$

$$KMnO_4 + H_2SO_4 + FeSO_4 \longrightarrow MnSO_4 + Fe_2(SO_4)_3 + K_2SO_4 + H_2O$$

（2）取试管 1 支，加 3 滴 0.1mol/L $K_2Cr_2O_7$ 溶液，5 滴 3mol/L H_2SO_4，摇匀，再加 H_2S 溶液数滴，观察现象。指出反应中的氧化剂和还原剂。

$$K_2Cr_2O_7 + H_2SO_4 + H_2S \longrightarrow Cr_2(SO_4)_3 + S(s) + KHSO_4 + H_2O$$

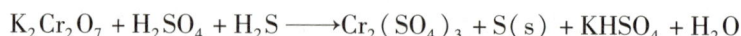

2. 确定锌、铅、铜在电极电势表中的顺序 在分别装有 3mL 1mol/L $Pb(NO_3)_2$ 溶液和 1mol/L $CuSO_4$ 溶液的两支试管中，各加入表面洁净的 Zn 粒，观察有何现象？

以表面洁净的 Pb 片代替 Zn 粒，分别与 1mol/L $ZnSO_4$ 和 $CuSO_4$ 溶液反应，观察有何变化？写出反应式，并确定 Zn 电极、Pb 电极、Cu 电极在电极电势表中的位置。解释之。

3. 浓度、酸度和温度对氧化还原反应的影响

（1）浓度对氧化还原反应的影响 往两支装有少数锌粒的试管中，分别加 2mL 浓 HNO_3 和 2mol/L HNO_3 溶液。观察所发生的现象，（a）它们的反应速度有何不同？（b）它们的反应产物有何不同？浓 HNO_3 被还原后的主要产物可通过观察它的颜色来判断。稀 HNO_3 的还原产物可用检验溶液中是否有

NH_4^+ 生成来确定。

检验 NH_4^+ 方法（气室法）：将 5 滴被检液置于一表面皿的中心，再加 3 滴 6mol/L NaOH 溶液，混匀，在另一块较小的表面皿中心黏附一小块红色石蕊试纸，把它盖在大的表面皿上做成气室。放置 10 分钟，如红色石蕊试纸变蓝，则表示有 NH_4^+ 存在。

（2）酸度对氧化还原反应的影响 在两支各盛有 0.5mL 1mol/L NaBr 溶液的试管中，分别加 10 滴 3mol/L H_2SO_4 和 6mol/L HAc 溶液，然后往两支试管中各加 1 滴 0.01mol/L $KMnO_4$ 溶液。观察并比较两个试管中紫色溶液褪色的快慢。写出反应式，并加以解释。

（3）温度对氧化还原反应的影响 在两支试管中分别加 2mL 0.1mol/L $Na_2C_2O_4$、0.5mL 3mol/L H_2SO_4 和 1 滴 0.01mol/L $KMnO_4$ 溶液，混匀。将其中一支试管放入 80℃ 的水浴中加热，另一支不加热，观察两管褪色的快慢。写出方程式，并加以解释。

（4）酸度、碱度对氧化还原反应产物的影响 取小试管 3 支，各加 10 滴 0.1mol/L Na_2SO_3 溶液，再分别加 10 滴 1mol/L H_2SO_4、蒸馏水和 40% NaOH 溶液，摇匀后，再各加 3 滴 0.01mol/L $KMnO_4$ 溶液，观察现象。$KMnO_4$ 在酸性、中性和强碱性介质中的还原产物分别为 Mn^{2+}、MnO_2 和 MnO_4^{2-}，试写出上述反应的反应式。

4. 催化剂对氧化还原反应的影响 取 5mL 0.2mol/L $MnSO_4$ 溶液和 1mL 3mol/L H_2SO_4 溶液在试管内充分振摇，并加入一小匙过二硫酸铵固体，充分振摇溶解后分成两份，往一份溶液中加 1～2 滴 0.1mol/L $AgNO_3$，静置片刻，观察溶液颜色有何变化？写出反应式，与没有加 $AgNO_3$ 溶液的那份溶液比较，反应情况有何不同？

5. 选择氧化剂 在含有 NaBr、NaI 的混合液中，要使 I^- 氧化为 I_2，又不使 Br^- 氧化，在常用的氧化剂 $Fe_2(SO_4)_3$ 和 $KMnO_4$ 中，选择哪一种能符合要求？

（1）取小试管 2 支，分别加 10 滴 1mol/L NaBr、NaI 溶液，各加 10 滴 3mol/L H_2SO_4，1mL $CHCl_3$，然后分别加入 2～3 滴 0.01mol/L $KMnO_4$，振摇。观察各试管中三氯甲烷层的变化。
[I_2 在三氯甲烷中呈粉红色（或紫红色），Br_2 在三氯甲烷中呈橙黄色]

（2）用 $Fe_2(SO_4)_3$ 溶液代替 $KMnO_4$ 溶液，操作同前，观察各试管中又有什么变化？写出反应式。从实验结果确定选择哪一种氧化剂？试解释之。

6. 利用原电池产生的电流电解 Na_2SO_4 溶液 往 1 只 50mL 的烧杯中加 25mL 1mol/L $ZnSO_4$ 溶液，在其中插入锌棒，往另 1 只 50mL 的烧杯中加 25mL 1mol/L $CuSO_4$ 溶液，在其中插入铜棒。按图 3-3 用盐桥连接构成原电池。把两根分别以导线连接在原电池锌棒和铜棒上的铜棒插入装有 50mL 0.1mol/L Na_2SO_4 溶液和 3 滴酚酞的烧杯中，观察连接锌棒的那根铜棒周围的溶液的变化。试解释之。

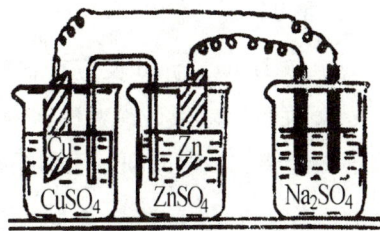
图 3-3 电解硫酸钠溶液的装置

实 验 指 导

1. 写出下列氧化剂和还原剂通常情况下被还原或被氧化的产物。

氧化剂 MnO_4^-（酸性），H_2O_2，CrO_4^{2-}，H_2SO_4（浓），HNO_3（2mol/L），MnO_4^-（中性），Fe^{3+}，Cu^{2+}，$S_2O_8^{2-}$，Cl_2，Br_2，MnO_4^-（碱性），$Cr_2O_7^{2-}$，Pb^{4+}，HNO_3（浓），I_2。

还原剂　H_2O_2，Zn，Br^-，$C_2O_4^{2-}$，Fe^{2+}，S^{2-}，I^-，SO_3^{2-}。

2. 怎样从标准电极电势判断金属的置换反应是否能够进行，并以锌、铅、铜为例说明。

3. 影响电极电势的因素有哪些？

4. 浓度、酸度对氧化还原产物及反应方向有何影响？

5. 怎样装置 $Cu-Zn$ 原电池，盐桥有什么作用？

6. 电解 Na_2SO_4 水溶液时，为什么阴极得不到金属钠？用石墨作电极和以铜作电极在阳极上的反应是否相同？为什么？

7. 完成下列反应

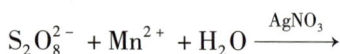

$MnO_4^- + H_2O_2 + H^+ \longrightarrow$

$MnO_4^- + Fe^{2+} + H^+ \longrightarrow$

$Cr_2O_7^{2-} + H_2S + H^+ \longrightarrow$

$Zn + HNO_3（浓）\longrightarrow$

$Zn + HNO_3（2mol/L）\longrightarrow$

$MnO_4^- + Br^- + H^+ \longrightarrow$

$MnO_4^- + C_2O_4^{2-} + H^+ \longrightarrow$

$MnO_4^- + SO_3^{2-} + H^+ \longrightarrow$

$MnO_4^- + SO_3^{2-} + H_2O \longrightarrow$

$MnO_4^- + SO_3^{2-} + OH^- \longrightarrow$

$S_2O_8^{2-} + Mn^{2+} + H_2O \xrightarrow{AgNO_3}$

基本操作

1. 掌握气室法检验 NH_4^+ 的方法。

2. 掌握利用原电池的装置来电解硫酸钠溶液的方法。

注意事项

1. 浓 HNO_3 与 Zn 反应有刺激性气体 NO_2 产生，必须在通风橱进行。

2. 确定锌电极、铅电极、铜电极在电极电势表中的顺序实验，由于置换反应速度较慢，应将试管放置试管架上一段时间，切勿振摇，然后再观察现象。

报告格式（参看"卤素"实验指导）

实验后思考

1. 什么情况下查酸表 $E_A^{\ominus}$？什么情况下查碱表 $E_B^{\ominus}$？

2. 在标准电势表上电势差值大的两电对，其反应速率是否一定很快？

3. 如何利用电极反应（半电池反应）来写氧化还原反应式？

4. 什么情况下用标准电势表来判断反应方向？什么情况下通过 Nernst 方程计算来判断？

Experiment 5　Redox Reaction

◆ Objectives

1. To understand the intrinsic properties of redox reaction,and be acquainted with some common oxidizing and reducing agents.

2. To learn about some metals' location in the table of electrode potentials,and enhance the knowledge about electrode potentials.

3. To compare the relative strengths of reducing and oxidizing agents.

4. To learn about the influence of concentration,acidity and temperature on redox reaction.

5. To learn to assemble a galvanic cell and electrolysis with the galvanic cell.

◆ Principles

A very large class of reactions can be regarded as occurring by the loss of electrons from one substance and gain by another substance. Since electron gain is called reduction and electron loss is oxidation,the joint process is called an oxidation-reduction reaction,or more simply,redox reaction. The substance that supplies electrons (the oxidation number of an atom increases) is the reducing agent and the substance that gains electrons(the oxidation number of an atom decreases) is the oxidizing agent. The reducing agent is to be oxidized while the oxidizing agent is to be reduced. Oxidation and reduction always occur together,and the number of electrons gained by the oxidizing agent must be identical to the number of electrons lost by the reducing agent.

Electrode potentials can be used to judge the relative intensity of oxidizing and reducing agents and determine the possible direction of a given redox reaction. The concentration,pH and emperature have effects on the magnitude of the electrode potential. The table of standard electrode potentials can be used as the summary of the redox reaction rules in aqueous solution. In the table of standard electrode potentials,generally,the reducing agent(written at the right of the reactions) in the upper of the table is astronger reducing agent,which can reduce the oxidizing agents below it. Conversely,the oxidizing agent(written at the left of the reactions) in the bottom of the table is astronger oxidizing agent,which can oxide the reducing agents listed above it.

A galvanic cell is a device designed to produce electron current basing on redox reactions.

Na_2SO_4 solution can be electrolyzed with the electron current produced by a galvanic cell.

◆ Equipment and Chemicals

Equipment

Two metallic Cu plate connected by a metallic wire

A Zn plate and a Cu plate connected by a metallic wire

Salt bridge (a U-shaped tube containing saturated solution of KCl in agar)

Chemicals

Solid： granulated plumbum; granulated zinc

Acid： HNO_3（2mol/L,conc.）; HCl 1mol/L;

　　　　H_2SO_4（3mol/L,1mol/L）; HAc 6mol/L; H_2S solution

Alkali： NaOH（40%,6mol/L）

Salt： $KMnO_4$ 0. 01mol/L； $FeSO_4$ 0. 5mol/L； Na_2SO_3 0. 1mol/L；

$Na_2C_2O_4$ 0. 1mol/L； NaBr 1mol/L； $K_2Cr_2O_7$ 0. 1mol/L；

NaI 1mol/L； $Fe_2(SO_4)_3$ 0. 025mol/L； $Pb(NO_3)_2$ 1mol/L；

$CuSO_4$ 1mol/L； Na_2SO_4 0. 5mol/L； $MnSO_4$ 0. 2mol/L；

$AgNO_3$ 0. 1mol/L

Others：litmus test paper（red）； H_2O_2 3%； $CHCl_3$

◆ Procedures

1. Oxidizing agents and reducing agents

（1）In each of two tubes，mix 5 drops of 0. 01mol/L $KMnO_4$ solution with 3 drops of H_2SO_4（3mol/L）solution. To one tube，add 1～2 drops of H_2O_2（3%），and to another tube，add 2～3 drops of $FeSO_4$（0. 5mol/L）solution. Observe the reactions and point out which are the oxidizing agents and reducing agents.

$$KMnO_4 + H_2SO_4 + H_2O_2 \rightarrow MnSO_4 + O_2 \uparrow + K_2SO_4 + H_2O$$

$$KMnO_4 + H_2SO_4 + FeSO_4 \rightarrow MnSO_4 + Fe_2(SO_4)_3 + K_2SO_4 + H_2O$$

（2）Add 3 drops of 0. 1mol/L $K_2Cr_2O_7$ solution to 5 drops of 3mol/L H_2SO_4 solution，then add a few drops of H_2S solution. Observe the reaction and point out which are the oxidizing agent and reducing agent.

$$K_2Cr_2O_7 + H_2SO_4 + H_2S \rightarrow Cr_2(SO_4)_3 + S \downarrow + KHSO_4 + H_2O$$

2. Locate the sequence of Zn,Pb,Cu in the table of electrode potentials

React granulated zinc with 3mL of 1mol/L $Pb(NO_3)_2$ and 1mol/L $CuSO_4$ solution respectively，and observe.

React granulated plumbum with 3mL of 1mol/L $ZnSO_4$ and $CuSO_4$ solution respectively，and observe.

Write out the reaction equations，and locate the relative sequence of Zn,Cu,Pb in the table of electrode potentials. Explain the relation between the relative sequence and the activities of the metals.

3. Factors influencing redox reactions

（1）The effect of concentration on redox reactions　React granulated zinc with 2mL concentrated HNO_3 and 2mol/L HNO_3 respectively. Observe the reactions，and point out the differences of reaction rate and products. The reduced product of concentrated HNO_3（NO_2）can be detected by its color，while the reduced product of diluted HNO_3（NH_4^+）can be detected by the following test.

Detection of NH_4^+（with a gas chamber）：Place 5 drops of the detected solution and 3 drops of 6mol/L NaOH solution on a watch glass，cover the watch glass with another smaller watch glass，attaching beforehand a piece of moist red litmus paper to the concave side. Then observe the litmus paper within 10 minutes. The blue color of red litmus paper indicates the presence of the NH_4^+ ion in the solution.

（2）The effect of pH on reaction rate　To two portions of 0. 5mL of 1mol/L NaBr solution acidified with 10 drops of 3mol/L H_2SO_4 and 6mol/L HAc solution respectively，add 1 drop of 0. 01mol/L $KMnO_4$ solution. Compare the rate of decolorization of both samples.

Write out the reaction equation，and explain the difference in reaction rate.

（3）The effect of temperature on reaction rate　Into each of two tubes，add 2mL of 0. 1mol/L $Na_2C_2O_4$ solution，0. 5mL of 3mol/L H_2SO_4 solution and 1 drop of 0. 01mol/L $KMnO_4$ solution，shake and mix. Heat one tube in 80℃ water bath，while keep another tube under room temperature. Compare the rate of decolorization of both samples. Write out the reaction equation，and explain the difference between the two solutions.

（4）The effect of pH on redox reaction products　Into three tubes containing 10 drops of $0.1mol/L$ Na_2SO_3, add 10 drops of $1mol/L$ H_2SO_4, distilled water, 40% NaOH solution respectively. Shake and add 3 drops of $0.01mol/L$ $KMnO_4$ solution to the tubes. Observe the reactions and write out the reaction equations. In acid solution MnO_4^- is reduced to the pale pink manganous ion, Mn^{2+}. In neutral solution MnO_4^- is reduced to black, insoluble, solid MnO_2, while in strongly alkaline solution it can be reduced to the +6 state, where it exists as the bright green manganate ion, MnO_4^{2-}.

4. The effect of catalyst on redox reactions　Mix 5mL of $0.2mol/L$ $MnSO_4$ solution and 1mL of $3mol/L$ H_2SO_4 solution, then add a spoon of solid $(NH_4)_2S_2O_8$. Shake constantly to dissolve the solid completely. Divide the solution into two portions. To one portion, add $1\sim2$ drops of $0.1mol/L$ $AgNO_3$ solution, while add nothing to another. Observe and write out the reaction equation. Explain the difference of the two portions of solution.

5. Selection of oxidizing agents　Between the commonly used oxidizing agents, $Fe_2(SO_4)_3$ and $KMnO_4$, choose a oxidizing agent which can oxidize I^- into I_2 while not oxidize Br^- into Br_2 in the solution of NaBr and NaI.

Tests：

（1）Into each of two tubes containing 10 drops of $1mol/L$ NaBr solution and 10 drops of $1mol/L$ NaI solution, add 10 drops of $3mol/L$ H_2SO_4, 1mL of $CHCl_3$ and $2\sim3$ drops of $0.01mol/L$ $KMnO_4$. Shake constantly to mix and observe the color of $CHCl_3$ layer in the two tubes.

The color of I_2 in $CHCl_3$ is pink (or purple), and Br_2 takes out orange color.

（2）Substitute $KMnO_4$ solution with $Fe_2(SO_4)_3$ solution and repeat the same operation above. Observe the color of $CHCl_3$ layer in the two tubes.

Write out the reaction equations and select the appropriate oxidizing agent. Explain your selection.

6. Electrolyze Na_2SO_4 solution with a galvanic cell　Assemble a galvanic cell with a salt bridge, a stripe of Zn in 25mL $1mol/L$ $ZnSO_4$ solution and a stripe of Cu in 25mL $1mol/L$ $CuSO_4$ solution. Immerse two metallic Cu electrodes into a beaker containing 50mL of $0.1mol/L$ Na_2SO_4 solution and 3 drops phenolphthalein. Link the two Cu electrodes with the electrodes of the galvanic cell respectively with metallic wire. Observe the solution around the Cu electrode linked to the stripe of Zn. Explain the phenomena.

◆ Instructions

1. Requirements

（1）Write out the usual reduced products of the following oxidants and oxidized products of the following reductants.

Oxidants：MnO_4^- (in acidic medium), H_2O_2, CrO_4^{2-}, H_2SO_4 (conc.), HNO_3 ($2mol/L$), MnO_4^- (in neutral medium), Fe^{3+}, Cu^{2+}, $S_2O_8^{2-}$, Cl_2, Br_2, MnO_4^- (in alkaline medium), $Cr_2O_7^{2-}$, Pb^{4+}, HNO_3 (concentrated), I_2.

Reductants：H_2O_2, Zn, Br^-, $C_2O_4^{2-}$, Fe^{2+}, S^{2-}, I^-, SO_3^{2-}.

（2）Taking Zn, Pb, Cu as examples to explain how to judge on basis of the standard electrode potentials whether a displace reaction between metals can be carried out?

（3）What factors will affect the electrode potential?

（4）What are the influence of concentration and pH on the redox reaction products and directions?

（5）How to assemble a Cu-Zn galvanic cell? What is the use of the salt bridge?

（6）Why can we not get solid sodium at the cathode during the electrolysis of Na_2SO_4 solution?

（7）Complete the following redox reactions.

$MnO_4^- + H_2O_2 + H^+ \longrightarrow$

$MnO_4^- + Fe^{2+} + H^+ \longrightarrow$

$Cr_2O_7^{2-} + H_2S + H^+ \longrightarrow$

$Zn + HNO_3(conc.) \longrightarrow$

$Zn + HNO_3(2mol/L) \longrightarrow$

$MnO_4^- + Br^- + H^+ \longrightarrow$

$MnO_4^- + C_2O_4^{2-} + H^+ \longrightarrow$

$MnO_4^- + SO_3^{2-} + H^+ \longrightarrow$

$MnO_4^- + SO_3^{2-} + H_2O \longrightarrow$

$MnO_4^- + SO_3^{2-} + OH^- \longrightarrow$

$S_2O_8^{2-} + Mn^{2+} + H_2O \xrightarrow{AgNO_3}$

2. Operation

（1）Master the method of detecting ammonium ion in a gas chamber.

（2）Master the method of electrolysis of Na_2SO_4 solution by using of a galvanic cell.

3. Notes

（1）The reaction between concentrated HNO_3 and Zn will produce poisonous gas NO_2, so it should be carried out in a hood.

（2）Owing to the low reaction rates, in the test of locating the sequence of metals, the tubes should be placed in the test tube rack for a while without any disturbance.

4. Report format

（1）Objectives

（2）Principle

（3）Procedure（illustrated with tables）

5. Questions

（1）At what conditions the table of electrode potentials in acidic medium or the table in alkaline medium should be consulted respectively?

（2）Can we get a conclusion that a redox reaction consisting two half-reactions will occur at a higher reaction rate with a larger $\Delta E^{\ominus}$?

（3）How to use the table of electrode potentials to form a redox reaction equation?

（4）At what conditions will the table of electrode potentials be used to judge the direction of a given redox reaction? And at what conditions will the Nernst equation be used?

Words

redox reaction	氧化还原反应	electrode	电极
oxidizing agent（oxidant）	氧化剂	electrode potential	电极电势
reducing agent（reductant）	还原剂	gas chamber	气室
galvanic cell	原电池	plumbum	铅

electrolysis	电解	zinc	锌
electrolyze	电解	catalyst	催化剂
oxidation	氧化	cathode	阴极
reduction	还原	salt bridge	盐桥
reduce	还原	agar	琼脂
oxidize	氧化	granulated	颗粒状的
oxidation number	氧化数		

实验六　配合物

PPT

目的要求 ●●●

通过实验了解配合物的形成条件和配合物的一些应用。

实验原理 ●●●

配合物的形成过程是一个可逆反应，例如 Cu^{2+} 离子与 NH_3 分子结合可形成四氨合铜配离子，反应式为：

$$Cu^{2+} + 4NH_3 \rightleftharpoons [Cu(NH_3)_4]^{2+}$$

1. 增加配位剂（如 NH_3）的浓度，上述平衡向生成配离子的方向移动；降低配位剂的浓度，平衡向配离子解离的方向移动。

2. 在一个配合物溶液中，加入一种可以与中心离子结合生成难溶物的沉淀剂，就会导致溶液中未配位的金属离子的浓度降低，促进配离子的解离。反之，一种配位剂若能与金属离子结合生成稳定的配合物，并且此配合物是易溶性的，则加入足量的配位剂可以使该金属离子的难溶盐溶解，若先加入配位剂而后加入沉淀剂，可以阻止沉淀的生成。沉淀剂及配位剂和金属离子的竞争反应结果，决定于相应的难溶物的 K_{sp} 和相应的配离子的 K_{st} 的相对大小。

3. 若配离子的配体是弱碱（如 NH_3）或者是弱酸根（如 CN^-、$S_2O_3^{2-}$、乙二胺四乙酸根），则加入强酸会促使配离子的解离。

4. 在同一金属离子的溶液中，同时存在两种配位剂，则此金属离子首先与能生成较稳定配合物的配位剂结合。例如，在含有 F^- 和 SCN^- 的溶液中加入 Fe^{3+}，则不能生成血红色的 $[Fe(SCN)_6]^{3-}$ 配离子，而主要生成 $[FeF_6]^{3-}$ 配离子。

5. 在配离子中，受配位体的影响，能够改变中心离子原来的电子结构，因而改变其氧化还原性质。例如，$[Co(CN)_6]^{4-}$ 的还原性要比 Co^{2+} 的还原性强得多。

6. 在 Cr^{3+} 与 EDTA 所形成的配合物溶液中加入过量 $NH_3 \cdot H_2O$ 可生成 $[Cr(OH)Y]^{2-}$ 的配离子。

7. 一些常数值

（1）难溶银盐的溶度积

　　AgCl 的 $K_{sp} = 1.8 \times 10^{-10}$

　　AgBr 的 $K_{sp} = 5.3 \times 10^{-13}$

　　AgI 的 $K_{sp} = 8.5 \times 10^{-17}$

（2）含银配离子的稳定常数值

$[Ag(NH_3)_2]^+$ 的 $K_{st} = 1.5 \times 10^7$

$[Ag(S_2O_3)_2]^{3-}$ 的 $K_{st} = 2.4 \times 10^{13}$

$[Ag(CN)_2]^-$ 的 $K_{st} = 1.3 \times 10^{21}$

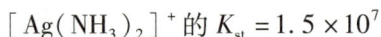

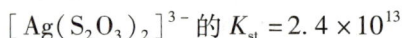

（3）标准电极电势

$E^{\ominus}_{Co^{3+}/Co^{2+}} = +1.842$ V

$E^{\ominus}_{I_2/I^-} = +0.5345$ V

仪器和药品 •••

1. 仪器 50mL 小烧杯，玻棒，10×100 小试管，滴管，15×150 试管。

2. 药品

（1）固体 $CuSO_4 \cdot 5H_2O$；NaF。

（2）酸 HCl 1mol/L。

（3）碱 氨水 2mol/L，浓；NaOH 1mol/L；Na_2CO_3 0.1mol/L。

（4）盐 Na_2S 0.1mol/L；$BaCl_2$ 0.1mol/L；KSCN 1mol/L；NaCl 0.1mol/L；KBr 0.1mol/L；EDTA-2Na 0.1mol/L；$K_4P_2O_7$；$Na_2S_2O_3$ 0.5mol/L；KCN 0.5mol/L；$AgNO_3$ 0.1mol/L；KI 0.1mol/L；枸橼酸钠 1mol/L；$CuSO_4$ 0.1mol/L；$CaCl_2$ 0.1mol/L；$CoCl_2$ 0.5mol/L；$CrCl_3$ 0.1mol/L。

3. 其他 乙醇 95%；H_2O_2 30%；淀粉溶液；丙酮；pH 试纸或红色石蕊试纸；酚酞指示剂；碘试液。

实验内容 •••

1. 配合物的形成——硫酸四氨合铜的制备及性质

（1）反应 $CuSO_4 + 4NH_3 \rightleftharpoons [Cu(NH_3)_4]SO_4$

（2）操作 在小烧杯中放入 2.5g 相当于 0.01mol $CuSO_4 \cdot 5H_2O$，加 10mL 水，搅拌至全部溶解，加 5mL 浓氨水（相当于 0.07mol NH_3），混匀，加等体积酒精，搅拌混匀，放置 2~3 分钟，滤得析出的结晶 $[Cu(NH_3)_4]SO_4 \cdot H_2O$，用少量乙醇洗 1~2 次，记录产品性状。

（3）性质

① 取少量产品，溶于几滴水，观察并记录溶液的颜色，再继续加水，观察溶液颜色有何变化？

② 取少量产品，溶于几滴水，逐滴加 1mol/L HCl 至过量，观察并记录溶液的颜色的变化，再加入过量浓氨水，观察溶液颜色的变化。

根据以上现象，讨论该配合物在溶液中的形成和解离。

③ 取少量产品，溶于几滴水，分在三个小试管中。

在第一管中加 0.1mol/L Na_2CO_3 溶液，观察有无碱式碳酸铜沉淀生成。

在第二管中加 0.1mol/L Na_2S 溶液，观察有无硫化铜沉淀生成。

根据这两个实验结果讨论 Cu^{2+} 离子浓度在溶液中的变化。

在第三管中加 0.1mol/L $BaCl_2$ 溶液，观察有无硫酸钡沉淀生成。

根据配合物的组成，讨论配合物中 Cu^{2+} 和 SO_4^{2-} 所处的位置。

④ 取少量产品（已干燥），闻一闻有无氨臭味，然后放在一支干试管里，管口挂一条润湿的 pH 试纸或红色石蕊试纸，微火加热。观察并记录：a. 试纸的颜色变化，b. 残余固体的颜色，c. 有无氨臭味，d. 写出反应式。

说明 NH_3 分子是否参与组成配合物，其结合是否牢固。

2. 配合物的稳定性　根据 $AgCl$、$AgBr$、AgI 的溶度积以及 $[Ag(NH_3)_2]^+$、$[Ag(S_2O_3)_2]^{3-}$、$[Ag(CN)_2]^-$ 的稳定常数估计在下列各步中应有什么现象，并用实验检验你的估计。

（1）取 $3 \sim 4$ 滴 $0.1mol/L$ $AgNO_3$ 于试管中，加等体积 $0.1mol/L$ $NaCl$ 溶液。

（2）再加 2 滴浓氨水。

（3）再加 2 滴 $0.1mol/L$ KBr。

（4）再加 2 滴 $0.5mol/L$ $Na_2S_2O_3$。

（5）再加 2 滴 $0.1mol/L$ KI。

（6）再加 2 滴 $0.5mol/L$ KCN。

观察并记录各步结果，并写出反应式。

3. 配合物的形成和酸碱性

（1）配合物形成时的 pH 变化　在两试管中分别放 $0.1mol/L$ $CaCl_2$ 和 $0.1mol/L$ 乙二胺四乙酸二钠（Na_2H_2Y）溶液各 2mL，各加 1 滴酚酞指示剂，都用 $2mol/L$ 氨水调到溶液刚刚变红。把两溶液混合后，溶液的颜色有何变化？写出反应式。在什么情况下，配合物形成时溶液的 pH 降低？

（2）溶液 pH 对配位平衡的影响

①枸橼酸与 Fe^{3+} 的配位。在试管中放 1mL $0.1mol/L$ $FeCl_3$ 溶液，加 1mL $1mol/L$ 枸橼酸钠溶液，观察并记录溶液的颜色变化。分成 3 份，一支试管作为对照，其余二支试管中分别加 $1mol/L$ $NaOH$ 及 $1mol/L$ HCl 使分别成碱性和酸性，比较三支试管中颜色有何不同（Fe^{3+} 呈橙黄色，Fe^{3+} 的枸橼酸配合物呈亮黄至黄绿色，$Fe(OH)_3$ 呈红棕色）。讨论枸橼酸铁在酸性和碱性溶液中的稳定性。

②pH 对 $[Fe(SCN)_6]^{3-}$ 形成的影响。在试管中放 $4 \sim 5$ 滴 $0.1mol/L$ $FeCl_3$ 溶液，加 1mL $1mol/L$ $KSCN$ 溶液，分成两份，一份中加数滴 $1mol/L$ HCl；一份中加 $1mol/L$ $NaOH$，观察并记录溶液颜色变化。讨论 $[Fe(SCN)_6]^{3-}$ 在酸性和碱性溶液中的稳定性。

4. 配合物的动力学稳定性　取 10 滴 $0.1mol/L$ $CrCl_3$ 溶液，加 2mL $0.1mol/L$ 乙二胺四乙酸二钠溶液，混匀，观察溶液颜色有无变化，有无配合物形成（Cr^{3+} 的 EDTA 配合物呈深紫色），将溶液煮沸几分钟，有无配合物形成？在其中加入 $1 \sim 2$ 滴 $2mol/L$ 氨水，是否生成绿色 $Cr(OH)_3$ 沉淀？讨论 Cr^{3+} 与乙二胺四乙酸不易形成配合物的原因。

5. 配合物的形成对氧化还原能力（热力学）的改变

（1）取 $CoCl_2$ 溶液 $4 \sim 5$ 滴，加 30% H_2O_2，观察并记录反应现象。H_2O_2 能否把 Co^{2+} 氧化成 Co^{3+}（Co^{3+} 棕色）？

（2）取 $CoCl_2$ 溶液 $4 \sim 5$ 滴，加过量浓氨水至沉淀溶解，观察颜色的变化。再加 H_2O_2 有无颜色变化？再在其中加 $6mol/L$ HCl 酸化，记录并观察颜色的变化。此时钴是几价的？形成氨配合物对 Co^{2+} 的还原性有什么影响？

（3）在试管里放 1mL $CoCl_2$ 溶液，加 5 滴淀粉指示剂，逐滴加入碘试液，观察并记录颜色的变化。在另一试管中放 2 滴 $CoCl_2$ 溶液，加过量 KCN 溶液至开始生成的沉淀全部溶解，再往其中加入淀粉指示剂 5 滴并逐滴加入碘试液，观察并记录颜色的变化，并写出反应方程式。讨论加氰化钾对 Co^{2+} 还原性的影响。

6. 配位掩蔽　F^- 对 Fe^{3+} 的掩蔽　在试管中加数滴 $0.1mol/L$ $FeCl_3$ 溶液，加数滴 $1mol/L$ $KSCN$ 溶液，再加入固体 NaF，摇匀，观察、记录颜色的变化并写出反应方程式。

在另一试管中，加数滴 $0.5mol/L$ $CoCl_2$ 溶液，加数滴 $KSCN$ 溶液，再加等体积丙酮，出现

$[Co(SCN)_4]^{2-}$ 的蓝色，可用以检定 Co^{2+}，加入固体 NaF 少许，蓝色是否褪去？

根据上述实验，设计一种在 Fe^{3+} 存在下检验 Co^{2+} 的方法。

7. 螯合物的生成　焦磷酸合铜螯合离子的生成　在试管中加 2 滴 0.1mol/L $CuSO_4$ 溶液，逐滴加 $K_4P_2O_7$ 溶液，直至沉淀溶解成深蓝色透明溶液。

$$2Cu^{2+} + P_2O_7^{4-} \Longrightarrow Cu_2P_2O_7(s)$$

$$Cu_2P_2O_7 + 3P_2O_7^{4-} \Longrightarrow 2[Cu(P_2O_7)_2]^{6-}$$

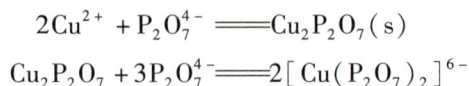

焦磷酸合铜配离子的结构为：

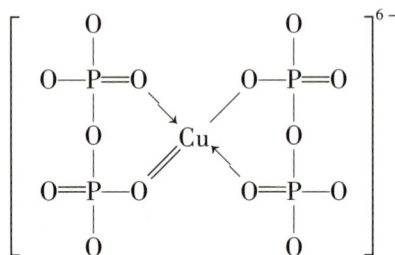

实 验 指 导

预习要求 ••••

1. 什么是配离子？什么是配合物？它们有哪些特性？

2. 形成配合物需要哪些条件？

3. 怎样用实验证明在溶液中形成了 $[Cu(NH_3)_4]^{2+}$？

4. 配合物与复盐有何区别？怎样证明？

5. 形成配合物后，原物质的哪些性质会发生改变？怎样改变？

6. 配合物中化学键的本质是什么？

7. 怎样判断配合物的相对稳定性？

8. 影响配合物稳定性的主要因素有哪些？

9. 螯合剂与配位剂有何区别和联系？

10. 在 Fe^{3+} 溶液中，先加 KSCN 溶液，再加 EDTA 溶液。会发生什么现象？为什么？

注意事项 ••••

1. KCN 为剧毒药品，使用时须加小心。做完实验的试液不能倒入水槽，要统一回收。

2. 制备 $[Cu(NH_3)_4]SO_4$ 时首先要将 $CuSO_4$ 固体全部溶解后才能加 NH_3 水，而且必须加浓 NH_3 水。

报告格式 ••••

1. 目的

2. 实验内容

（1）按讲义中"实验内容"的标题，尽量用化学方程式、流程图来表示。

（2）结论要简要明确。

实验后思考 ••••

1. 配位平衡与一般的化学平衡是否相同？

2. 配离子的解离与配离子之间的相互转化和沉淀的溶解与转化有哪些相似的地方?

3. 在配位平衡中，是否也存在同离子效应?

4. 哪些元素较容易形成配合物?

Experiment 6　Coordination Compounds

◆ Objectives

To learn about the set of conditions to form coordination compounds and the application of coordination compounds.

◆ Principles

The reactions of forming coordination compounds are reversible. For example:

$$Cu^{2+} + 4NH_3 \Longleftrightarrow [Cu(NH_3)_4]^{2+}$$

1. Increasing the concentration of ligand (NH_3) will cause a shift of the equilibrium to the side of complex ion. Conversely, decreasing the concentration of ligand will result in a reverse shift of the equilibrium.

2. Adding a precipitant to precipitate the central metal ion will make the complex ion decompose. Adding a ligand, which reacts with the central metal ion to form stable complex ion, can dissolve the slightly soluble salts of the metal ion. To the solution of the metal ion, adequate ligands added prior to the precipitant will prevent the precipitation.

3. If the ligand is a weak base or the anion of weak acid, adding strong acid will make the complex ion decompose.

4. Adding two kinds of ligands to the solution of a metal ion, a more stable complex ion will form. For instance, adding Fe^{3+} to the solution of F^- and CN^-, we will get $[FeF_6]^{3-}$ instead of $[Fe(SCN)_6]^{3-}$.

5. The outer electron structure of the central metal ion will change by adding ligands to form complex ion. Consequently, its redox property will also change. An example is that $[Co(CN)_6]^{4-}$ has stronger reductive ability than Co^{2+}.

6. By adding excess $NH_3 \cdot H_2O$ to the complex ion of Cr^{3+} and EDTA, $[Cr(OH)Y]^{2-}$ will be formed.

7. There are some constants needed in this experiment:

K_{sp} of AgCl:　　　1.8×10^{-10}

K_{sp} of AgBr:　　　5.3×10^{-13}

K_{sp} of AgI:　　　8.5×10^{-17}

K_{st} of $[Ag(NH_3)_2]^+$:　1.5×10^7

K_{st} of $[Ag(S_2O_3)_2]^{3-}$:　2.4×10^{13}

K_{st} of $[Ag(CN)_2]^-$:　1.3×10^{21}

$E^{\ominus}_{Co^{3+}/Co^{2+}} = +1.842$ V

$E^{\ominus}_{I_2/I^-} = +0.5345$ V

◆ Equipment and Chemicals

1. Equipment　Small beaker(50mL), glass rod, small test tube(10×100), pipette, test tube(16×150)

2. Chemicals

（1）Solid $CuSO_4 \cdot 5H_2O$；NaF

（2）Acid   HCl 1mol/L

（3）Alkali   ammonium solution（2mol/L,conc. ）；NaOH 1mol/L；Na_2CO_3 0.1mol/L

（4）Salt Na_2S 0.1mol/L；$BaCl_2$ 0.1mol/L；KSCN 1mol/L；

NaCl 0.1mol/L；KBr 0.1mol/L；

EDTA −2Na 0.1mol/L；$K_4P_2O_7$；

$Na_2S_2O_3$ 0.5mol/L；KCN 0.5mol/L

$AgNO_3$ 0.1mol/L；KI 0.1mol/L

Sodium citrate 1mol/L；$CuSO_4$ 0.1mol/L；

$CaCl_2$ 0.1mol/L；$CoCl_2$ 0.5mol/L；$CrCl_3$ 0.1mol/L；

3. Others   ethanol（95%）；H_2O_2（30%）；starch solution；acetone；pH test paper or red litmus paper；phenolphthalein；iodine solution

◆ Procedures

1. Formation of coordination compounds

Preparation and properties of $[Cu(NH_3)_4]SO_4$

Reaction：$CuSO_4 + 4NH_3 \rightleftharpoons [Cu(NH_3)_4]SO_4$

Operation：Dissolve 2.5g $CuSO_4 \cdot 5H_2O$（0.01mol）with 10mL of water in a beaker, add 5mL of concentrated ammonia solution（0.07mol）, mix the solution, then add 15mL of ethanol and stir. Waiting 2 ~ 3 minutes, filter and collect the crystal of $[Cu(NH_3)_4]SO_4 \cdot H_2O$, wash the crystal with ethanol twice. Record the appearance of the crystal.

Property

（1）Dissolve a portion of the product in several drops of water, observe and record the color of the solution. Add more drops of water and observe the color change.

（2）Dissolve a portion of the product in several drops of water, add excess 1mol/L HCl solution drop by drop, observe and record the color change of the solution. Then add excess concentrated ammonia solution to observe the color change. Discuss the formation and dissociation of $[Cu(NH_3)_4]SO_4$ in solution based on the tests.

（3）Dissolve a portion of the product with several drops of water, divide the solution to three tubes.

Add 0.1mol/L Na_2CO_3 solution to the first tube, and observe whether $Cu_2(OH)_2CO_3$ is precipitated from the solution.

Add 0.1mol/L Na_2S solution to the second tube, and observe whether CuS is precipitated from the solution.

Discuss the concentration change of Cu^{2+} in the two portions of solution based on the tests above.

Add 0.1mol/L $BaCl_2$ solution to the third tube, and observe whether $BaSO_4$ is precipitated from the solution.

Discuss the position of Cu^{2+} and SO_4^{2-} in $[Cu(NH_3)_4]SO_4$.

（4）Smell a portion of dried product to determine whether there is the odor of NH_3, then put it in a dry test tube, and hang a piece of moist pH paper or red litmus paper near the mouth of the tube and heat the crystal. Observe and record（a）the color change of the test paper,（b）the color of the residue in the tube,（c）the odor from the tube. Write out the reaction equation.

Conclude that whether NH_3 takes part in composing the coordination compound and whether the bond between NH_3 and Cu^{2+} is stable.

2. The stability of coordination compounds Estimate the phenomena of the following tests based on the involved solubility product constants and stability constants.

Perform the consecutive operations, observe and record the result of each step, then write out the reaction equations.

(1) Put 3~4 drops of 0. 1mol/L $AgNO_3$ solution in a tube, then add equal volume of 0. 1mol/L NaCl solution.

(2) Add 2 drops of concentrated ammonia solution to solution (1).

(3) Add 2 drops of 0. 1mol/L KBr solution to solution (2).

(4) Add 2 drops of 0. 5mol/L $Na_2S_2O_3$ solution to solution (3).

(5) Add 2drops of 0. 1mol/L KI solution to solution (4).

(6) Add 2 drops of 0. 5mol/L KCN solution to solution (5).

3. Relation between the formation of complex ion and solution pH

(1) The pH change during formation of coordination compounds.

Put 2mL of 0. 1mol/L $CaCl_2$ solution into one test tube, and 2mL of 0. 1mol/L EDTA-2Na solution to another tube, add 1 drop of phenolphthalein to the tubes respectively, acidify the solution with 2mol/L ammonia solution to be red. Mix the two portions of solution together, and observe the color change. Write out the reaction equation and answer: At what conditions the formation of coordination compounds will result in the decreasing of the solution pH?

(2) The effect of pH on complexation equilibria.

(a) The coordination between citrate and Fe^{3+}: Add 1mL of 1mol/L sodium citrate solution to 1mL of 0. 1mol/L $FeCl_3$ solution, observe and record the color change. Divide the solution into three portions. One portion is used for blank test. Acidify the second portion with 1mol/L HCl solution and basify the third portion with 1mol/L NaOH. Compare the different color of the three portions (Fe^{3+} is orange, its coordination compound with citrate is bright yellow to yellow-green, $Fe(OH)_3$ is red brown). Discuss the stability of coordination compound between citrate and Fe^{3+} in acidic and alkaline solution.

(b) The effect of pH on formation of $[Fe(SCN)_6]^{3-}$: Add 1mL of 1mol/L KSCN solution to 4~5 drops of 0. 1mol/L $FeCl_3$ solution. Divide the solution into two portions. Add several drops of 1mol/L HCl solution to one portion, and several drops of 1mol/L NaOH solution to another, compare the different color in two portions and record. Discuss the stability of $[Fe(SCN)_6]^{3-}$ in acidic and alkaline solution.

4. The activity of coordination compounds Mix 10 drops of 0. 1mol/L $CrCl_3$ solution with 2mL of 0. 1mol/L EDTA-2Na solution, observe whether the color of solution will change and conclude whether there is coordination compound formed (the EDTA complex ion of Cr^{3+} is dark purple). Boil the solution for a few minutes to see whether there is coordination compound formed. Add 1~2 drops of 2mol/L ammonia solution to see whether there are green $Cr(OH)_3$ precipitated. Does the difficulty of forming the EDTA complex ion of Cr^{3+} result from its low stability?

5. The effect of coordination on redox property

(1) Add 30% H_2O_2 solution to 4~5 drops of $CoCl_2$ solution, observe and record the phenomenon. Can

H_2O_2 solution oxide Co^{2+} to brown Co^{3+}?

（2）Add excess concentrated ammonia solution to 4 ~ 5 drops of $CoCl_2$ solution until the precipitate dissolves. Observe the color change. Add H_2O_2 solution to see whether the color will change. Acidify the solution with 6mol/L HCl solution, observe and record the color change. Explain the effect of formation of complex ion on the reductive property of Co^{2+}.

（3）Put 1mL of $CoCl_2$ solution and 5 drops of starch solution in a test tube, then add iodine solution drop by drop, observe and record the color change. Put 2 drops of $CoCl_2$ solution in another test tube, add excess KCN solution until the formed precipitate dissolves, then add 5 drops of starch solution and add iodine solution drop by drop, observe and record the color change. Write out the reaction equations and discuss the effect of KCN on reductive property of Co^{2+}.

6. Masking Masking Fe^{3+} with F^-: Into a tube containing a few drops of 0.1mol/L $FeCl_3$ solution, add several drops of 1mol/L KSCN solution, then add solid NaF, shake the tube to dissolve NaF. Record the color of the solution and write out the reaction equation.

Into another tube containing a few drops of 0.5mol/L $CoCl_2$ solution, add several drops of KSCN solution, then add equal volume of acetone, blue $[Co(SCN)_4]^{2-}$ is formed which indicates the presence of Co^{2+}. Add solid NaF to the blue solution and observe whether the blue color will disappear.

Design a test to detect Co^{2+} in the presence of Fe^{3+}.

7. Formation of chelate Forming a chelate of $[Cu(P_2O_7)_2]^{6-}$: Add $K_4P_2O_7$ solution dropwise to two drops of 0.1mol/L $CuSO_4$ solution until the precipitate dissolves to produce a dark blue solution.

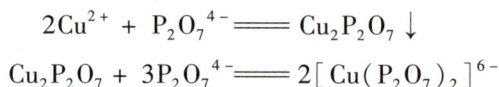

$$2Cu^{2+} + P_2O_7^{4-} =\!=\!= Cu_2P_2O_7 \downarrow$$
$$Cu_2P_2O_7 + 3P_2O_7^{4-} =\!=\!= 2[Cu(P_2O_7)_2]^{6-}$$

The structure of $[Cu(P_2O_7)_2]^{6-}$ is showed below:

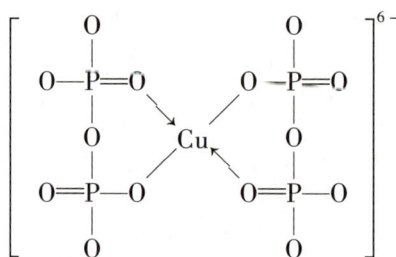

◆ Instructions

1. Requirements

（1）What are complex ions and coordination compounds?

（2）What set of conditions should be needed to produce coordination compounds?

（3）How to prove that $[Cu(NH_3)_4]^{2+}$ is formed in the experiment?

（4）Show the differences between coordination compounds and double salts with laboratory methods.

（5）What properties of the substance will change when coordination compounds are formed?

（6）Explain the intrinsic property of the bonds in coordination compounds.

（7）How to judge the relative stability of various coordination compounds?

（8）What factors will influence the stability of coordination compounds?

（9）Compare the similarity and differences between chelant and complexant.

（10）Add KSCN solution to ferric solution prior to the addition of EDTA solution, explain what will happen to the mixture.

2. Notes

（1）Special care should be taken with the highly poisonous KCN, and the solution should be recollected after the experiment.

（2）During the preparation of $[Cu(NH_3)_4]SO_4$, ammonia solution should be added to the solution only when the solid $CuSO_4$ dissolves completely. And the ammonia solution should be concentrated.

3. Report format

（1）Objectives

（2）Procedure

（a）Show the contents with reaction equations and flow charts.

（b）Give concise conclusions to each test.

4. Questions

（1）Does the complexation equilibria is same to other chemical equilibria?

（2）Show the similarity between the complexation equilibrium of coordination compound and the dissolve equilibrium of precipitate.

（3）Does common ion effect in the complexation equilibrium exist?

（4）Which elements are prone to form coordination compounds?

Words

coordination compound	配合物	chelate	螯合物
complex ion	配离子	chelant	螯合剂
sodium citrate	枸橼酸钠	complexant	配位剂
mask	掩蔽	complexation equilibrium	配位平衡

第四章　元素的化学性质实验

实验一　卤　素

PPT

目的要求 •••

1. 验证卤素、卤化氢和卤素的含氧酸及其盐的物理性质和化学性质。
2. 掌握实验室中制备卤素的一般原理和方法。
3. 掌握卤素离子的一般鉴别方法。

实验原理 •••

氯、溴、碘是周期系第Ⅶ主族元素。它们的原子最外电子层上有 7 个电子，容易得到一个电子生成卤化物，因此卤素都是很活泼的非金属，其氧化数通常是 -1。卤素还能生成含氧酸，在其含氧酸中氧化数表现为 $+1$、$+3$、$+5$、$+7$。

氯、溴、碘都可以用氧化剂从其卤化物中制取。

卤素都是氧化剂，它们的离子都是还原剂。作为氧化剂的卤素分子的化学活泼性按下列顺序变化：

$$F_2 > Cl_2 > Br_2 > I_2$$

而作为还原剂的卤素阴离子的化学活泼性则按相反的顺序变化：

$$I^- > Br^- > Cl^- > F^-$$

卤素分子都是非极性分子，故易溶于非极性溶剂（有机溶剂）中。碘还易溶于碘化钾溶液中，生成 KI_3。

卤化银不溶于水和稀硝酸，而 CO_3^{2-}、PO_4^{3-}、CrO_4^{2-} 等阴离子形成的银盐溶于硝酸，所以可在硝酸溶液中使卤素阴离子形成卤化银沉淀以防止其他阴离子的干扰。

卤化银在氨水中溶解度不同，可以控制氨的浓度来分离混合的卤离子。实验中，常用 $(NH_4)_2CO_3$，使 AgCl 沉淀溶解，与 AgBr、AgI 分离。反应式如下：

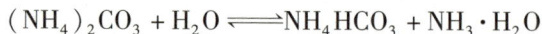

$$(NH_4)_2CO_3 + H_2O \rightleftharpoons NH_4HCO_3 + NH_3 \cdot H_2O$$

$$AgCl + 2NH_3 \cdot H_2O \rightleftharpoons [Ag(NH_3)_2]^+ + Cl^- + 2H_2O$$

卤素的含氧酸根都具有氧化性，次氯酸的氧化能力是氯的含氧酸中最强的，因此它具有漂白、杀菌的作用，它的盐如次氯酸钠常用作为漂白剂与消毒剂。

仪器和药品 •••

1. 仪器　离心管；离心机。

2. 药品

（1）固体　I_2，红磷，锌粉，KCl，KBr，KI，MnO_2，$KClO_3$。

（2）酸　H_2SO_4 3mol/L，18mol/L；HCl 12mol/L；HNO_3 6mol/L。

（3）盐　KBr 0.1mol/L；KCl 0.1mol/L；KI 0.1mol/L；$AgNO_3$ 0.1mol/L；NaClO 溶液；I_2 溶液；三氯甲烷。

3. 试纸 蓝色石蕊试纸，$Pb(Ac)_2$ 试纸，滤纸片，KI 淀粉试纸。

实验内容 ●●●

1. 碘与金属、非金属的反应

(1) 碘溶液与锌粉的作用 将一小匙锌粉加入盛有 1mL 碘溶液的试管中，不断振荡（另取一支试管加 1mL 碘溶液作对照）。观察反应过程中碘溶液的颜色变化（如现象不明显可微热），写出反应式，并加以解释。

(2) 碘和红磷的作用 取少许碘和红磷于试管中混合，滴入 1~2 滴水（如果红磷潮湿就可不加水），在水浴中加热片刻后反应猛烈发生，用湿润的蓝色石蕊试纸在管口试验 HI 的生成，记录观察到的现象，并写出反应式（先写生成 PI_3 的反应式、再写 PI_3 的水解反应式）。

2. 氯、溴、碘氧化性的比较

(1) 氯与碘氧化性的比较 在试管中加 1 滴 0.1mol/L KI 溶液，加蒸馏水稀释至 1mL，逐滴加入氯水，观察溶液颜色的变化，再加 1mL 三氯甲烷，振摇，观察三氯甲烷层的颜色。然后再向此溶液中加入过量的氯水（或通氯气）至三氯甲烷层的颜色消失为止。解释现象并写出反应式。

(2) 溴和碘氧化性的比较 在盛有约 1mL 0.1mol/L KI 溶液的试管中，加数滴溴水，再加入数滴淀粉溶液，记录观察到的现象并写出反应式。

(3) 氯与溴氧化性的比较 向盛有约 1mL 0.1mol/L KBr 溶液的试管中加入数滴氯水，观察溶液颜色的变化，再加 1mL 三氯甲烷振摇，观察三氯甲烷层的颜色，解释现象并写出反应式。

综合以上实验结果列出 Cl_2、Br_2、I_2 氧化性大小的递变顺序，并用标准电极电位来说明。

3. 卤素的制备 取 3 支干试管分别加入少许 KCl、KBr、KI 晶体，向各试管中加入 2mL 3mol/L H_2SO_4，再各加入少量 MnO_2，用碘化钾淀粉试纸在装有 KCl 的试管口检查证明放出的气体是 Cl_2，在其余两个试管中分别加入 1mL 三氯甲烷，观察三氯甲烷层中的颜色，写出有关反应式。

4. 比较卤化氢的还原性

(1) 在一支干燥试管中加入几小粒 KCl 晶体，加 2~3 滴浓 H_2SO_4，观察试管中的变化，并用蓝色石蕊纸在试管口检查证明所逸出的气体是 HCl。

(2) 在一支干燥试管中加入几小粒 KBr 晶体，加 2~3 滴浓 H_2SO_4，观察试管中的变化，并用沾有 I_2 试液的试纸在试管口检查，证明所逸出的气体是 SO_2。

(3) 在一支干燥试管中加入几小粒 KI 晶体，加 2~3 滴浓 H_2SO_4，观察试管中的变化，并用醋酸铅试纸在试管口检查证明所逸出的气体是 H_2S。

综合比较三支试管的反应产物，列出 Cl^-、Br^-、I^- 还原性的强弱递变顺序，写出有关反应式，并用标准电极电势解释实验结果。

5. 次氯酸盐和氯酸盐的氧化性

(1) 在试管中加 0.1mol/L KI 溶液 1mL 和三氯甲烷 1mL，再加 1~2 滴次氯酸钠溶液，振摇，观察三氯甲烷层的颜色，再逐滴加入过量的次氯酸钠，不断振摇直至三氯甲烷层颜色消失，记录现象，写出有关反应式。

(2) 在试管中加入少量 $KClO_3$ 晶体，用 1~2mL 水溶解后，加入 10 滴 0.1mol/L KI 溶液，把得到的溶液分成两份，一份用 1mol/L H_2SO_4 酸化，一份留作对照。稍等片刻，观察有何变化。试比较氯酸盐在中性和酸性溶液的氧化性。

6. Cl⁻、Br⁻、I⁻混合溶液的分离和检出

（1）AgX 沉淀的生成　于离心管中加入 3 滴 0.1mol/L NaCl、0.1mol/L KBr 和 0.1mol/L KI，混合后，加 2 滴 6mol/L HNO₃ 酸化，再滴加 0.1mol/L AgNO₃ 溶液至沉淀完全，离心沉淀，弃去溶液，沉淀用蒸馏水洗涤 2 次，每次用水 4~5 滴，搅拌后离心沉淀，弃去洗液（用毛细吸管吸取）得卤化银沉淀。

（2）AgCl 的溶解及 Cl⁻ 的检出　往上面所得卤化银沉淀上加 2mL 12%（NH₄）₂CO₃ 溶液充分搅拌后，离心，将清液 [Ag(NH₃)₂]Cl 移于试管中，用 6mol/L HNO₃ 酸化，如有白色 AgCl 沉淀生成，表示有 Cl⁻ 存在，将沉淀用作 Br⁻、I⁻ 的检定。

（3）Br⁻ 和 I⁻ 的检出　在（2）中所得的沉淀中加 5 滴水和少量锌粉，充分搅拌，待卤化银被还原完全后（沉淀全变黑色），离心沉降，吸取上清液于另一支离心管（或小试管）中，加 10 滴三氯甲烷再滴加氯水，每加 1 滴均要充分摇动试管，并观察三氯甲烷层颜色变化，如三氯甲烷层显紫红色则表示有 I⁻ 存在（生成 I₂）。继续加入氯水至红紫色退去（被氧化成无色 NaIO₃），而三氯甲烷层呈橙色或金黄色，表示有 Br⁻ 存在。

有关反应式：

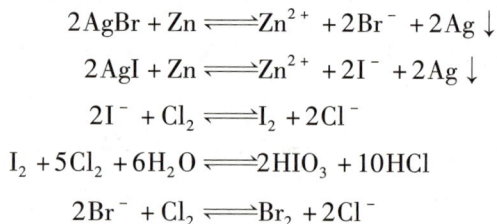

$$2AgBr + Zn \Longrightarrow Zn^{2+} + 2Br^- + 2Ag \downarrow$$
$$2AgI + Zn \Longrightarrow Zn^{2+} + 2I^- + 2Ag \downarrow$$
$$2I^- + Cl_2 \Longrightarrow I_2 + 2Cl^-$$
$$I_2 + 5Cl_2 + 6H_2O \Longrightarrow 2HIO_3 + 10HCl$$
$$2Br^- + Cl_2 \Longrightarrow Br_2 + 2Cl^-$$

实　验　指　导

预习要求 ●●●

1. 为什么卤素在化合物中的氧化数常是单数？

2. 从电极电势说明为什么作为氧化剂的活泼性 $F_2 > Cl_2 > Br_2 > I_2$，而作为还原剂的活泼性 $F^- < Cl^- < Br^- < I^-$？

3. 在实验室中制备少量 Cl_2、Br_2、I_2 可利用什么反应？

4. 卤素单质有哪些主要化学性质？

5. 卤化氢的还原性有什么递变规律？实验中如何验证 HCl、SO_2、H_2S 气体？

6. 如何证明次氯酸盐的氧化性？

7. 水溶液中，氯酸盐的氧化性与介质有何关系？

8. 如何分离和检出 Cl⁻、Br⁻、I⁻？

9. 实验前查出本实验中有关的标准电极电势。

10. 完成下列反应式，指出氧化剂和还原剂。

（1）$Zn(s) + I_2(sol) \longrightarrow$

（2）$Mg(s) + I_2(s) \longrightarrow$

（3）$P(s) + I_2(s) \xrightarrow[H_2O]{\Delta}$

（4）$I^- + Cl_2(g) \longrightarrow$

（5）$I_2 + Cl_2 + H_2O \longrightarrow$

（6）$I^- + Br_2 \longrightarrow$

（7）$Br^- + Cl_2 \longrightarrow$

（8）$Cl^- + MnO_2 + H^+ \longrightarrow$

（9）$Br^- + MnO_2 + H^+ \longrightarrow$

（10）$Cl^- + H_2SO_4(Conc) \longrightarrow$

（11）$Br^- + H_2SO_4(Conc) \longrightarrow$

（12）$I^- + H_2SO_4(Conc) \longrightarrow$

（13）$I^- + HClO + H^+ \longrightarrow$

（14）$I_2 + HClO + H_2O \longrightarrow$

（15）$I^- + ClO_3^- + H^+ \longrightarrow$

（16）$AgCl(s) + NH_3（过量）\longrightarrow$

（17）$AgBr(s) + Zn \longrightarrow$

（18）$AgI(s) + Zn \longrightarrow$

基本操作 •••

1. 离心分离。

2. 检查沉淀完全与否。

3. 试管反应中少量气体逸出的检别。

4. 试管中少量液体的萃取。

注意事项 •••

1. 本实验中，PI_3、Cl_2 都是有毒气体，并有刺激性，吸入人体会刺激喉管，引起咳嗽和喘息。因此，在做产生 PI_3 和 Cl_2 气实验时，须在通风橱内进行，室内也要注意通风换气。

2. 当检验反应中所逸出的气体时，必须把所用的试纸用水润湿后放在试管口，不能投入试管内。

3. 关于离心机的使用方法与溶液和沉淀的分离操作，可参看"沉淀平衡"实验指导中的注意事项。

报告格式 •••

1. 目的

2. 实验内容（示例）

（1）

试样	加入试剂	现　象	反应方程式
结论			

（2）Cl^-、Br^-、I^- 离子混合溶液的分离和检出

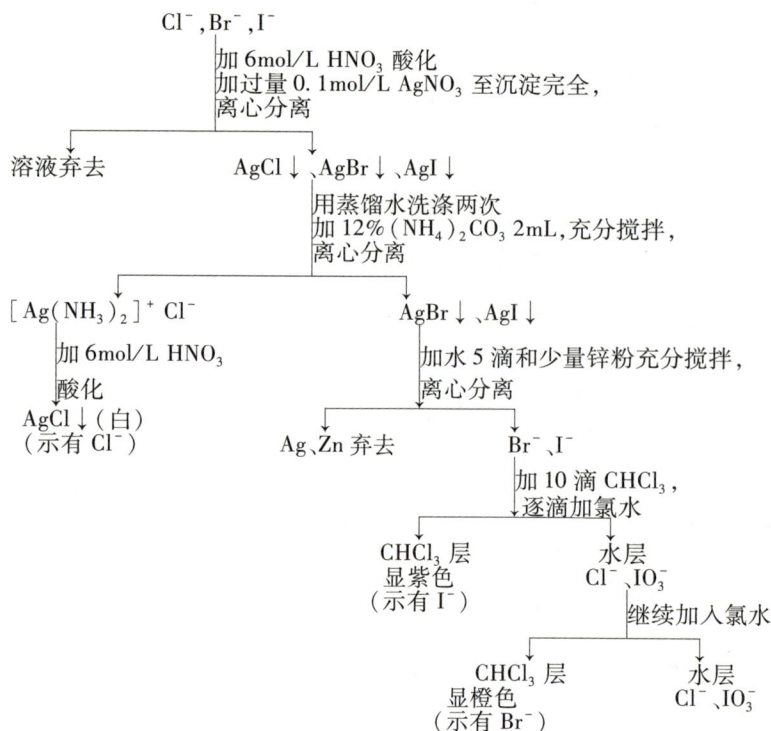

$$Cl^-, Br^-, I^-$$

加 6mol/L HNO_3 酸化
加过量 0.1mol/L $AgNO_3$ 至沉淀完全，
离心分离

溶液弃去 $AgCl\downarrow$、$AgBr\downarrow$、$AgI\downarrow$

用蒸馏水洗涤两次
加 12% $(NH_4)_2CO_3$ 2mL，充分搅拌，
离心分离

$[Ag(NH_3)_2]^+Cl^-$ $AgBr\downarrow$、$AgI\downarrow$

加 6mol/L HNO_3 加水 5 滴和少量锌粉充分搅拌，
酸化   离心分离

$AgCl\downarrow$（白）
（示有 Cl^-）

Ag、Zn 弃去 Br^-、I^-

加 10 滴 $CHCl_3$，
逐滴加氯水

$CHCl_3$ 层   水层
显紫色 Cl^-、IO_3^-
（示有 I^-）

继续加入氯水

$CHCl_3$ 层   水层
显橙色 Cl^-、IO_3^-
（示有 Br^-）

实验后思考 •••

1. 为什么 Ag 的卤化物（除氟化银）及 CO_3^{2-}，PO_4^{3-}，CrO_4^{3-} 等阴离子形成的 Ag 盐难溶于水，用分子结构的相关知识解释。

2. 用电极电势说明不同氧化数氯的含氧酸盐氧化能力强弱的顺序。"氧化数愈高，氧化能力愈强"这句话对吗？

3. 为什么 AgF 易溶于水？为什么 HF 的沸点在卤化氢中为最高？

4. 如何检验溶液中是否含 F^- 离子？

Experiment 1   Halogen

◆ Objectives

1. Validate（Test）the physical and chemical properties of halogen，haloid oxyacid，haloid oxysalt and hydrogen halid.

2. Master the methods of how to prepare halogen and the general principles.

3. Learn how to identify the halogen ions.

◆ Principles

1. Group Ⅶ elements comprise of fluorine F，chlorine Cl，bromine Br，iodine I and astatine At，which have similar structure of the outer and penultimate electron layers（ns^2np^5）. The presence of one unpaired electron determines its likeness to get one electron，forming corresponding halogenide. Halogens all have active nonmetallic characters with a usual oxidation state of -1，but the oxidation states in the haloid oxyacid are displayed as：$+1$，$+3$，$+5$，$+7$.

2. Free halogen molecule (such as Cl_2, Br_2, I_2) each is obtained by oxidizing its corresponding halogenide.

3. All the halogen molecules are oxidants with different chemical activity according to the sequence as $F_2 > Cl_2 > Br_2 > I_2$. Whereas all halogen ions are reducers with their reducing activity decreasing in the following series: $I^- > Br^- > Cl^- > F^-$.

4. Halogen molecules are non-polar so that they are readily soluble in some non-polar solvents (organic solvents). Especially iodine is soluble in KI solution due to the form of KI_3 molecules.

5. Most ionic halogenides are readily soluble in water with the exception of silver halide (AgX), which is also insoluble in dilute nitric acid (HNO_3) solution. But compounds of Ag^+ with CO_3^{2-}, PO_4^{3-}, CrO_4^{2-} all readily dissolve in HNO_3 solution. So silver halide (AgX) can be precipitated in HNO_3 solution as to prevent the disturbance from the other anions.

6. Diverseness of the solubility of AgX in ammonia solution ($NH_3 \cdot H_2O$) determines that controlling the density of NH_3 is useful to separate the mixed haloid ions. For example, $(NH_3)_2CO_3$ is added to dissolve AgCl, which then can be separated from AgBr and AgI. The relative reaction equations are given below:

$$(NH_4)_2CO_3 + H_2O = NH_4HCO_3 + NH_3 \cdot H_2O$$

$$AgCl + 2NH_3 \cdot H_2O = [Ag(NH_3)_2]^+ + Cl^- + 2H_2O$$

7. All the haloid oxyacid radicals have oxidizing activity. Among them hypochlorous acid (HClO) is the strongest oxidant. It is usually used as bleacher and antiseptic because of its capacity of bleaching and sterilizing.

◆ Equipment and Chemicals

1. Equipment　　Centrifugal tube, centrifugal machine

2. Chemicals

(1) Soild　I_2(s), red phosphor, zinc powder, KCl, KBr, KI, MnO_2, $KClO_3$

(2) Acid　H_2SO_4 3mol/L, 18mol/L; HCl 12mol/L; HNO_3 6mol/L.

(3) Salt　KBr 0.1mol/L; KCl 0.1mol/L; KI 0.1mol/L; $AgNO_3$ 0.1mol/L;

　　　　　NaClO solution; I_2 solution; Chloroform ($CHCl_3$)

3. Test paper　　Blue litmus test paper, Pb(Ac)$_2$ test paper, filter paper, KI/amylum test paper

◆ Procedures

1. Iodine reaction with metallic and nonmetallic elements

(1) Iodine solution reaction with zinc powder

Add a spoon of zinc powder to a test tube containing 1mL iodine solution. Shake sufficiently (in contrast to another test tube with only 1mL iodine solution). Observe the color change of iodine solution (heated if the phenomenon is not obvious). Write out the chemical reaction equations and explain the result above.

(2) Iodine reaction with red phosphor (P)

Mix some I_2(s) with red P into a test tube, and then add $1 \sim 2$ drops of water (when P is dry). You may find strong reaction occurs after the test tube is heated in water bath. Then put a piece of wet litmus paper near the top of the tube to detect the product (HI). Record the phenomenon, write out the chemical reaction equations and explain the result above.

2. Comparing the oxidizing activity of Cl_2, Br_2 and I_2

(1) Comparing the oxidizing activity of Cl_2 and I_2

Add a drop of 0.1mol/L KI solution to a test tube, and then dilute it with about 1mL of distilled water. Continuously add aqueous solution of chlorine drop by drop, observe the color change. Then add 1mL chloroform ($CHCl_3$), shake, and observe the color of $CHCl_3$ layer. Go on to add aqueous solution of chlorine until the color of $CHCl_3$ layer disappears. Write out the chemical reaction equations and explain the result above.

（2） Comparing the oxidizing activity of Br_2 and I_2

Add drops of aqueous solution of bromine to a test tube containing 1mL 0.1mol/L KBr solution, and then add drops of amylum solution. Record the phenomenon and write out the chemical reaction equations and explain the result above.

（3） Comparing the oxidizing activity of Cl_2 and Br_2

Add drops of aqueous solution of chlorine to a test tube containing 1mL 0.1mol/L KBr solution, observe the color change of the solution. Then add 1mL $CHCl_3$ while shaking, observe the color of the $CHCl_3$ layer again. Write out the chemical reaction equations and explain the result above.

Summarize the experimental results above, then range the halogen molecules (Cl_2, Br_2, I_2) conforming to their oxidizing activity increasing or decreasing. Illustrate that rule with standard electrode potentials.

3. Preparation of the halogens Add some KCl, KBr and KI crystals to three dry tubes respectively, then add 2mL 3 mol/L H_2SO_4 and some MnO_2 to each tube. Put a piece of KI/starch paper near the top of the tube containing KCl to detect the emitted gas (Cl_2). Finally add 1mL $CHCl_3$ to the other two tubes and observe the color of $CHCl_3$ layer carefully. Write out the chemical reaction equations and explain the result above.

4. Comparing the reducing activity of the hydrogen halogenide

（1） Add a few KCl particles to a dry test tube, then 2~3 drops of concentrated sulphuric acid (H_2SO_4) are added. Observe what occurs in the tube, put a piece of blue litmus test paper near the top of tube to detect the emitted gas (HCl).

（2） Add a few KBr particles to a dry test tube, and then 2~3 drops of concentrated sulphuric acid (H_2SO_4) are added. Observe what occurs in the tube, put a piece of test paper moistened with solution of I_2 near the top of tube to detect the emitted gas (SO_2).

（3） Add a few KI particles to a dry test tube, then 2~3 drops of concentrated sulphuric acid (H_2SO_4) are added. Observe what occurs in the tube, put a piece of blue $Pb(Ac)_2$ test paper near the top of tube to detect the emitted gas (H_2S).

Summarize the three different products above, then range the haloid ions (Cl^-, Br^-, I^-) according to their reducing activity increasing or decreasing. Illustrate that rule with standard electrode potential.

5. The oxidizing activity of hypochlorites and chlorates

（1） Add 0.1mol/L 1mL KI solution and $CHCl_3$ 1mL, then add 1~2 drops of natrium hypochlorite (NaClO) solution and shake sufficiently. Observe the color of the $CHCl_3$ layer. After that add superfluous NaClO solution drop by drop while shaking until the color of the $CHCl_3$ layer disappears. Write out the chemical reaction equations and explain the result above.

（2） Add some potassium chlorates ($KClO_3$) crystals into a test tube, and then dissolve them in 1~2mL water. Continuously add 10 drops of 0.1mol/L KI solution. Then mixture solution above are divided into two tubes, one of which is acidified by concentrated H_2SO_4, another is held for comparison. Wait a minute, observe and try to conclude the difference of the oxidizing activity of the chlorates when they are in the different medium, e.g. in a neutral solution or in an acidic solution.

6. Separation and detection of the solution mixed with Cl⁻, Br⁻, and I⁻

（1）How to produce the silver halogenide

Add 3 drops of 0.1mol/L NaCl(aq), 0.1mol/L KBr(aq) and 0.1mol/L KI(aq) to a centrifugal tube respectively, then add 2 drops of 6mol/L HNO₃ to acidify the mixed solution. After that dribble down 0.1mol/L AgNO₃ solution until the precipitates are completely produced. With the centrifugal machine you can separate these silver halid precipitates from the solution, and wash the precipitates twice with distilled water(4~5 drops once and stirring at the same time). Finally discard the washing solution (with a burette) after subside the precipitates using centrifugal machine.

（2）How to dissolve AgCl and how to detect Cl⁻

Add 2mL 12% (NH₄)₂CO₃ solution to the precipitates mentioned above, after stirring sufficiently, separate the precipitates from the solution again with centrifugal machine, remove the transparent solution ([Ag(NH₃)₂]Cl) to a test tube, and acidify it with 6mol/L HNO₃. If a white substance AgCl appears, it proves Cl⁻ existing. Take note of keeping the precipitates (AgBr and AgI) for the next process to detect the presence of Br⁻ and I⁻.

（3）How to detect Br⁻ and I⁻

Add 5 drops of water and some zinc powder to the left precipitates mentioned above. Stirring until the silver halide is completely reduced (that is the precipitate turns black). Separate the left precipitates form the solution again with centrifugal machine, imbibe the transparent solution to another centrifugal tube, and then add 10 drops of CHCl₃ and the aqueous solution of chlorine while shaking sufficiently at the same time. Observe the color change in CHCl₃ layer. If the red violet color appears, it proves I⁻ existing. Finally add the aqueous solution of chlorine until such color disappears again because the oxidation product NaIO₃ that is colorless is formed. Then if the orange or the golden color appears, it proves Br⁻ existing.

Some relevant reaction equations are as follows:

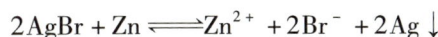

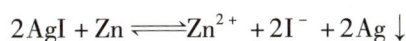

$$2AgBr + Zn \rightleftharpoons Zn^{2+} + 2Br^- + 2Ag \downarrow$$
$$2AgI + Zn \rightleftharpoons Zn^{2+} + 2I^- + 2Ag \downarrow$$
$$2I^- + Cl_2 \rightleftharpoons I_2 + 2Cl^-$$
$$I_2 + 5Cl_2 + 6H_2O \rightleftharpoons 2HIO_3 + 10HCl$$
$$2Br^- + Cl_2 \rightleftharpoons Br_2 + 2Cl^-$$

◆ Instructions

1. Requirements

（1）Why are the oxidation states of the halogens in their compounds singulars? Please correlate the atom structure to explain it.

（2）Explain why the oxidizing activity is decreasing as follows: $F_2 > Cl_2 > Br_2 > I_2$, but the reducing activity is increasing as follows: $F^- < Cl^- < Br^- < I^-$.

（3）Which reaction can be selected to prepare a small quantity of Cl_2, Br_2, I_2 in the lab?

（4）Enumerate the main chemical properties of the free halogen molecules.

（5）What rules do the reducing activity of the hydrogen halogenides obey? How to detect HCl(g), SO_2 and H_2S?

（6）How to validate the oxidizing activity of ClO⁻?

（7）In aqueous water, what's the relationship between the oxidizing activity of ClO_3^- and the medium?

（8）How to separate and detect Cl^-, Br^- and I^-?

（9）Please consult the relational standard electrode potential before the experiment?

（10）Complete the following reaction equations, discriminating the reducer and the oxidant in each reaction.

（a）$Zn(s) + I_2(sol) \longrightarrow$

（b）$Mg(s) + I_2(s) \longrightarrow$

（c）$P(s) + I_2(s) \xrightarrow[H_2O]{\Delta}$

（d）$I^- + Cl_2(g) \longrightarrow$

（e）$I_2 + Cl_2 + H_2O \longrightarrow$

（f）$I^- + Br_2 \longrightarrow$

（g）$Br^- + Cl_2 \longrightarrow$

（h）$Cl^- + MnO_2 + H^+ \longrightarrow$

（i）$Br^- + MnO_2 + H^+ \longrightarrow$

（j）$Cl^- + H_2SO_4(Conc) \longrightarrow$

（k）$Br^- + H_2SO_4(Conc) \longrightarrow$

（l）$I^- + H_2SO_4(Conc) \longrightarrow$

（m）$I^- + HClO + H^+ \longrightarrow$

（n）$I_2 + HClO + H_2O \longrightarrow$

（o）$I^- + ClO_3^- + H^+ \longrightarrow$

（p）$AgCl(s) + NH_3(superfluous) \longrightarrow$

（q）$AgBr(s) + Zn \longrightarrow$

（r）$AgI(s) + Zn \longrightarrow$

2. Operation

（1）Separate the precipitate with the centrifugal machine.

（2）Examine whether the precipitate is completely subsided.

（3）Detect the small quantity of the emitted gas from the test tube.

（4）Extract a small quantity of liquid in the test tube.

3. Report format

（1）Objectives

（2）Procedures

Samples	Chemicals	Phenomenon	Reaction equation

Conclusion

（3）Separation and detection of the mixed solution containing Cl^-, Br^-, I^-

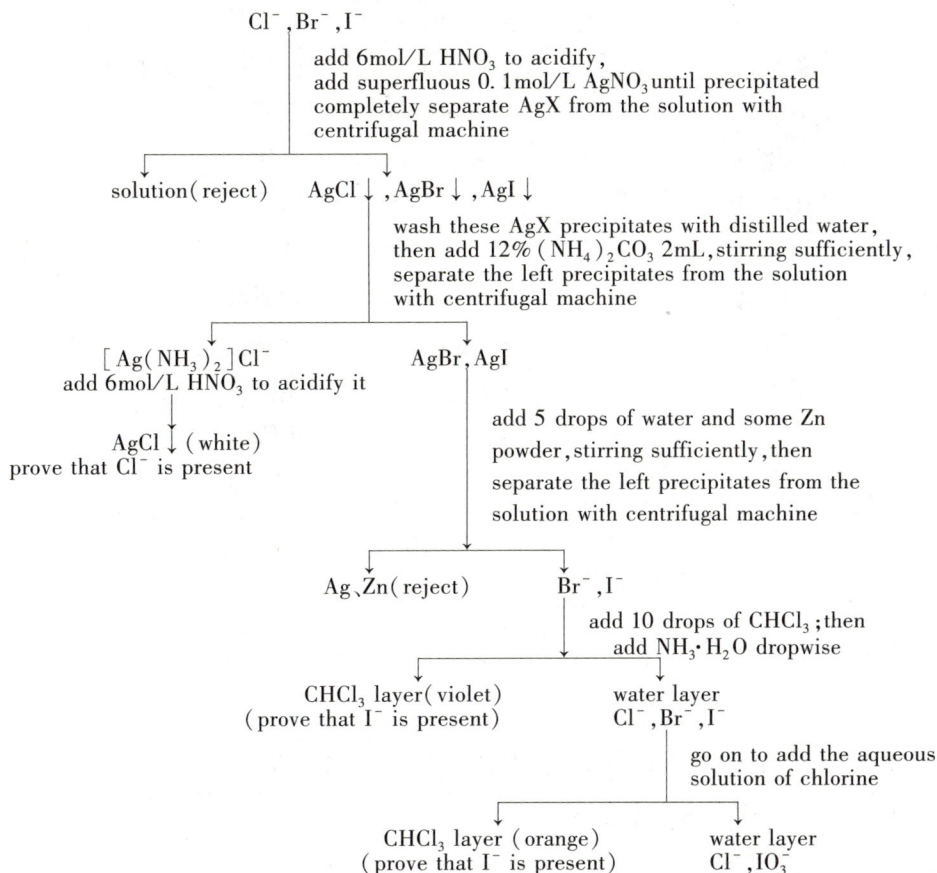

Cl^-,Br^-,I^-

add 6mol/L HNO_3 to acidify,
add superfluous 0.1mol/L $AgNO_3$ until precipitated
completely separate AgX from the solution with
centrifugal machine

solution(reject) AgCl↓,AgBr↓,AgI↓

wash these AgX precipitates with distilled water,
then add 12% $(NH_4)_2CO_3$ 2mL,stirring sufficiently,
separate the left precipitates from the solution
with centrifugal machine

$[Ag(NH_3)_2]^+Cl^-$
add 6mol/L HNO_3 to acidify it

AgBr,AgI

AgCl↓(white)
prove that Cl^- is present

add 5 drops of water and some Zn
powder,stirring sufficiently,then
separate the left precipitates from the
solution with centrifugal machine

Ag、Zn(reject) Br^-,I^-

add 10 drops of $CHCl_3$;then
add $NH_3 \cdot H_2O$ dropwise

$CHCl_3$ layer(violet)
(prove that I^- is present)

water layer
Cl^-,Br^-,I^-

go on to add the aqueous
solution of chlorine

$CHCl_3$ layer (orange)
(prove that I^- is present)

water layer
Cl^-,IO_3^-

4. Questions

(1) Explain the cause of insolubility of Ag_2CO_3, Ag_3PO_4, Ag_2CrO_4 and the silver halides (exception of AgF) in the water using Molecule Structure Theory.

(2) Illuminate the difference of oxidizing activity of oxychlorates with the different oxidation states of chlorine. What do you think of "Increasing oxidation state means increasing oxidizing activity"?

(3) Why is AgF readily soluble in water? Why is the boiling point of HF highest among the hydrogen haloid?

(4) How to detect the solution containing I^-?

Words

halogen	卤素	chloroform	三氯甲烷
halide	卤化物,卤化物的	litmus	石蕊
haloid	卤素的,卤素盐	aqueous	水的
oxychloride	氯氧化物	phosphor	磷
chlorine	氯	sulphuric acid	硫酸
bromine	溴	illustrate	说明
iodine	碘	amylum	淀粉
fluorine	氟	concentrated	浓的
hypochlorite	次氯酸盐	precipitate	沉淀

chlorate	氯酸盐	centrifugal	离心的
identify	鉴别	pungent	(气味)刺激性的
halogenide	卤化物	equilibrium	平衡
oxidant	氧化剂	medium	介质
reducer	还原剂	natrium	钠
dilute	稀释的	kalium	钾
oxidizing activity	氧化性	oxyacid	含氧酸
dissolve	溶解	oxysalt	含氧酸盐
ammonia	氨	stir	搅拌
radical	根	subside	沉降,下沉

实验二 氧、硫

PPT

目的要求 •••

1. 验证 H_2O_2、H_2S、H_2SO_3、$Na_2S_2O_3$、$(NH_4)_2S_2O_8$ 的化学性质。

2. 掌握 H_2O_2、S^{2-}、$S_2O_3^{2-}$ 的检查方法。

3. 了解金属硫化物的难溶性。

实验原理 •••

1. 氧、硫是周期系第Ⅵ主族元素，氧是人类生存必需的气体。氢和氧的化合物，除了水以外，还有 H_2O_2。过氧化氢是强氧化剂，但和更强的氧化剂作用时，它又是还原剂。

2. H_2S 是有毒气体，能溶于水，其水溶液呈弱酸性。在 H_2S 中，S 的氧化数是 -2，H_2S 是强还原剂。S^{2-} 可与多种金属离子生成不同颜色的金属硫化物沉淀，例如 ZnS（白色）、CuS（棕黑色）、HgS（黑色）、CdS（黄色）。

3. 在碱性溶液中，亚硝酰铁氰化钠 $Na_2[Fe(CN)_5NO]$ 与 S^{2-} 生成红紫色的配合物，这是鉴定 S^{2-} 离子的灵敏反应，其反应可能是：

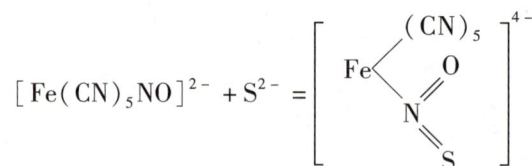

$$[Fe(CN)_5NO]^{2-} + S^{2-} = \left[Fe \underset{N}{\overset{(CN)_5}{\diagdown}} \begin{matrix} O \\ \| \\ N \\ \| \\ S \end{matrix} \right]^{4-}$$

4. SO_2 和 H_2SO_3 是还原剂，但与强还原剂作用时，又表现为氧化剂。

5. $Na_2S_2O_3$ 是一个还原剂，I_2 可以将它氧化成 $Na_2S_4O_6$。$Na_2S_2O_3$ 在酸性溶液中不稳定，会分解成 S 和 SO_2。

$$Na_2S_2O_3 + 2HCl \longrightarrow 2NaCl + SO_2 + S\downarrow + H_2O$$

$S_2O_3^{2-}$ 与 Ag^+ 的反应是它的特征反应。可用来鉴定 $S_2O_3^{2-}$ 离子。

$$S_2O_3^{2-} + 2Ag^+ \longrightarrow Ag_2S_2O_3$$
$$Ag_2S_2O_3 + H_2O \longrightarrow Ag_2S\downarrow + H_2SO_4$$

6. 浓 H_2SO_4 是一个强氧化剂，在加热时，它能氧化许多金属，而本身则随条件不同而被还原成 SO_2、S 或 H_2S。

H_2SO_4 的还原产物可按如下判断：如溶液呈白色混浊，示析出 S；如产生具有还原性的刺激性气体，则为 SO_2；如产生具有还原性或使 $Pb(Ac)_2$ 试纸变黑的，具有臭鸡蛋气味的气体则为 H_2S。

7. 过二硫酸盐系强氧化剂，能将 I^- 氧化成 I_2。

药品

1. 固体　MnO_2，$(NH_4)_2S_2O_8$，Zn 粒，Cu 粒，Na_2SO_3，$PbCO_3$ 或 $CdCO_3$。

2. 酸　H_2SO_4 浓，2mol/L；HCl 6mol/L；H_2S 溶液。

3. 碱　NaOH 40%；$NH_3 \cdot H_2O$ 2mol/L。

4. 盐　KI 0.1mol/L；$KMnO_4$ 0.01mol/L；$AgNO_3$ 0.1mol/L；$K_2Cr_2O_7$ 0.1mol/L；

　　$Pb(NO_3)_2$ 0.1mol/L；$CuSO_4$ 0.1mol/L；$Cd(NO_3)_2$ 0.1mol/L；

　　$FeSO_4$ 0.1mol/L；Na_2S 0.1mol/L；$Na_2S_2O_3$ 0.1mol/L；$ZnSO_4$ 0.1mol/L；

　　$MnSO_4$ 0.2mol/L；$Sr(NO_3)_2$ 0.5mol/L。

5. 其他　淀粉溶液，乙醇 95%；SO_2 水溶液；H_2O_2 3%；I_2 0.1mol/L；蓝色石蕊试纸，$Pb(Ac)_2$ 试纸。

实验内容

1. 过氧化氢

（1）H_2O_2 的氧化、还原性

①在试管中加 10 滴 0.1mol/L KI 溶液，用 2mol/L H_2SO_4 酸化后，加 2 滴 3% H_2O_2 溶液，观察有何变化。再加 5 滴淀粉试液，有何现象出现？写出反应式。

②取一小片 $Pb(Ac)_2$ 试纸，加 1 滴 H_2S 水溶液，则有黑色的 PbS 生成，然后再向 PbS 上滴 1 滴 3% H_2O_2 溶液，观察有何现象？写出反应式。

（2）H_2O_2 的酸性——过氧化钠的生成　往试管中加 0.5mL 40% NaOH 溶液和 2 滴 3% H_2O_2 溶液，再加 1mL 乙醇以降低生成物的溶解度，振荡，观察反应情况与产物的颜色和状态，写出反应式，并加以解释。

（3）H_2O_2 的催化分解　往盛有 2mL 3% H_2O_2 溶液的试管中，加少量 MnO_2 固体，观察 H_2O_2 的催化分解。试验有 O_2 生成（管口用火柴余烬检验）。写出反应式。

2. 硫化氢（H_2S）的性质

（1）H_2S 的还原性　取 2 支试管，分别加 5 滴 0.1mol/L $K_2Cr_2O_7$ 和 0.01mol/L $KMnO_4$ 溶液，再加 2mol/L H_2SO_4 酸化，再分别加入数滴 H_2S 溶液，观察现象，写出反应式。

（2）H_2S 与金属离子的反应　取 6 支试管，分别加 0.1mol/L $AgNO_3$、$Pb(NO_3)_2$、$CuSO_4$、$FeSO_4$①、$ZnSO_4$ 和 $Cd(NO_3)_2$ 溶液 2～3 滴，滴加 H_2S 饱和溶液，观察各试管中有无沉淀生成，若无沉淀，继续加 2mol/L 氨水至碱性，观察各试管中沉淀的颜色或生成沉淀的情况。

（3）S^{2-} 离子的鉴定　取数滴 0.1mol/L Na_2S 溶液，滴入新配制的 1% $Na_2[Fe(CN)_5NO]$（亚硝酰铁氰化钠）溶液，观察产生的现象。

---

①　由于 Fe^{2+} 在空气中易被氧化，所以在 $FeSO_4$ 溶液中难免存在 Fe^{3+}，当加入 H_2S 水溶液时，$2Fe^{3+} + H_2S \Longrightarrow 2Fe^{2+} + S\downarrow + 2H^+$ 溶液中将出现白色浑浊。

3. 亚硫酸（H_2SO_3）的性质

（1）H_2SO_3 的氧化性　取 H_2SO_3 溶液 2mL（即 SO_2 水溶液），先试酸性，然后滴加 H_2S 溶液，观察现象，写出反应式。

（2）H_2SO_3 的还原性　取 0.01mol/L $KMnO_4$ 溶液 10 滴，加 5 滴 2mol/L H_2SO_4 溶液，再滴加 H_2SO_3 溶液，观察现象，写出反应式。

4. 硫代硫酸钠（$Na_2S_2O_3$）的性质

（1）与酸反应　取 1mL 0.1mol/L $Na_2S_2O_3$ 溶液，滴加 6mol/L HCl 溶液，观察现象，并写出反应式。

（2）还原性　取 0.1mol/L $Na_2S_2O_3$ 溶液 5 滴，然后逐滴加入 0.1mol/L I_2 溶液，不断振摇试管，观察现象并写出反应式。

（3）$S_2O_3^{2-}$ 的鉴定　取 0.1mol/L $Na_2S_2O_3$ 溶液 5 滴，加 1mL 水，滴加 0.1mol/L $AgNO_3$ 溶液，即有白色 $Ag_2S_2O_3$ 沉淀生成，振摇后沉淀溶解，再加入过量的 $AgNO_3$ 溶液，又有白色沉淀生成，但迅速转变为黄色、红棕色，最后转变为棕黑色。写出反应式。

5. 浓 H_2SO_4 与金属作用

（1）浓 H_2SO_4 与活泼金属作用　取浓 H_2SO_4 约 2mL，小心放入 Zn 粒，微热，观察现象，如何用化学方法判断所产生的气体是什么。写出反应方程式。

（2）浓 H_2SO_4 与不活泼金属的作用　如上法做浓 H_2SO_4 和金属铜的反应，观察溶液颜色的变化，用化学方法判断所产生的气体，写出反应方程式。

6. 过二硫酸铵 $[(NH_4)_2S_2O_8]$ 的氧化性

（1）取 $(NH_4)_2S_2O_8$ 晶体少许，加入 2mL 水溶解后，滴加 0.1mol/L KI 溶液，观察溶液颜色的变化，再加入淀粉试液有何现象出现，写出反应方程式。

（2）取 2mL 2mol/L H_2SO_4 和 4～5 滴 0.2mol/L $MnSO_4$ 溶液，混合均匀后，加 1 滴 0.1mol/L $AgNO_3$（作催化剂）和少量 $(NH_4)_2S_2O_8$ 固体，微热，观察溶液的颜色有何变化。

$$5S_2O_8^{2-} + 2Mn^{2+} + 8H_2O === 10SO_4^{2-} + 2MnO_4^- + 16H^+$$

7. S^{2-}、SO_3^{2-}、$S_2O_3^{2-}$ 混合离子的分离和检出

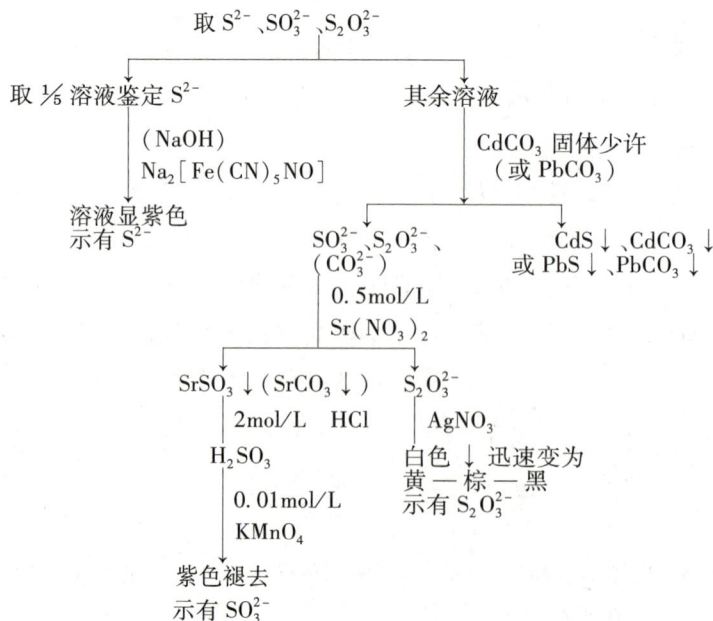

实 验 指 导

预习要求 ●●●

1. 说明过氧化氢在氧化还原性、酸碱性、稳定性等方面的性质，并根据实验内容用反应式表示。

2. 本实验中怎样试验硫化氢的还原性？

3. 实验室为何不能长期保存硫化氢水溶液？

4. 为何不能用硝酸同 FeS 作用来制备 H_2S？

5. 为什么在 $FeSO_4$ 溶液中滴加 H_2S 饱和溶液无沉淀生成，当加入 2mol/L 氨水至碱性后有 FeS 沉淀生成？

6. 试比较浓硫酸与 Zn 粒和 Cu 粒作用的产物。

7. 亚硫酸、硫代硫酸钠、过二硫酸铵有哪些主要性质？怎样用实验加以验证。

8. 如何分离和鉴定 S^{2-}、SO_3^{2-}、$S_2O_3^{2-}$？

9. $Na_2S_2O_3$ 分别与 I_2 和 Cl_2 反应，产物有何不同？

10. 完成下列反应：

$$KI + H_2O_2 + H_2SO_4 \longrightarrow$$

$$PbS + H_2O_2 \longrightarrow$$

$$H_2O_2 + NaOH \longrightarrow$$

$$H_2O_2 \xrightarrow{\ MnO_2\ }$$

$$Cr_2O_7^{2-} + H_2S + H^+ \longrightarrow$$

$$MnO_4^- + H_2S + H^+ \longrightarrow$$

$$MnO_4^- + SO_3^{2-} + H^+ \longrightarrow$$

$$Na_2S_2O_3 + HCl \longrightarrow$$

$$Na_2S_2O_3 + I_2 \longrightarrow$$

$$Na_2S_2O_3（过量）+ AgNO_3 \longrightarrow$$

$$Na_2S_2O_3 + AgNO_3（过量）\longrightarrow$$

$$Ag_2S_2O_3 + H_2O \longrightarrow$$

$$H_2SO_4（Conc）+ Zn \longrightarrow$$

$$H_2SO_4（Conc）+ Cu \longrightarrow$$

$$(NH_4)_2S_2O_8 + KI \longrightarrow$$

$$(NH_4)_2S_2O_8 + MnSO_4 + H_2O \xrightarrow{\ AgNO_3\ }$$

基本操作 ●●●

掌握用气体发生器制备 H_2S 气体的基本操作。

注意事项 ●●●

1. 本实验中 H_2S、SO_2 都是有毒气体，并有刺激性气味，须在通风橱进行。

2. $FeSO_4$ 溶液中，Fe^{2+} 在空气中易被氧化成 Fe^{3+}，当加入 H_2S 水溶液时，由于

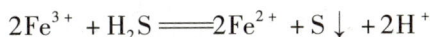

$$2Fe^{3+} + H_2S \Longrightarrow 2Fe^{2+} + S\downarrow + 2H^+$$

而出现白色沉淀。

3. 在 $Na_2S_2O_3$ 溶液中逐滴加入 $AgNO_3$，由于 $AgNO_3$ 局部浓度过大即产生白色沉淀，（沉淀是什么？）但一振摇沉淀即消失，当 $AgNO_3$ 过量时，白色沉淀产生且迅速转为黄色、红棕色最后为棕黑色。必须仔细观察。

4. 浓硫酸与活泼金属 Zn 粒作用产物可能有 SO_2、S、H_2S，注意观察与鉴别。

5. 在 S^{2-}、SO_3^{2-}、$S_2O_3^{2-}$ 混合离子的分离与鉴定实验中注意沉淀完全与沉淀分离。

报告格式 •••

1. 目的
2. 实验内容

试 样	加入试剂	现 象	反应方程式

结论

实验后思考 •••

1. 试另行设计分离和检出 S^{2-}、SO_3^{2-}、$S_2O_3^{2-}$ 混合离子的方案。

2. 有一白色固体 A，加入油状液体 B，可得紫黑色固体 C，C 微溶于水，加入 A 后溶解度增大，成棕色溶液 D。把 D 分成两份。一份中加一种无色溶液 E，另一份通入气体 F，都褪色成无色透明溶液。E 溶液遇酸有刺激性气体 G 产生同时有黄色沉淀 H 析出，气体 G 能使 $KMnO_4$ 褪色，将气体 F 通入溶液 E 中，在所得的溶液中加入 $BaCl_2$ 溶液有白色沉淀产生，且沉淀难溶于 HNO_3，问 A、B、C、D、E、F、G、H 各为何物？写出各步反应式。

3. 根据硫化物溶解度的差别如何分离 Na^+，Cu^{2+}，Zn^{2+}？

Experiment 2　　Oxygen and Sulphur

◆ Objectives

1. Test the chemical properties of H_2O_2, H_2S, H_2SO_3, $Na_2S_2O_3$ and $(NH_4)_2S_2O_8$.

2. Grasp the methods of detecting H_2O_2, S^{2-}, $S_2O_3^{2-}$.

3. Understand the insolubility of the sulphides of metals.

◆ Principles

1. Group Ⅵ elements comprise the typical elements oxygen (O) and sulphur (S). Oxygen is the most important gas to human being's life. Because of its abundance and high oxidizing activity. Oxygen predetermines the form of existence of all other elements in the Earth. The most important oxide is water (hydrogen oxide H_2O), but another important substance is hydrogen peroxide (H_2O_2), which is widely used as a powerful oxidizing agent. Its reducing properties are only displayed with such stronger oxidants as the MnO_4^- ion.

2. Hydrogen sulphide (H_2S) is very poisonous with offensive odor. The aqueous solution of H_2S is a weak acid known as hydrosulphuric acid. Hydrogen sulphide is a strong reducer with the lowest oxidation state (S^{2-}).

When sulphides of metals are formed, they shows characteristic colors, e. g. , ZnS is white, CuS is brown black, HgS is black and CdS is yellow.

3. When in the alkaline solution, a red violet complex is produced by $Na_2[Fe(CN)_5NO]$ reacting with S^{2-}, which is used in detecting S^{2-}:

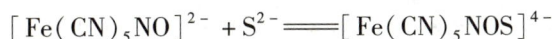

$$[Fe(CN)_5NO]^{2-} + S^{2-} = [Fe(CN)_5NOS]^{4-}$$

4. Sulphur dioxide (SO_2) and its aqueous solution known as sulphurous acid (H_2SO_3) are all strong reductants unless they react with other stronger reducers, the oxidizing properties are also displayed.

5. $Na_2S_2O_3$ displays reducing properties in reaction with I_2 and the final product is $Na_2S_4O_6$. In acid medium $Na_2S_2O_3$ undergoes decomposition:

$$Na_2S_2O_3 + 2HCl = 2NaCl + SO_2 + S + H_2O$$

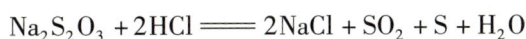

The characteristic reaction between thiosulphate ($S_2O_3^{2-}$) and silver salt (Ag^+) is employed to detect $S_2O_3^{2-}$:

$$S_2O_3^{2-} + 2Ag^+ \longrightarrow Ag_2S_2O_3$$
$$Ag_2S_2O_3 + H_2O \longrightarrow Ag_2S \downarrow + H_2SO_4$$

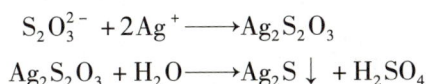

6. The concentrated sulphuric acid (H_2SO_4) is a strong oxidant. When heated, it always oxidizes many metals with the reduction products such as sulphur dioxide SO_2, sulphur S or hydrogen sulphide H_2S depending on the diverse conditions. Sulphur may be produced if the solution turns white and turbid; SO_2 may be produced if an offensive odor gas is emitted; H_2S may be produced if $Pb(Ac)_2$ test paper turns black and an offensive gas smells like rancid eggs is emitted.

◆ Chemicals

1. Solid MnO_2, $(NH_4)_2S_2O_8$, zinc particles, cuprum particles, Na_2SO_3, $PbCO_3$ or $CdCO_3$

2. Acid   Concentrated H_2SO_4 2mol/L, HCl 6mol/L; aqueous solution of H_2S

3. Alkali   NaOH 40%; $NH_3 \cdot H_2O$ 2mol/L

4. Salt   KI 0. 1mol/L; KMnO_4 0. 01mol/L; AgNO_3 0. 1mol/L; K_2Cr_2O_7 0. 1mol/L;
   Pb(NO_3)_2 0. 1mol/L; CuSO_4 0. 1mol/L; Cd(NO_3)_2 0. 1mol/L; FeSO_4 0. 1mol/L;
   Na_2S 0. 1mol/L; Na_2S_2O_3 0. 1mol/L; ZnSO_4 0. 1mol/L; MnSO_4 0. 2mol/L
   Sr(NO_3)_2 0. 5mol/L

5. Rest   Asylum solution, ethanol 95%; the aqueous solution of SO_2, H_2O_2 3%, I_2 0. 1mol/L; Blue litmus test paper, $Pb(Ac)_2$ test paper

◆ Procedures

1. Hydrogen peroxides (H_2O_2)

(1) The oxidizing and reducing activity of H_2O_2

(a) Add 10 drops of 0. 1mol/L KI solution to a test tube, then acidify it with 2mol/L H_2SO_4, then add 2 drops of 3% H_2O_2 solution. Observe the change occurred. Then add 5 drops of starch solution, and observe the color change. Write out the chemical reaction equations.

(b) Add drops of the aqueous solution of H_2S to a piece of $Pb(Ac)_2$ test paper, a black PbS must be formed. Then add a drop of 3% H_2O_2 solution to PbS, observe what occurs. Write out the chemical reaction equations.

(2) The acidity of H_2O_2 – to form Na_2O_2

Add 0.5mL 40% NaOH solution and 2 drops of 3% H_2O_2 solution to a test tube, then 1mL C_2H_5OH is added to decrease the solubility of the product, shake, observe the color and the state of the product. Write out the chemical reaction equations.

(3) The decomposition of H_2O_2 under catalyst

Add a small quantity of $MnO_2(s)$ to a test tube containing 2mL 3% H_2O_2 solution, observe the decomposition. Put a match with spark near the outlet of the tube to validate oxygen is emitted. Write out the chemical reaction equations.

2. The properties of hydrogen sulphide (H_2S)

(1) The reducing activity of H_2S

Add 5 drops of 0.1mol/L $K_2Cr_2O_7$ and 0.01mol/L $KMnO_4$ to two tubes respectively, then add 2mol/L H_2SO_4 to acidify them. Finally add drops of the aqueous solution of H_2S and then observe the phenomenon. Write out the chemical reaction equations.

(2) Reaction with metal ions

Add drops of 0.1mol/L $AgNO_3$, $Pb(NO_3)_2$, $CuSO_4$, $FeSO_4$, $ZnSO_4$ and $Cd(NO_3)_2$ to each of the six test tubes respectively, then add saturated solution of H_2S. Observe carefully whether the precipitate is formed in each tube. If in some tube there is no precipitate, please add 2mol/L $NH_3 \cdot H_2O$ to create a alkaline environment. Take attention to the color of these precipitates.

(3) Identification of S^{2-}

Add drops of 0.1mol/L Na_2S solution to 1% $Na_2[Fe(CN)_5NO]$ solution just prepared. Observe the phenomenon.

3. The Properties of sulphurous acid (H_2SO_3)

(1) The oxidizing activity of H_2SO_3

Take 2mL of H_2SO_3 solution (aqueous solution of SO_2) and test its acidity. Then add H_2S solution, please observe the phenomenon and write out the reaction equation.

(2) The reducing activity of H_2SO_3

Add 5 drops of H_2SO_4 solution to 10 drops of 0.01mol/l $KMnO_4$. Then add H_2SO_3 solution, please observe the phenomenon and write out the reaction equation.

4. The properties of natrium thiosulphate ($Na_2S_2O_3$)

(1) Reaction with acid Add 6mol/L hydrochloric acid (HCl) solution to a test tube containing 1mL 0.1mol/L $Na_2S_2O_3$ solution, observe the phenomenon and write out the chemical reaction equations.

(2) Reducing activity Add drops of 0.1mol/L I_2 solution drop by drop to a test tube containing 5 drops of 0.1mol/L $Na_2S_2O_3$ solution while shaking constantly, observe the phenomenon and write out the chemical reaction equations.

(3) Identification of $S_2O_3^{2-}$ Add 5 drops of 0.1mol/L $Na_2S_2O_3$ solution and 1mL H_2O to a test tube, and then add 0.1mol/L $AgNO_3$ solution. If a white precipitate ($Ag_2S_2O_3$) is produced (but disappear after shaking), add superfluous $AgNO_3$ solution again, then a white precipitate reappears. But it promptly turns from yellow to red brown to final brown black.

5. Concentrated H_2SO_4 reacting with metals

(1) Concentrated H_2SO_4 reaction with active metals Cautiously add some zinc particles into a test tube containing 2mL concentrated H_2SO_4, heated a little, observe the phenomenon. Estimate the emitted gases with

chemical method and write out the reaction equation.

（2）Concentrated H_2SO_4 reaction with inactive metals　Reaction between Concentrated H_2SO_4 and copper particles occurs at the same condition mentioned above. Estimate the emitted gases with chemical method and write out the reaction equation.

6. The oxidizing activity of $(NH_4)S_2O_8$

（1）Add a small quantity of $(NH_4)_2S_2O_8$ crystals to a test tube, then 2mL H_2O is added to dissolve them. After that titrate 0.1mol/L KI solution to the tube, observe the color change. Then add drops of starch solution observe the phenomenon and write out the chemical reaction equations.

（2）Mix 2mL 2mol/L H_2SO_4 and 4~5 drops of 0.2mol/L $MnSO_4$ solution well-proportionally, then a drop of 0.1mol/L $AgNO_3$（as a catalyst）and a small quantity of $(NH_4)_2S_2O_8$ crystals are added, little heated. Observe the color change of the solution.

$$5S_2O_8^{2-} + 2Mn^{2+} + 8H_2O === 10SO_4^{2+} + 2MnO_4^- + 16H^+$$

7. Separation and detection of the mixed solution containing $S^{2-}, SO_3^{2-}, S_2O_3^{2-}$

15 drops of the mixture solution containing $S^{2-}, SO_3^{2-}, S_2O_3^{2-}$

20% of the solution kept for identifying S^{2-}

（NaOH）
$Na_2[Fe(CN)_5NO]$

Solution appears violet（prove that S^{2-} is present）

the left solution a few $CdCO_3$（s）or $PbCO_3$

$SO_3^{2-}, S_2O_3^{2-}$（CO_3^{2-}）

$CdS, CdCO_3$ or $PbS, PbCO_3$

0.5mol/L $Sr(NO_3)_2$

$S_2O_3^{2-}$

$AgNO_3$

White but promptly turn yellow-brown-black（prove that $S_2O_3^{2-}$ is present）

$SrSO_3$（$SrCO_3$）

2mol/L HCl

H_2SO_3

0.01mol/L $KMnO_4$

Violet color disappears（prove SO_3^{2-} is present）

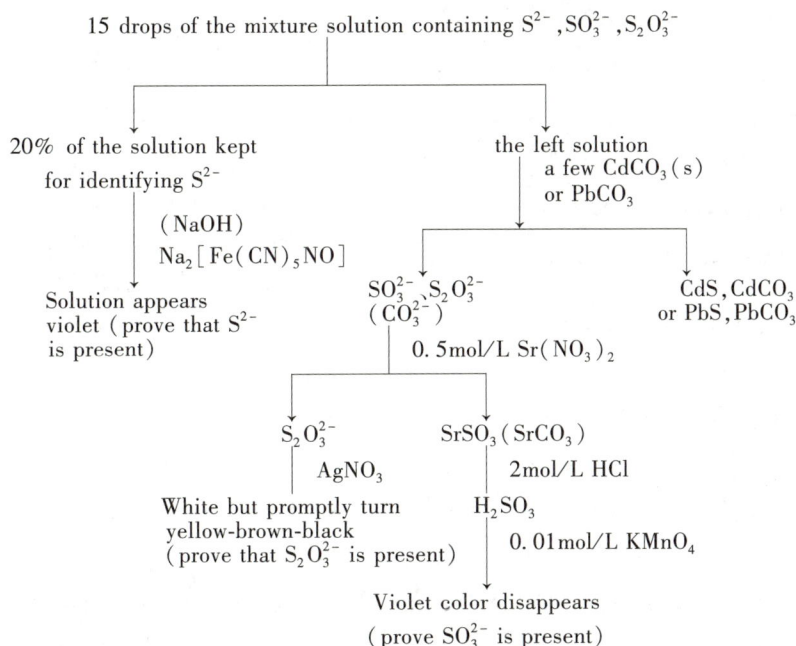

◆ Instructions

1. Requirements

（1）Illustrate the oxidizing and reducing activity, acidity and solubility of H_2O_2, write the relevant reaction equations.

（2）How to test the reducing activity of H_2S in this experiment?

（3）Why do the aqueous solution of H_2S not be kept immutable for a long period time in the lab?

（4）Why cannot H_2S be produced by HNO_3 reacting with FeS?

（5）Why is no precipitate formed when adding the saturated solution of H_2S to $FeSO_4$ solution? But if 2mol/L $NH_3 \cdot H_2O$ is added more?

（6）Try to compare the product of concentrated H_2SO_4 reacting with zinc particles and reacting with copper particles.

（7）Enumerate the main properties of $H_2SO_3, Na_2S_2O_3, (NH_4)_2S_2O_8$ and validate them with experiments.

（8）How to separate and detect S^{2-}, SO_3^{2-}, $S_2O_3^{2-}$?

（9）What different product is formed when $Na_2S_2O_3$ reacts with I_2 and with Cl_2?

（10）Complete the following reaction equations：

$$KI + H_2O_2 + H_2SO_4 \longrightarrow$$

$$PbS + H_2O_2 \longrightarrow$$

$$H_2O_2 + NaOH \longrightarrow$$

$$H_2O_2 \xrightarrow{\quad MnO_2 \quad}$$

$$Cr_2O_7^{2-} + H_2S + H^+ \longrightarrow$$

$$MnO_4^- + H_2S + H^+ \longrightarrow$$

$$MnO_4^- + SO_3^{2-} + H^+ \longrightarrow$$

$$Na_2S_2O_3 + HCl \longrightarrow$$

$$Na_2S_2O_3 + I_2 \longrightarrow$$

$$Na_2S_2O_3\,(superfluous) + AgNO_3 \longrightarrow$$

$$Na_2S_2O_3 + AgNO_3\,(superfluous) \longrightarrow$$

$$Ag_2S_2O_3 + H_2O \longrightarrow$$

$$H_2SO_4\,(Conc) + Zn \longrightarrow$$

$$H_2SO_4\,(Conc) + Cu \longrightarrow$$

$$(NH_4)_2S_2O_8 + KI \longrightarrow$$

$$(NH_4)_2S_2O_8 + MnSO_4 + H_2O \xrightarrow{\quad AgNO_3 \quad}$$

2. Operation

Grasp the basic operation with the equipment used to prepare H_2S.

3. Notes

（1）H_2S and SO_2 are both poisonous gases with offensive odor. Be careful and you should work in a ventilated cabinet.

（2）Fe^{2+} is easily oxidized to Fe^{3+} in the air, so when H_2S solution is added to $FeSO_4$ solution, a white precipitate is produced：

$$2Fe^{3+} + H_2S =\!=\!= 2Fe^{2+} + S + 2H^+$$

（3）When trickle down drops of $AgNO_3$ to $Na_2S_2O_3$ solution, a white precipitation is formed immediately because of the partly superfluous density of $AgNO_3$. After shaking the tube, it disappears. If abundant $AgNO_3$ solution is added more, the white precipitate is produced again but at once turn yellow, then red brown, last brown black. So pay attention to the color change in this process.

（4）Be careful to observe and detect the variance products when the concentrated H_2SO_4 reacting with active metal（Zinc particle in this experiment）.

（5）Be sure of the completeness of forming precipitates and the exhaustiveness of separating them from the solution mixed with S^{2-}, SO_3^{2-}, $S_2O_3^{2-}$.

4. Report format

（1）Objectives

（2）Procedure

Samples	Chemicals	Phenomenon	Reaction equation

Conclusion

5. Questions

（1）Please design another experiment to separate and detect S^{2-}, SO_3^{2-}, $S_2O_3^{2-}$.

（2）A white solid, if added with oiled liquid B, a violet black solid C is formed, which is soluble in the water. Then if added with A, C is melt into D, a brown solution. Then D is divided into two tubes, one of which is added with a colorless transparent solution E, another is added with gas E. Both of them in two tubes turn to colorless transparent solution. When E reacts with some acid, gas G that makes $KMnO_4$ fade is emitted and a yellow precipitate H is produced. After gas F is added to E, $BaCl_2$ solution is added more, a white precipitate insoluble in HNO_3 is produced. Please write out the concrete substances corresponding to A, B, C, D, E, F, G and H and all the reaction equations.

（3）How to separate Na^+, Cu^{2+}, Zn^{2+}.

实验三　氮、磷、砷、锑、铋

PPT

目的要求 •••

1. 了解氮、磷、砷、锑、铋重要化合物的性质。
2. 学会砷的鉴别方法。

实验原理 •••

1. 氮族元素为周期系第 V 主族元素，它们原子的最外层上有 5 个电子，所以它们的氧化数最高为 +5，最低为 -3。

2. 氨的水溶液呈弱碱性，铵盐加热时易分解。

3. 亚硝酸（HNO_2）可用酸分解亚硝酸盐而得到，但不稳定，易分解成 NO 和 NO_2。HNO_2 具有氧化性，但遇强氧化剂时可呈还原性。

4. 硝酸是强酸，亦是强氧化剂，它与非金属作用时，常被还原为 NO_2 或 NO，与金属作用时被还原的产物决定于硝酸的浓度和金属的活泼性。

5. 磷酸是三元酸，故可形成酸式盐和正盐，磷酸的钙盐在水中的溶解度是不同的。$Ca_3(PO_4)_2$ 和 $CaHPO_4$ 难溶于水，而 $Ca(H_2PO_4)_2$ 则易溶于水。

6. 氮族元素从 N 到 Bi 由于离子半径渐增，其氧化物的酸性渐减，碱性渐增，故 N_2O_3、P_2O_3 为酸性氧化物。AsO_3、Sb_2O_3 为两性氧化物，Bi_2O_3 为碱性氧化物。

从 As 到 Bi 三价化合物的还原性逐渐减弱。

从 As 到 Bi 五价化合物的氧化性逐渐增强。

7. 砷的鉴定主要利用 AsO_3^{3-} 或 AsO_4^{3-} 还原为 AsH_3，然后利用 AsH_3 的还原性还原 Ag^+ 离子为 Ag。

反应式为：

$$AsO_3^{3-} + 9H^+ + 3Zn \Longrightarrow AsH_3 \uparrow + 3Zn^{2+} + 3H_2O$$

$$AsO_4^{3-} + 11H^+ + 4Zn \Longrightarrow AsH_3 \uparrow + 4Zn^{2+} + 4H_2O$$

$$6AgNO_3 + AsH_3 \Longrightarrow Ag_3As \cdot 3AgNO_3 \downarrow （黑） + 3HNO_3$$

$$6Ag^+ + AsH_3 + 3H_2O \Longrightarrow 6Ag \downarrow （黑） + H_3AsO_3 + 6H^+$$

试样中有硫化物存在时，遇酸发生 H_2S 能使 $AgNO_3$ 变为黑色的 Ag_2S。为了消除 H_2S 的干扰，需用 $Pb(Ac)_2$ 棉花吸收 H_2S。

试样中若有锑化物存在，亦可产生 SbH_3，同样使 $AgNO_3$ 变黑，故有干扰，此时可设法在碱性溶液中进行反应，锑化物则不会产生 SbH_3（用碱式还原法没有 H_2S、SbH_3 的干扰，但 AsO_4^{3-} 必须先还原成 AsO_3^{3-} 形式）。

8. Sb^{3+} 的鉴定主要取用 Sb^{3+} 与 $Na_2S_2O_3$ 生成了橘红色硫氧化锑 Sb_2OS_2 沉淀。反应式为：

$$2Sb^{3+} + 3S_2O_3^{2-} \longrightarrow 4SO_2 \uparrow + Sb_2OS_2 \downarrow$$
$$（橘红色）$$

溶液中酸性不宜过强，否则 $Na_2S_2O_3$ 将分解为 SO_2 和 S，有碍鉴定反应的进行。

9. Bi^{3+} 的鉴定主要利用 Bi^{3+} 与亚锡酸钠溶液作用，试剂使 Bi^{3+} 还原为黑色金属铋，是铋离子的主要定性反应之一。反应式为：

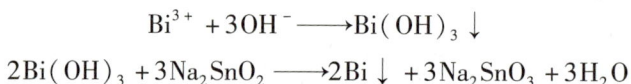

$$Bi^{3+} + 3OH^- \longrightarrow Bi(OH)_3 \downarrow$$

$$2Bi(OH)_3 + 3Na_2SnO_2 \longrightarrow 2Bi \downarrow + 3Na_2SnO_3 + 3H_2O$$

反应进行时，必须避免加入浓碱和加热，否则亚锡酸钠将分解生成黑色金属锡的沉淀。

$$2Na_2SnO_2 + H_2O \longrightarrow Na_2SnO_3 + Sn \downarrow + 2NaOH$$

仪器和药品

1. 仪器 大试管（装有软木塞及导气管），离心机，离心管。

2. 药品

（1）固体 NH_4Cl，$AgNO_3$，$Pb(NO_3)_2$，$NaNO_2$，Cu，Na_3BiO_3，As_2O_3（剧毒），Zn 粒，$Na_2S_2O_3$。

（2）酸 H_2SO_4 1mol/L；HNO_3 浓，6mol/L；HCl 6mol/L，2mol/L。

（3）碱 $NaOH$ 6mol/L；2mol/L；$NH_3 \cdot H_2O$ 6mol/L。

（4）盐 NH_4NO_3 0.1mol/L，$CaCl_2$ 0.1mol/L，$AgNO_3$ 0.1mol/L，KI 0.1mol/L，$AsCl_3$ 0.1mol/L，$SnCl_2$ 溶液，$NaNO_2$ 1mol/L，$SbCl_3$ 0.1mol/L，$KMnO_4$ 0.01mol/L，$Bi(NO_3)_3$ 0.1mol/L，Na_3PO_4 0.1mol/L，Na_2S 2mol/L，Na_2HPO_4 0.1mol/L，NaH_2PO_4 0.1mol/L，$MnSO_4$ 0.05mol/L。

3. 其他 奈氏试剂，红色石蕊试纸，淀粉溶液，pH 试纸，H_2S 溶液，I_2 溶液 0.01mol/L，$Pb(Ac)_2$ 棉花。

实验内容

1. 铵离子的鉴定（气室法） 取一块较大的表面皿，其上覆盖一块小表面皿制成气室。在小表皿的内壁贴一小块用水湿润的红色石蕊试纸和一小块用奈氏试剂湿润的试纸，在大表皿上滴加铵离子试液及 6mol/L NaOH 各 2 滴，立即将小表面皿盖上，稍待片刻，必要时在水浴上微热，即见红色石蕊试纸变蓝，奈氏试剂显棕褐色，写出反应式（同时做空白试验）。

2. 亚硝酸的性质

（1）亚硝酸的氧化性 在试管中加入两滴 0.1mol/L KI 溶液，加水稀释至 1mL。用 1mol/L H_2SO_4 酸化后，滴加 1mol/L NaNO₂ 溶液 2 滴，观察现象。再加入淀粉溶液一滴，观察现象。写出反应式。

（2）亚硝酸的还原性　　在试管中加入 5 滴 0.01mol/L KMnO₄ 溶液。用 1mol/L H₂SO₄ 酸化，然后滴加 1mol/L NaNO₂ 溶液，观察溶液颜色的变化。写出反应式。

3. 硝酸和硝酸盐的性质

（1）硝酸的氧化性　　分别取一颗小铜粒于两试管中，一管加入浓 HNO₃ 10 滴；另一管中加入 6mol/L硝酸 10 滴，必要时可微微加热，比较两试管反应有何不同？如何说明硝酸是氧化性的酸（本实验在通风橱中进行）？

（2）硝酸盐受热分解　　在 3 支试管中分别加少量硝酸银、硝酸铅和硝酸钠固体，用喷灯直火灼热，观察放出气体的颜色。取带有火星的火柴插入试管，检验气体产物。管中残渣是什么？写出反应式。

4. 磷酸盐的性质　　取 3 支试管，分别加 2 滴 0.1mol/L Na₃PO₄、Na₂HPO₄ 和 NaH₂PO₄ 溶液及 1mL 蒸馏水，用 pH 试纸检验它们的 pH，然后在每支试管中加入 10 滴 0.1mol/L CaCl₂ 溶液，振摇。观察何者有沉淀产生？加入氨水至碱性，观察有何变化。最后各加入 6mol/L 盐酸至酸性，沉淀是否溶解？比较磷酸钙、磷酸氢钙与磷酸二氢钙的溶解度，说明它们之间相互转化的条件。写出反应式。

5. 砷、锑、铋的性质

（1）硫化物和硫代酸盐　　在 3 支离心管中分别加 2 滴 0.1mol/L AsCl₃、SbCl₃ 和 Bi(NO₃)₃ 溶液，各加 2 滴 6mol/L HCl，再加 H₂S 饱和溶液至沉淀完全，观察硫化物的颜色，离心分离，弃去清液，然后向各管再加 20 滴 2mol/L Na₂S 溶液，搅拌之，哪种沉淀溶解，哪种沉淀不溶解？在沉淀已溶解后的清液中加 2mol/L HCl 酸化，又会发生什么变化？写出反应式。

（2）氧化物或氢氧化物的酸碱性

①三氧化二砷的性质　　取两支干试管，各加少量 As₂O₃（极毒，向老师领取）。在一支试管内加少量 2mol/L NaOH，振荡，As₂O₃ 是否溶解（保留溶液，供下面 Na₃AsO₃ 的还原性实验用）？在另一试管内加 6mol/L HCl，振荡，As₂O₃ 是否溶解？加热，As₂O₃ 是否溶解？根据实验结果说明 As₂O₃ 的酸性和碱性相对强弱。

②氢氧化亚锑的生成与性质　　向盛有 10 滴 0.1mol/L SbCl₃ 溶液的试管中滴加 2mol/L NaOH 溶液，观察现象，把沉淀分成两份，分别试验它们与 6mol/L NaOH 溶液和 6mol/L HCl 的作用，沉淀是否溶解？写出反应式。

③氢氧化铋（Ⅲ）的生成与性质　　向盛有 10 滴 0.1mol/L Bi(NO₃)₃ 溶液的试管中，滴加 2mol/L NaOH 溶液，观察反应产物的颜色和状态。把沉淀分成两份，分别试验它们与 6mol/L HCl 和 6mol/L NaOH 溶液的作用，沉淀是否溶解？写出反应式。

（3）三价砷（锑、铋）的还原和五价砷（锑、铋）的氧化性

①取少量（5~6 滴）由本实验得到的亚砷酸钠弱碱性溶液，滴加碘溶液，有何现象？然后将溶液用浓盐酸酸化，又有何变化？写出反应式并解释之。

②铋酸钠的氧化性。往盛有 5 滴 0.05mol/L MnSO₄ 试液和 10 滴 6mol/L HNO₃ 试管中加入少量铋酸钠固体，振摇试管并微热，观察到什么变化？写出反应式并解释之。

6. 离子鉴定

（1）砷的鉴定　　如图 4-4 所示，在试管中放含有砷的试液 1 滴，锌粒少许，滴加 6mol/L HCl 10 滴，在试管上半部放 Pb(Ac)₂ 棉花一小团，在棉花上放一小片沾有 AgNO₃ 溶液的滤纸，管口用纸套罩住，如图所示，数分钟后，使蘸有 AgNO₃ 的滤纸变为黑褐色或黑色，证明砷存在。

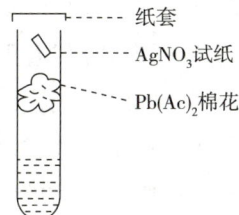

（2）Sb³⁺ 离子的鉴定　　取 Sb³⁺ 离子试液 10 滴，加热至沸，趁热加入

纸套
AgNO₃试纸
Pb(Ac)₂棉花

图 4-4　砷的鉴定

固体 $Na_2S_2O_3$ 少许，生成红色硫氧化锑（Sb_2OS_2）沉淀，示有 Sb^{3+} 离子。

（3）Bi^{3+} 离子的鉴定　亚锡酸钠溶液的配制。取 2 滴 $SnCl_2$ 溶液，滴入 2mol/L NaOH 溶液，使最初生成的 $Sn(OH)_2$ 沉淀溶解生成亚锡酸钠溶液。取 Bi^{3+} 离子试液 2 滴，加入新鲜配制的亚锡酸钠碱性溶液 4～5 滴，生成黑色金属铋沉淀，示有 Bi^{3+} 离子。

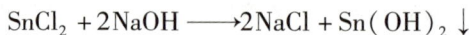

$$SnCl_2 + 2NaOH \longrightarrow 2NaCl + Sn(OH)_2 \downarrow$$

$$Sn(OH)_2 + 2NaOH \longrightarrow 2H_2O + Na_2SnO_2 \downarrow$$

$$2Bi^{3+} + 6OH^- + 3SnO_2^{2-} \longrightarrow 2Bi \downarrow + 3SnO_3^{2-} + 3H_2O$$

7. As^{3+}、Sb^{3+}、Bi^{3+} 混合离子的分离和检出

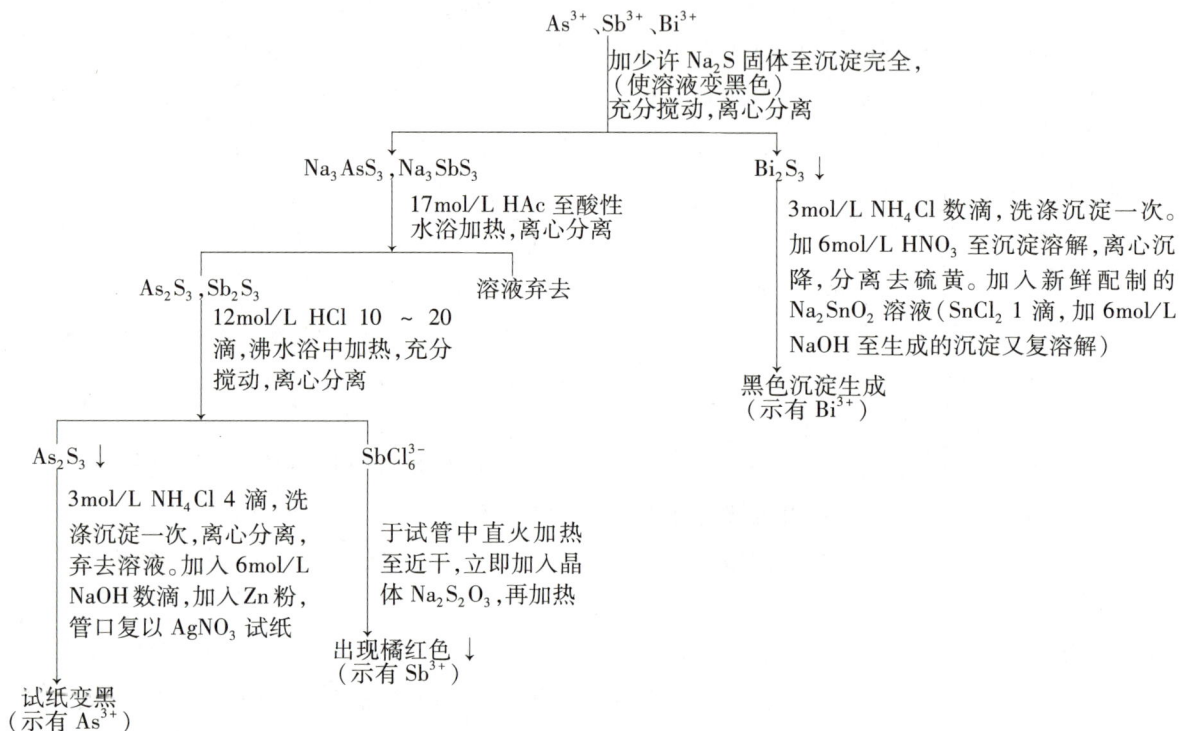

注：加 NH_4Cl 溶液洗涤沉淀，目的是防止少量沉淀以胶体状悬浮在溶液中。

实　验　指　导

预习要求

1. 为什么单质氮在常温下有很高的化学稳定性？

2. 为什么一般情况下不用 HNO_3 作为酸性反应介质，稀硝酸和金属作用与稀 H_2SO_4 或稀 HCl 和金属作用有何不同？

3. 试以 NaH_2PO_4 和 Na_2HPO_4 为例，说明酸式盐是否都呈酸性？

4. 如何配制 $SbCl_3$ 和 $Bi(NO_3)_3$ 水溶液？它们有何特性？

5. 本实验中，溶液的酸、碱性影响氧化还原反应方向的实例有哪些？

6. 氮族元素的金属性和非金属性有什么变化规律？这些元素最常见的氧化数有哪些？

7. 亚硝酸与亚硝酸盐为什么既具有氧化性又具有还原性？试举例说明。

8. 怎样鉴定 NH_4^+ 离子？

9. 磷酸的各种钙盐的溶解性有什么不同？

10. 如何分离 Sb^{3+} 和 Bi^{3+}？

基本操作 •••

气室法鉴定 NH_4^+ 离子。

注意事项 •••

1. 本实验中，NO_2、AsH_3、H_2S 均为有毒气体，有刺激性，吸入后能刺激神经与肺泡。因此在做这一类实验时，必须在通风橱内进行，实验室也要注意通风。

2. As_2O_3（俗称砒霜）是剧毒的白色固体，致死量仅为 0.1g。其他可溶性的砷化合物也有剧毒，切勿进入口内或与伤口接触，用毕要洗手，废液要妥善处理。锑、铋的化合物也有毒，使用中要注意。

3. 在硝酸盐热分解时，反应产物中伴有大量有毒气体 NO_2，所以预先要准备好带火星的纸捻或火柴，当夹有 NO_2（红棕色）气体出现时，立刻用以检验。实验结束后，应立即处理。

报告格式 •••

1. 目的
2. 实验内容

试　样	加入试剂	现　象	反应方程式
结论			

实验后思考 •••

化合物 A 是无色液体，在它的水溶液中加入 HNO_3 和 $AgNO_3$ 时生成白色沉淀 B，B 能溶于氨水得一溶液 C，C 中加入 HNO_3 时 B 重新沉淀，将 A 的水溶液以 H_2S 饱和，得黄色沉淀 D，D 不溶于稀 HNO_3，但能溶于 KOH 和 KHS 的混合液，得到溶液 E。酸化 E 时，D 重新沉淀，试鉴别字母所表示的物质。

Experiment 3　Nitrogen, Phosphorus, Arsenic, Antimony and Bismuth

◆ Objectives

1. Understand the properties of important compounds of Nitrogen, Phosphorus, Arsenic, Antimony and Bismuth.

2. Learn the identification of arsenic.

◆ Principles

1. The group VA elements, Nitrogen, Phosphorus, Arsenic, Antimony and Bismuth, have the valence shell configuration ns^2np^3. So they have the highest oxidation state of +5, the lowest oxidation state of -3.

2. The solution of ammonia is alkali, salt of ammonium is easily decomposed.

3. Nitrous acid is obtained by decomposing nitrite salts in acid, but unstable. It is easily converted into nitric

oxide and nitrogen dioxide. Nitric acid is oxidizing agent, but is reducing agent while meeting stronger oxidant.

4. Nitric acid is a strong acid and powerful oxidant. HNO_3 is generally reduced to NO_2 or NO while reacting with nonmetal. Depending on the concentration of nitric acid and the activity of metal, the reducing products are obtained while HNO_3 reacts with metal.

5. Phosphoric acid is a tribasic acid and converted into acidic salt and orthophosphate. The solubility calcium of phosphoric acids is different in water. $Ca_3(PO_4)_2$ and $CaHPO_4$ is insoluble, but $Ca(H_2PO_4)_2$ is soluble.

6. Because the group VA elements increase in radius from nitrogen to bismuth, the acid strength of their oxides is gradually weakening, but the alkali strength is gradually strengthening. So dinitrogen trioxide N_2O_3 and phosphorus trioxide P_2O_3 are acidic oxides, As_2O_3 and Sb_2O_3 is amphoteric oxides. The reducing properties of the state (+3) compounds decrease from As to Bi, but the oxidizing properties of the state (+5) compounds increase from As to Bi.

7. Identification of arsenic

Small amounts of arsenic are detected by their reduction to arsine by Zn with HCl. The reactions is shown below:

$$AsO_3^{3-} + 9H^+ + 3Zn = AsH_3 + 3Zn^{2+} + 3H_2O$$
$$AsO_4^{3-} + 11H^+ + 4Zn = AsH_3 + 4Zn^{2+} + 4H_2O$$
$$6AgNO_3 + AsH_3 = Ag_3As \cdot 3AgNO_3(black) + 3HNO_3$$
$$6Ag^+ + AsH_3 + 3H_2O = 6Ag(black) + H_3AsO_3 + 6H^+$$

Sulphide, thiosulphate, thioclyanate and antimony (+3) interfere because the formation of hydrogen sulphide, stibine, and mercury inhibit the test. If the gas is first passed through cotton-wool impregnated with silver nitrate, hydrogen sulphide released by the sulphur anions mentioned above is trapped, so that they no longer interfere.

8. Identification of antimony (+3) sodium thiosulphate solution precipitates and orange sulphide, Sb_2O_2 from a very warm slightly acidic test solution

$$2Sb^{3+} + 3S_2O_3^{2-} \longrightarrow 4SO_2 + Sb_2OS_2$$

9. Identification of bismuth (+3) by reaction with sodium stannite solution bismuth (+3) is reduced to the black metal bismuth the reaction is below

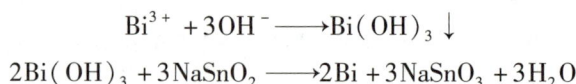

$$Bi^{3+} + 3OH^- \longrightarrow Bi(OH)_3 \downarrow$$
$$2Bi(OH)_3 + 3NaSnO_2 \longrightarrow 2Bi + 3NaSnO_3 + 3H_2O$$

Avoiding to add concentrated base and heat, otherwise sodium stannite decompose to black precipitate, Sn.

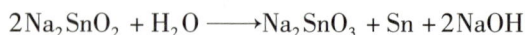

$$2Na_2SnO_2 + H_2O \longrightarrow Na_2SnO_3 + Sn + 2NaOH$$

◆ Chemicals

Solids: NH_4Cl, $AgNO_3$, $Pb(NO_3)_2$, Cu, Na_3BiO_3, As_2O_3 (very toxic), Zn, $Na_2S_2O_3$.

Acids: H_2SO_4 1mol/L; HNO_3, 6mol/L; HCl 6mol/L, 2mol/L

Bases: NaOH 6mol/L, 2mol/L; $NH_3 \cdot H_2O$ 6mol/L

Salts: NH_4NO_3 0.1mol/L, $CaCl_2$ 0.1mol/L, $AgNO_3$ 0.1mol/L, KI 0.1mol/L,

　　　 $AsCl_3$ 0.1mol/L, $SnCl_2$ solution, $NaNO_2$ 1mol/L, $SbCl_3$ 0.1mol/L, $KMnO_4$ 0.01mol/L,

　　　 $Bi(NO_3)_3$ 0.1mol/L, Na_3PO_4 0.1mol/L, Na_2S 2mol/L, Na_2HPO_4 0.1mol/L,

　　　 $MnSO_4$ 0.05mol/L.

Others：Nessler's reagent，starch solution，pH test paper，H_2S solution，I_2 solution（0.01mol/L），$Pb(Ac)_2$ cotton.

◆ Procedures

1. Identification of ammonium，NH_4^+　Cover with a small glass plate on a big one to make the gas room. A small moist red litmus paper and a test paper impregnated with Nessler's reagent are stuck in the interior walls of the small one，Drip respectively 2 drops of ammonium ion and NaOH solution in the big one，cover the little one immediately，leave for a moment，warming on water bath if required. Then the red litmus paper change to be blue，the test paper gives a brown colour. Write out the reactions equations and the blank test is carried out.

2. The properties of nitrite salts

（1）Oxidation of nitrous acid　Add 2 drops 0.1mol/L KI solution to a test tube，dilute to 1mL with water. Acidified with 1mol/L H_2SO_4，and add 2 drops of 1mol/L $NaNO_2$，observe the phenomena. Then add a drop of starch solution again，observe the phenomena. Write out the reactions equations.

（2）Reduction of nitrous acid　Add 5 drops of 0.01mol/L $KMnO_4$ solution，acidified with 1mol/L H_2SO_4，then dropwise add 1mol/L $NaNO_2$ solution，observe the colour change in the solution. Write out the reaction equation.

3. The properties of nitric acid and nitrates

（1）Oxidation of nitric acid　Add copper chips particle to two test tubes respectively. Then add 10 drops of concentrated HNO_3 to one test tube. 10 drops of 6mol/L HNO_3 to the other（warming if required）. Compare the reactions in the two test tubes. How to explain that nitric acid is a strong oxidizing agent.

（2）The thermal decomposition of nitrates　Add respectively an amount of $AgNO_3$，$Pb(NO_3)_2$，and $NaNO_3$ to three different test tubes，burning with a Bunsen flame. Observe the colour of gas emitted. Insert the match with slightly fire to the test tube，detect the gas product. What is the residue in the test tube. Write out the reactions equations.

4. The properties of phosphates　Add 2 drops of 0.1mol/L Na_3PO_4，Na_2HPO_4 and NaH_2PO_4 respectively and 1mL water to three different test tubes，measure their hydrogen ion index with pH test paper. Then add 10 drops of 0.1mol/L $CaCl_2$ solution to each one，shaking. Observe which forms precipitate. Add ammonia to make it alkaline. Observe the change. At last add 6mol/L HCl to make it acidic whether the precipitate dissolves. Compare the solubility of $Ca_3(PO_4)_2$，$CaHPO_4$ and $Ca(H_2PO_4)_2$，illustrate the condition of transform between each other. Write out the reactions equations.

5. The properties of arsenic，antimony and bismuth

（1）Sulphides　Add 2 drops of 0.1mol/L $AsCl_3$，$SbCl_3$ and $Bi(NO_3)_3$ respectively and 6mol/L HCl to three different centrifugal tubes，In addition the saturated solution of hydrogen sulphurate to precipitate completely，observe the colour of sulphides，centrifuge the solid and discard the liquid，then add respectively 20 drops of 2mol/L Na_2S and stir. Which kind of is dissolved? Which kind of precipitate is undissolved? Something takes place. While the liquid Precipitate is acidified with hydrochloric acid after dissolving. Write out the reactions equations.

（2）Oxides and oxyacids

（a）Properties of arsenic（Ⅲ）oxide（very poisonous, obtained from a teaching assistant）respectively to two different dry test tubes. To one is added small amounts of 2mol/L NaOH, stirring. View whether arsenic（Ⅲ）oxide dissolves（retaining the solution to be used for the lab of reduction of sodium arsenic acid）. To the other is added 6mol/L HCl, shaking. Whether arsenic（Ⅲ）oxide dissolves. If heating what happens? State the acid and alkaline relative power of the arsenic trioxide according to the result of the experiment.

（b）Properties and formation of antimony（Ⅲ）hydroxide

To 10 drops of 0.1mol/L antimony（Ⅲ）halide, add 2mol/L NaOH. Observe the phenomenon in the test tube. Divide the precipitate into two portions. Respectively react with 6mol/L sodium hydroxide and 6mol/L hydrochloric acid to detect whether the precipitate dissolve. Write out the reactions equations.

（c）Properties and formation of bismuth（Ⅲ）hydroxide

To 10 drops of 0.1mol/L Bi（NO$_3$）$_3$ solution, drop 2mol/L sodium hydroxide. Observe the shape and colour of the products. Divide the precipitate into two portions. Respectively react with 6mol/L sodium hydroxide and 6mol/L hydrochloric acid to detect whether the precipitate dissolve Write out the reactions.

（3）Reduction of arsenic（antimony, bismuth）in state and oxidation of arsenic（antimony, bismuth）in state（+5）

To small amounts of sodium arsenic acid（weakly alkaline）from this experiment, drop iodine solution. What changes take place? Then acidified with concentrated hydrochloride acid. What changes take place? Write out the reaction and explain all of them.

Oxidation of sodium bismuthate: Take 5 drops of 0.05mol/L MnSO$_4$ and 10 drops of 6mol/L nitric acid, add small amounts of sodium bismuthate solid, shake and heat the test tube. Observe what takes place, and explain it.

6. Ion identification

（1）Identification of arsenic

To 1 drops of the solution containing arsenic and small amounts of zinc chip in a test tube, drop 10 drops of 6mol/L hydrochloric acid. Put a small group cotton containing lead acetate in the upper half of the test tube, the small filter paper which is impregnated with silver nitrate solution is put on the cotton, cover the test tube with the paper. Wait a moment. Reappearance of the black colour in the filter paper confirms the presence of arsenic.

（2）Identification of antimony

Add 10 drops of antimony（+3）ion solutim, heat until boiling, then add small amounts of sodium thiosulphate solid. Formation of red precipitate, antimony sulphide proves the presence of antimony（+3）ions.

（3）Identification of bismuth（+3）

Preparation of sodium stannite solution: Add sodium hydroxide to a tin（+2）chloride solution until the initial white precipitate of tin（+2）oxide dissolves. Then add 4 or 5 drops of fresh sodium stannite solution to 2 drops of bismuth（+3）ions solution. Formation of black precipitate, metal bismuth, is evidence for the presence of bismuth（+3）ions.

7. Separation and identification of a mixture of arsenic（+3）,antimony（+3）and bismuth（+3）ions

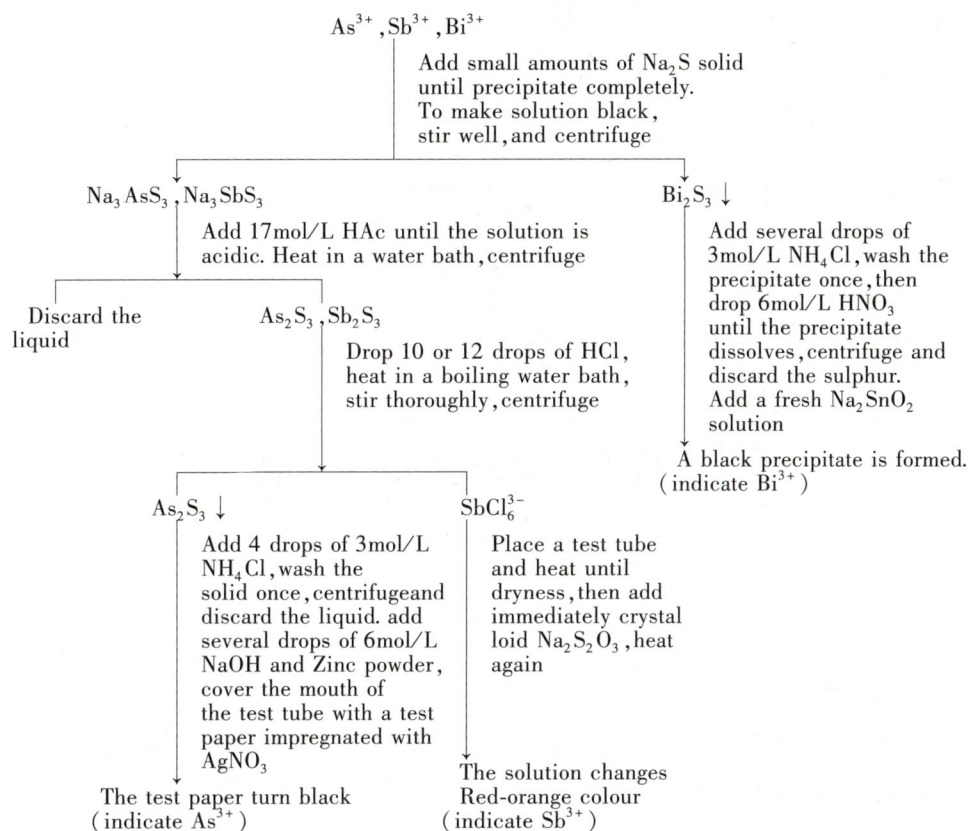

As^{3+},Sb^{3+},Bi^{3+}

Add small amounts of Na$_2$S solid until precipitate completely. To make solution black, stir well,and centrifuge

Na_3AsS_3,Na_3SbS_3

$Bi_2S_3\downarrow$

Add 17mol/L HAc until the solution is acidic. Heat in a water bath,centrifuge

Add several drops of 3mol/L NH$_4$Cl,wash the precipitate once,then drop 6mol/L HNO$_3$ until the precipitate dissolves,centrifuge and discard the sulphur. Add a fresh Na$_2$SnO$_2$ solution

Discard the liquid

As_2S_3,Sb_2S_3

Drop 10 or 12 drops of HCl, heat in a boiling water bath, stir thoroughly,centrifuge

A black precipitate is formed. (indicate Bi^{3+})

$As_2S_3\downarrow$

$SbCl_6^{3-}$

Add 4 drops of 3mol/L NH$_4$Cl,wash the solid once,centrifugeand discard the liquid. add several drops of 6mol/L NaOH and Zinc powder, cover the mouth of the test tube with a test paper impregnated with AgNO$_3$

Place a test tube and heat until dryness,then add immediately crystal loid Na$_2$S$_2$O$_3$,heat again

The test paper turn black (indicate As^{3+})

The solution changes Red-orange colour (indicate Sb^{3+})

◆ **Instructions**

1. Requirement

（1）Why does nitrogen has very high chemical stability at the normal atmospheric temperature?

（2）Why is nitric acidnot generally regarded as the acid reaction medium?

（3）How to prepare SbCl$_3$ and Bi(NO$_3$)$_3$ solution? Discuss the characteristic features of them?

（4）Why can nitrous acid and nitric acid be used as oxidizing agents and reducing agents? Giving examples.

（5）How to identify ammonium ion?

（6）What is the difference of solubility between the different oxyacids of phosphorus containing calcium?

（7）How to separate Sb^{3+} and Bi^{3+}?

（8）How do you detect arsenate in a mixture?

（9）Discuss the thermal stability of nitrates taking examples of nitrates of Na,Mg,Pb.

（10）Describe the gradation in the properties of sulphides of group ⅤA elements. How do they compare with those of ⅣA elements?

2. Operation　Test for presence of ammonium.

3. Note

（1）In this experiment,arsine,nitrogen dioxide and hydrogen sulphide,which are all the poisonous and irritating gases,can stimulate nerve and alveolus after sucking. So you must work with them in the fume hood during the experiment.

（2）Arsenic（+3）oxide,As$_2$O$_3$,is a white and very poisonous solid. The lerhal dose of As$_2$O$_3$ is only 0. 1 grams. Don't eat or touch wound. Wash your hands after finishing your work. The waste liquid should be dealt with carefully.

4. Reprt Format

（1）Objectives

（2）Procedures

Samples	Chemicals	Phenomenon	Reaction equation

Conclusion

实验四　碱金属、碱土金属

PPT

目的要求 •••

1. 试验金属钠的强还原性。

2. 掌握钠、钾、镁、钙、钡的鉴定方法。

3. 比较镁、钙、钡的氢氧化物、硫酸盐、铬酸盐、草酸盐、碳酸盐的溶解性。

4. 了解对阳离子未知液的分析方法。

实验原理 •••

1. 碱金属是周期系第Ⅰ主族元素、原子最外层的电子构型为 ns^1，它们容易失去这一个电子而表现强还原性。

2. 碱金属的盐类一般都易溶于水，只有少数几种盐难溶，如钴亚硝酸钾（二钾），醋酸铀酰锌钠，可以利用它们的难溶性来检验钠离子和钾离子。

碱土金属的硝酸盐、氯化物都易溶于水。碳酸盐、硫酸盐、磷酸盐等难溶。可利用难溶盐的生成，如磷酸铵镁、草酸钙、硫酸钡、铬酸钡沉淀以检验镁离子、钙离子和钡离子。

3. 碱金属、碱土金属及其挥发性的化合物，在无色火焰中灼烧时，原子中外层电子接收能量被激发到较高能级上。处于激发态的电子是不稳定的，当这些电子跃回低能级时，便将多余的能量以光子形式放出、产生特征的焰色。

仪器和药品 •••

1. 仪器　钴玻片，镊子，酒精喷灯。

2. 药品

（1）固体　金属钠。

（2）酸　HCl 2mol/L，浓；HAc 2mol/L；HNO₃ 2mol/L；H₂SO₄ 3mol/L。

（3）碱　NaOH 2mol/L；氨水 2mol/L。

（4）盐　NaCl 0.5mol/L；CaSO₄ 饱和溶液；Na₂SO₄ 0.5mol/L；KCl 0.5mol/L；Na₂CO₃ 0.5mol/L；KMnO₄ 0.01mol/L；CaCl₂ 0.5mol/L；K₂CrO₄ 0.5mol/L；NH₄Cl 10%；BaCl₂ 0.5mol/L；Na₂HPO₄ 0.5mol/L；（NH₄)₂C₂O₄ 3%；SrCl₂ 0.5mol/L；（NH₄)₂HPO₄ 0.5mol/L；Na₃[Co(NO₂)₆] 20%

$NH_3 \cdot H_2O\text{-}NH_4Cl\text{-}(NH_4)_2CO_3$ 混合溶液；$Na[B(C_6H_5)_4]$ 0.1%；醋酸铀酰锌溶液。

实验内容

1. 碱金属

（1）金属钠的性质、与氧的作用　用镊子取一小块金属钠，迅速用滤纸吸干其表面的煤油，用刀削去外层，使露出新鲜面，立即放入坩埚中加热。当开始燃烧时，停止加热，观察反应情况和产物的颜色、状态。

将反应物转入干试管中，加入少许水，即发生反应（反应放热，必须将试管放在冷水中）。

检验管口是否有氧放出（怎样试验？）。

检验水溶液是否呈碱性（用 pH 试纸试验）。

检验水溶液是否有 H_2O_2 生成（将溶液用 1mol/L H_2SO_4 酸化，加 1 滴 0.01mol/L $KMnO_4$，观察紫色是否褪去）。写出氧化产物与水作用的反应式。

（2）钠盐、钾盐的鉴定

①生成醋酸铀酰锌钠鉴定 Na^+ 离子　于一小试管中，加 1 滴 Na^+ 离子试液（0.5mol/L NaCl 溶液），加 2 滴 2mol/L HAc 和约 10 滴醋酸铀酰锌试液，用玻棒摩擦试管内壁，即有黄绿色醋酸铀酰锌钠沉淀生成。写出离子反应式。

②生成钴亚硝酸钠钾鉴定 K^+ 离子　于一小试管中，加 2 滴 K^+ 离子试液（0.5mol/L KCl 溶液），再加 3~4 滴钴亚硝酸钠试液，即有黄棕色沉淀生成。写出离子反应式。

③生成四苯硼钾鉴定 K^+ 离子　于一小试管中，加 2 滴 K^+ 离子试液（0.5mol/L KCl 溶液），加入 3~4 滴四苯硼钠试剂，即有白色四苯硼钾沉淀生成。写出离子反应式。

2. 碱土金属

（1）氢氧化镁的生成和性质　在 3 支小试管中，各加入约 5 滴 0.5mol/L $MgCl_2$ 溶液，再向各试管中滴加 2 滴 2mol/L NaOH 溶液，观察生成的氢氧化镁沉淀的颜色和状态，然后再分别滴加 3~4 滴 2mol/L NaOH、2mol/L HCl 和 10% NH_4Cl 溶液，观察现象，并比较 3 个试管中沉淀量的多少。写出反应式，并解释之。

（2）难溶盐的生成和性质、硫酸盐的溶解度比较　在 3 支试管中，分别加 5 滴 0.5mol/L $CaCl_2$、$SrCl_2$、$BaCl_2$ 溶液，然后各加 10 滴 0.5mol/L $CaCl_2$、$SrCl_2$、$BaCl_2$ 溶液，然后各加 10 滴 0.5mol/L Na_2SO_4 溶液，观察反应产物的颜色和状态。

比较 $CaSO_4$、$SrSO_4$、$BaSO_4$ 的溶解度大小。

（3）钙、锶、钡碳酸盐的生成和性质

①取 3 支试管，分别加 5 滴 0.5mol/L $CaCl_2$、$SrCl_2$、$BaCl_2$ 溶液，再加 6~7 滴 0.5mol/L Na_2CO_3 溶液，观察现象，再向各管中约加 10 滴 2mol/L HAc，观察现象并写出反应式。

②取 1 支试管，加 5 滴 0.5mol/L $MgCl_2$ 溶液，5 滴氨水 - 氯化铵 - 碳酸铵混合溶液 [含 1mol/L $NH_3 \cdot H_2O\text{-}NH_4Cl$ 和 0.5mol/L $(NH_4)_2CO_3$]，观察现象，并解释之。

（4）钙、钡铬酸盐的生成和性质　在 2 支试管中，各加 5 滴 0.5mol/L $CaCl_2$、0.5mol/L $BaCl_2$ 溶液，再加 10 滴 0.5mol/L K_2CrO_4 溶液，观察现象。试验产物对 2mol/L HAc、2mol/L HCl 溶液的作用。写出反应式。

（5）钙离子的鉴定　生成草酸钙鉴定 Ca^{2+} 离子：在 1 支试管中，加 5 滴 Ca^{2+} 离子试液（0.5mol/L $CaCl_2$ 溶液）和 10 滴 $(NH_4)_2C_2O_4$ 溶液，观察反应现象，试验产物对 2mol/L HAc 和 2mol/L HCl 溶液的作用，写出反应式。

（6）镁离子的鉴定   生成磷酸铵镁鉴定 Mg^{2+} 离子：在 1 支试管中加 10 滴 Mg^{2+} 离子试液（0.5mol/L $MgCl_2$ 溶液），加 5 滴 $NH_3 \cdot H_2O$-NH_4Cl 溶液，再加 10 滴 Na_2HPO_4 溶液，振荡试管，有白色磷酸铵镁沉淀生成。写出反应式。

3. 钠、钾、钙、锶、钡盐的焰色试验   取铂丝棒（或镍丝棒）反复蘸以浓 HCl，在酒精喷灯上灼烧至无色。分别蘸以 0.5mol/L NaCl、KCl、$CaCl_2$、$SrCl_2$、$BaCl_2$ 溶液在氧化焰中灼烧，观察并比较它们的焰色有何不同。

（观察钾盐的焰色时，需用钴玻璃滤光）。

4. 未知液的分离和检出   取可能含 Na^+、K^+、NH_4^+、Mg^{2+}、Ca^{2+}、Ba^{2+} 的混合溶液 20 滴，于一离心管中混合均匀后，先按（1）中的步骤检验 NH_4^+ 离子。

（1） NH_4^+ 的检出——气室法   取 3 滴混合溶液于一块表面皿上，再滴加 6mol/L NaOH 溶液至显碱性为止。另取一块较小表面皿，在凹面上贴一块湿的 pH 试纸、和一块以奈氏试剂润湿的滤纸，将此表面皿迅速覆盖在大表面皿上。如果 pH 试纸变成蓝紫色并使蘸有奈氏试剂的滤纸变成红褐色，表示试液中有 NH_4^+ 离子（同时做空白试验）。

检出 NH_4^+ 以后，再按下列步骤进行分离和检出。

（2） $BaCO_3$、$CaCO_3$ 的沉淀   在试液中加 6 滴 3mol/L NH_4Cl 溶液，并加 2mol/L 氨水使溶液呈碱性，再多加 3 滴氨水。在搅拌下加 10 滴 1mol/L $(NH_4)_2CO_3$ 溶液，离心管放在 60℃ 的热水浴中加热几分钟，然后离心沉降，分离，把清液移到另 1 支离心管中，按（5）中操作处理，沉淀供（3）用。

（3） Ba^{2+} 的分离和检出   在（2）中所得的沉淀用 10 滴热水洗涤，离心沉降，分离弃去洗涤液，加 3mol/L HAc 溶解沉淀（需加热，并不断搅拌）。然后加 5 滴 4mol/L NH_4Ac 溶液，加热后，滴加 1mol/L K_2CrO_4 溶液数滴，如有黄色沉淀产生即表示有 Ba^{2+} 存在，如清液呈橘黄色时，表明 Ba^{2+} 已沉淀完全，否则需要加 1mol/L K_2CrO_4 溶液使 Ba^{2+} 沉淀完全，离心沉降，分离，清液留做检查 Ca^{2+}。

（4） Ca^{2+} 的检出   向（3）所得的清液中加 1 滴 2mol/L 氨水和 1 滴 2mol/L $(NH_4)_2C_2O_4$ 溶液，加热后，如有白色沉淀产生，表示有 Ca^{2+}。

（5）残留 Ba^{2+}、Ca^{2+} 的除去   向（2）所得的清液内加 2mol/L $(NH_4)_2C_2O_4$ 和 2mol/L $(NH_4)_2SO_4$ 各 1 滴。加热几分钟，如果溶液浑浊，离心分离，弃去沉淀，把清液移到坩埚中。

（6） Mg^{2+} 的检出   取几滴（5）中的清液，加到试管中，再加 1 滴 2mol/L $(NH_4)_2HPO_4$ 溶液，摩擦试管内壁。如果产生白色结晶，表示有 Mg^{2+}。

另取 1 滴（5）中的清液，加在点滴板的穴中，再加 2 滴 2mol/L NaOH 溶液使呈碱性，然后加 1 滴镁试剂，如产生蓝色沉淀，表示有 Mg^{2+} 离子存在。

（7）铵盐的除去   将（5）中已经移在坩埚中的清液，小心地蒸发至只剩下几滴为止，再加 8～10 滴浓硝酸，然后蒸发至干，为了防止溅出，应在蒸到最后 1 滴时，借石棉网上的余热把它蒸发至干，最后用大火灼烧至不再冒白烟。冷却后，往坩埚中加入 8 滴蒸馏水，使溶解。从坩埚中取出此溶液 1 滴，加在点滴板的穴中、再加 2 滴奈氏试剂，如果不产生红褐色沉淀，表明铵盐已被除尽，否则需重复上述除铵盐的操作。铵盐除尽后，溶液供（8）、（9）检出 Na^+、K^+ 离子。

（8） Na^+ 的检出   取（7）中的溶液 2 滴，加 10 滴醋酸铀酰锌试剂，并用玻璃棒摩擦试管内壁，如有黄绿色晶体生成，示有 Na^+ 离子。

（9） K^+ 的检出   将（7）中剩余的溶液加到试管中，加 2 滴 $Na_3[Co(NO_2)_6]$ 溶液，如产生黄色沉淀，表示有 K^+ 离子。

实　验　指　导

预习要求 ●●●

1. 预习有关碱金属、碱土金属元素的内容，掌握有关元素及其化合物的重要性质。

2. 实验前查出与本实验有关的难溶盐的溶度积常数。

3. 通过计算，说明 $MgCl_2$ 溶液中，滴入氨水时，会生成 $Mg(OH)_2$ 和 NH_4Cl，而 $Mg(OH)_2$ 沉淀又能溶于饱和的 NH_4Cl 溶液。

4. 试从理论上说明，碱土金属碳酸盐为什么可以溶于 HAc 溶液。

5. 为什么从能否溶于 HAc 或 HCl，就可以比较出 CaC_2O_4、BaC_2O_4 或 $CaCrO_4$、$BaCrO_4$ 溶解度的相对大小？

6. 为什么在氨－氯化铵－碳酸铵的混合溶液中，Mg^{2+} 离子不会析出沉淀，而依此作为分离 K^+、Na^+、NH_4^+、Mg^{2+} 与 Ba^{2+}、Ca^{2+} 的方法？

7. 完成下列反应式：

(1) $Na + O_2 \longrightarrow$

(2) $Na_2O_2 + H_2O \longrightarrow$

(3) $H_2O_2 \xrightarrow{\text{分解}}$

(4) $H_2O_2 + MnO_4^- + H^+ \longrightarrow$

(5) $Na^+ + Zn^{2+} + 3UO_2^{2+} + 9Ac^- \longrightarrow$

(6) $K^+ + Na^+ + [Co(NO_2)_6]^{3-} \longrightarrow$

(7) $K^+ + [B(C_6H_5)_4]^- \longrightarrow$

(8) $Mg^{2+} + OH^- \longrightarrow$

(9) $Mg(OH)_2\downarrow + NH_4^+ \longrightarrow$

(10) $\begin{matrix} Ca^{2+} \\ Sr^{2+} \\ Ba^{2+} \end{matrix} + SO_4^{2-} \longrightarrow$

(11) $\begin{matrix} Ca^{2+} \\ Sr^{2+} \\ Ba^{2+} \end{matrix} + CO_3^{2-} \longrightarrow$

(13) $Ca^{2+} + C_2O_4^{2-} \longrightarrow$

(14) $Mg^{2+} + HPO_4^{2-} + NH_4^+ + OH^- \longrightarrow$

基本操作 ●●●

1. 金属钠的加热。

2. 酒精喷灯的使用。

3. 练习用焰色反应作为鉴定阳离子的辅助试验。

注意事项 ●●●

1. 金属钠、钾遇水会引起爆炸，在空气中也会立即被氧化。所以通常把它们保存在煤油中，置于阴凉处。使用时应在煤油中切割成小块，用镊子夹取，并用滤纸把煤油吸干，切勿与皮肤接触。未用完的金属钠碎屑不能乱丢，可加入少量无水乙醇，使其缓慢分解。

2. 使用酒精喷灯须注意，在开启开关、点燃管口气体以前，必须充分灼热灯管，否则流出的酒精不能全部气化，会有液态酒精由管口喷出，可能形成"火雨"，甚至引起火灾。

3. 以焰色反应检查有关金属离子时，必须用洗净的铂丝蘸上被试离子液在氧化焰中灼烧。注意观察描述各离子的特征焰色。

报告格式

1. 目的

2. 实验内容（以表格形式列出）

3. Na^+、K^+、NH_4^+、Mg^{2+}、Ba^{2+}、Ca^{2+} 混合未知液的分离和检出

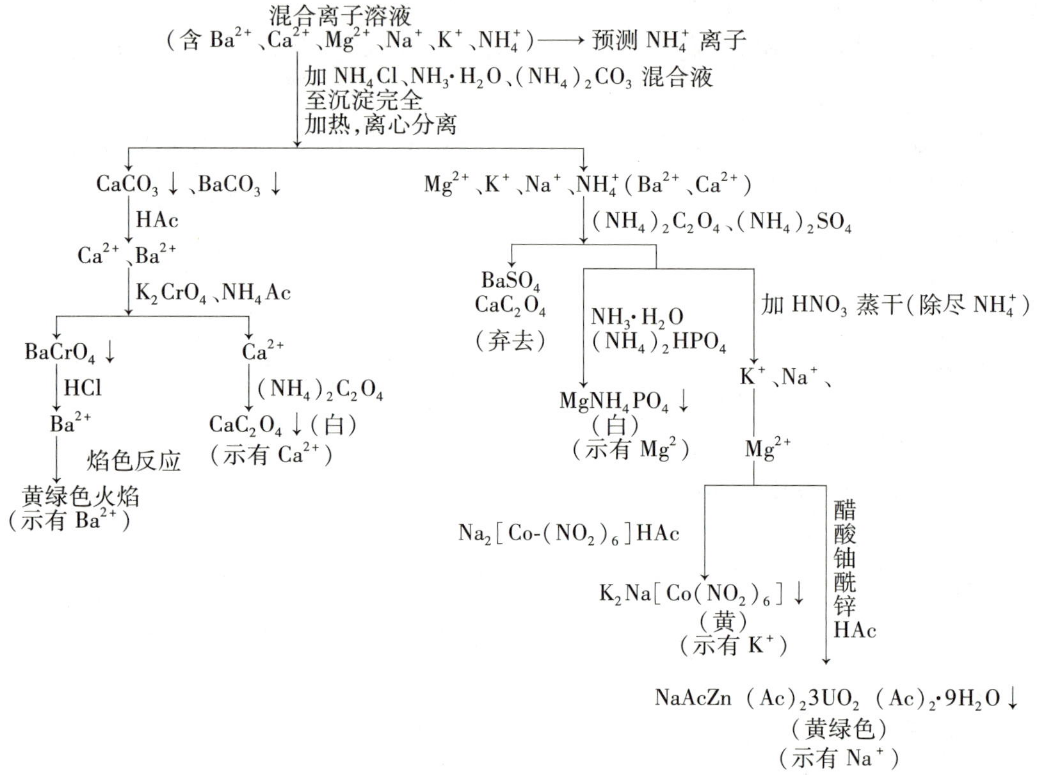

混合离子溶液
（含 Ba^{2+}、Ca^{2+}、Mg^{2+}、Na^+、K^+、NH_4^+）$\longrightarrow$ 预测 NH_4^+ 离子

加 NH_4Cl、$NH_3\cdot H_2O$、$(NH_4)_2CO_3$ 混合液
至沉淀完全
加热，离心分离

$CaCO_3\downarrow$、$BaCO_3\downarrow$ 　　　　Mg^{2+}、K^+、Na^+、NH_4^+（Ba^{2+}、Ca^{2+}）

HAc 　　　　　　　　　　　$(NH_4)_2C_2O_4$、$(NH_4)_2SO_4$

Ca^{2+}、Ba^{2+}

K_2CrO_4、NH_4Ac 　　　　$BaSO_4$
　　　　　　　　　　　　　　CaC_2O_4　$NH_3\cdot H_2O$　加 HNO_3 蒸干（除尽 NH_4^+）
$BaCrO_4\downarrow$　　Ca^{2+}　（弃去）　$(NH_4)_2HPO_4$

HCl　　　$(NH_4)_2C_2O_4$　　　　　　　　K^+、Na^+、

Ba^{2+}　　$CaC_2O_4\downarrow$（白）　$MgNH_4PO_4\downarrow$（白）

焰色反应　（示有 Ca^{2+}）　（示有 Mg^2）　Mg^{2+}

黄绿色火焰
（示有 Ba^{2+}）　　　　　$Na_2[Co\text{-}(NO_2)_6]HAc$　　醋酸铀酰锌 HAc

$K_2Na[Co(NO_2)_6]\downarrow$（黄）
（示有 K^+）

$NaAcZn(Ac)_2 3UO_2(Ac)_2\cdot 9H_2O\downarrow$（黄绿色）
（示有 Na^+）

实验后思考

1. 设计一个分离 K^+、Mg^{2+}、Ba^{2+} 离子的方法。

2. 现有五种溶液，它们是：$NaOH$、$NaCl$、$MgSO_4$、KOH、K_2CO_3，选用合适的试剂，将它们逐一鉴别。

3. 有一白色固体，初步试验，它不溶于水，用盐酸处理，则产生气泡，得一澄清溶液，如果用硫酸处理，也产生气泡，但不能形成澄清的溶液，这一白色固体是什么化合物？

Experiment 4　Alkali Metals, Alkali Earth Metals

◆ Objectives

1. Test the strong reduction of sodium.

2. Master the method of identifying sodium, potassium, magnesium, calcium.

3. Compare the solubility of hydroxides, sulphates, chromates, oxalic acid salts, carbonates of the magnesium, calcium, barium.

4. Understand the analytical method to an unknown cation solution.

◆ Principles

1. The alkali metals are members of group 1 of the periodic table, and each has a ground state valence electronic configuration ns^1. The elements show strongly reduction by loss of one electron.

2. Salts of the alkali metals generally dissolve in water, only a few salts being insoluble. For example, the inferior potassium nitrate of cobalt, sodium zinc uranylacetate etc. Make use of their insolubility to identify sodium and potassium ions.

The nitrates, chlorides of the alkaline-earth metals are easy to dissolve in water. Carbonates, sulphates and phosphate salts are insoluble. Magnesium, calcium, barium ions are identified by the formation of their insoluble salts, for example, precipitates with magnesium ammonium phosphate ($MgNH_4PO_4$), calcium oxalate, barium sulphate, barium chromium.

3. While alkali metals, alkaline earth metals and thein compounds are heated strongly in the colorless flame, the outer electrons are excited to relatively high energy in the atom, but unstable. When these electrons jump back to low energy and emit surplus energy in photon form, a characteristic flame color is observed.

◆ Equipment and Chemicals

1. Equipment　Bunsen flame, forceps, blue cobalt glass

2. Chemicals

(1) Solids　sodium metal

(2) Acid　HCl 2mol/L, concentrated; HAc 2mol/L; HNO_3 2mol/L; H_2SO_4 3mol/L

(3) Bases　NaOH 2mol/L; $NH_3 \cdot H_2O$ 2mol/L

(4) Salts　NaCl 0.5mol/L; $CaSO_4$ saturation solution; Na_2SO_4 0.5mol/L; KCl 0.5mol/L;

　　　　　Na_2CO_3 0.5mol/L; $KMnO_4$ 0.01mol/L; $CaCl_2$ 0.5mol/L;

　　　　　NH_4Cl 10%; $BaCl_2$ 0.5mol/L; Na_2HPO_4 0.5mol/L; $(NH_4)_2C_2O_4$ 3%

　　　　　$Na_3[Co(NO_2)_6]$ 20%; $NH_3 \cdot H_2O$-NH_4Cl-$(NH_4)_2CO_3$ mixed solution,

　　　　　$Na[B(C_6H_5)_4]$ 0.1%; Zinc uranylacetate

◆ Procedures

1. Alkali metals

(1) The properties of metal sodium and reaction of sodium with water

Small piece of metal sodium is cut with tweezers blot. Its superficial kerosene is absorbed.

With the filter paper removing the fresh superficial membrane in sodium metal, immediately put it into a crucible to heat; when the sodium begin burning, stop heating. Observe the situation of reaction and the color, shape of products.

Then the products are transferred to a dry test tube, adding some water.

The reaction occur (because the reaction emit heat, the test tube must placed in cold water).

Detect whether oxygen will be given out (how to detect).

Detect whether the solution is alkaline with pH test paper.

Detect whether H_2O_2 is yielded in the aqueous (add a drop of 0.01mol/L $KMnO_4$ to the solution with

1mol/L H_2SO_4, then observe whether pink will disappear). Write out the reaction of the oxide with water.

(2) Identify of sodium salts and potassium salts

(a) Identify sodium ion by yielding sodium zinc uranylacetate.

Add a little drop of aqueous solution of Na^+ ion (0.5mol/L NaCl solution), two drops of 2mol/L HAc and about ten drops of zinc uranylacefate to a test tube. If necessary, rub the inside wall of the test tube with a glass rod, a yellow precipitate is formed. Write out the ion equation of the reaction.

(b) Identification of the potassium ion K^+ by forming sodium hexanitrocobaltate, $Na_3[Co(NO_2)_6]$.

Add two drops of the potassium ion K^+ ions solution (0.5mol/L KCl solution) to a test tube, then add several drops of $Na_3[Co(NO_2)_6]$ solution. A brown yellow precipitate is obtained. Write out the ion equation of the reaction.

(c) Identification of the potassium ion K^+ by forming potassium tetraphenylboron.

Add two drops of the potassium ion K^+ solution (0.5mol/L KCl solution) to a test tube, then add several drops of sodium tetraphenylboron solution. A white precipitate is obtained. Write out the ionic equation of the reaction.

2. Alkali Metals

(1) The properties, formation of magnesium hydroxide

Add 5 drops of 0.5mol/L $MgCl_2$ solution to three test tube respectively, then add 2 drops of 2mol/L NaOH solution to each one. Observe the color, state of the precipitate of $Mg(OH)_2$, then add 3 ~ 4 drops of 2mol/L NaOH, 2mol/L HCl, 10% NH_4Cl solution respectively. Observe the phenomena, and compare the quantity of the precipitate in the three test tubes. Write out the reaction. Giving reasons.

(2) Formation of insoluble salts and their properties, comparing the solubility of sulphate salts

Add 5 drops of 0.5mol/L $CaCl_2$, $SrCl_2$, $BaCl_2$ solution to three test tubes respectively, then 10 drops of Na_2SO_4 solution, observe the color and state of the products.

Compare the solubility of sulphates.

(3) The properties and formation of carbonates of calcium, strontium, barium

(a) Add 5 drops of 0.5mol/L $CaCl_2$, $SrCl_2$, $BaCl_2$ solution to three test tubes respectively, and then 6 or 7 drops of 0.5mol/L Na_2CO_3 solution, observe the phenomena and write out the reaction.

(b) Add 5 drops of 0.5mol/L $MgCl_2$ solution, and a ammonia-ammonium chloride-ammonium carbonate mixed solution (a combination of 1mol/L $NH_3 \cdot H_2O$-NH_4Cl and 0.5mol/L $(NH_4)_2CO_3$) to a test tube. Observe and explain the phenomena.

(4) The properties and formation of chromate of calcium, barium

Add 5 drops of each 0.5mol/L $CaCl_2$, $BaCl_2$ to a different clean test tube respectively, and then 10 drops of 0.5mol/L potassium chromate solution to each one, observe the phenomena. Test that the products react with 2mol/L HAc and 2mol/L HCl solution. Write out the reactions.

(5) Identification of calcium ion Ca^{2+} (calcium is precipitated as the oxalates to identify the presence of calcium)

Add 5 drops of the calcium ion Ca^{2+} solution (0.5mol/L $CaCl_2$ solution) and 10 drops of ammonium oxalate. Observe the phenomena of the reaction, test that products react with 2mol/L HAc and 2mol/L HCl solution. Write out the reactions.

（6）Identification of the magnesium ion Mg^{2+} by forming the white precipitate of $MgNH_4PO_4$

The white precipitate of $MgNH_4PO_4$ can be obtained by adding Na_2HPO_4 to a solution containing Mg^{2+}, $NH_3 \cdot H_2O$, NH_4Cl, and then raising the pH to more than 12 by add NaOH. Write our the reaction.

3. Flame tests of the salts of Na, K, Ca, Sr, Ba

Add 15 drops of each 0.5mol/L solution to a different clean test tube. To clean the wire, dip it into the test tube of concentrated HCl and heat the wire in the hottest part of the flame until no color shows. When the platinum wire is clean, dip the wire in the test tube containing a 0.5mol/L solution and hold it in the hottest part of the flame. Record your observation of the color of the flame. (when observing the color of the flame of the potassium ion, a blue cobalt glass should be used)

4. Separation and detection of an unknown solution

Add 20 drops of the mixed solution possibly containing Na^+, K^+, NH_4^+, Mg^{2+}, Ca^{2+}, Ba^{2+} to a clean centrifugal test tube, mix the test solution. Firstly detect the ammonium ion NH_4^+ in term of the following step:

Step 1. Detection of the ammonium ion NH_4^+

Add 3 drops of the mixed solution to a watch glass, and 6mol/L NaOH dropwise until the solution is basic. Stick a pH test paper with water and filter paper impregnated with Nessler's reagent on the concave surface of another smaller watch glass. Cover this one on the big one rapidly. If pH test paper become blue-purple, and the filter paper of impregnated with Nessler's reagent become reddish brown, it means ammonium ion is present in the solution (Make the blank test at the same time).

After identifying the ammonium ion NH_4^+, then separate and detect the other ions according to the following.

Step 2. Precipitates formation of $BaCO_3$, $CaCO_3$

To the unknown solution containing 6 drops of 3mol/L NH_4Cl, add 2mol/L ammonia water dropwise, until the solution is basic, add 3 drops more of ammonia. Add 10 drops of 1mol/L $(NH_4)_2SO_4$ dropwise and stir. Place the solution in a hot water bath at 60℃ for a few minutes. Centrifuge and separate the precipitate. Transfer the liquid to another centrifugal tube and then treat with it. The precipitate for step 3.

Step 3. Separation and detection of Ba^{2+}

Wash the solid from step 2 with 10 drops of hot water. Centrifuge the precipitate and discard the liquid. Dissolve the precipitate with 3mol/L HAc (heat, stir well). Then add 5 drops of 4mol/L NH_4Ac, heat, a few drops of 1mol/L K_2CrO_4 dropwise. You should see a yellow precipitate if Ba^{2+} is present. The yellow orange liquid indicates that calcium ions precipitate incompletely, otherwise barium ions precipitate completely by adding 1mol/L K_2CrO_4, centrifuge and separate.

The liquid is used for detecting for Ca^{2+}.

Step 4. Test for presence of calcium

To the liquid retained from step 3, add 1 drop of 2mol/L ammonia and 2mol/L $(NH_4)_2C_2O_4$. You should see a white precipitate after heating if Ca^{2+} is present.

Step 5. Removing Ca^{2+}, Ba^{2+} from the remaining liquid

To the liquid retained from step 2, add 1 drop of 2mol/L $(NH_4)_2C_2O_4$ and 2mol/L $(NH_4)_2SO_4$, heat for a few minutes. If the solution turns turbidness, centrifuge and discard the precipitate, transfer the liquid to a crucible.

Step 6. Test for presence of magnesium

Add a few drops of the liquid from step 5 to a clean test tube, then add 1 drop of 2mol/L $(NH_4)_2HPO_4$, rub the inside of the test tube with a glass rod. Appearance of the white crystalline precipitate confirms the presence of magnesium.

Add other one drop of the liquid from step 5 to a spot plate, then add 2 drops of 2mol/L NaOH to make the solution become basic, and 1 drop of magnesium reagent. Appearance of the blue precipitate indicates the presence of magnesium.

Step 7. Removing of ammoniumion, NH_4^+

Heat carefully the liquid in the crucible from step 5 until the volume is reduced to about several drops. Then add 8 or 10 drops of concentrated nitric acid. Evaporate the solution to dryness (While steaming to the last drop, take advantage of remaining energy of the wire gauze and steam it to prevent spattering out). At last burn with the fire until the white smoke disappears.

Dissolve the residue in 8 drops of water after cooling. Add 1 drop of the solution to a spot plate, then add 2 drops of Nessler's reagent. Once a redish brown precipitate is not obtained, this indicates the solution does not have ammonium ions. If not, you have to repeat the operation of the removing of ammonium in the above. After eliminating the ammonium salt, the retained solution is used for step 8 and 9 to detect the presence of Na^+, K^+.

Step 8. Detection of Na^+

To 2 drops of the solution from step 7, add 10 drops of zinc uranyl acetate reagent, and scratch the inside of the test tube with a glass rod. Appearance of the green yellow precipitate confirms the presence of Na^+.

Step 9. Detection of K^+

To the solution retained from step 7 in a test tube, add 2 drops of $Na_3[Co(NO_2)_6]$.

Appearance of the yellow precipitate confirms the presence of K^+.

◆ Instructions

1. Requirements

(1) Preview the relevant contents about alkali metals and alkali earth metals in the textbook of inorganic chemistry, Grasp the important properties of the elements and chemical compounds.

(2) Consult solubility constants of insoluble salts related to the experiment before the lab.

(3) Why does the $MgCl_2$ solution produce $Mg(OH)_2$ and NH_4Cl when ammonia is added, but the $Mg(OH)_2$ precipitate is soluble in a saturated NH_4Cl solution?

(4) Explain why the alkali earth metals is soluble in a HAc solution.

(5) Compare the solubility of CaC_2O_4 ($CaCrO_4$) and BaC_2O_4 ($CaCrO_4$).

2. Operation

(1) Heat the metal sodium

(2) Use of the Bunsen flame

(3) Identification of cations with flame test

3. Notes

(1) Metal sodium, potassium can be caused to explode while meeting water, will be oxidized immediately in the air, so usually keep them in the kerosene, lay them in the shady and cool place. Should cut into small one in the kerosene, insert and fetch with the tweezers, and absorb the kerosene with the filter paper. It is sure not to

keep in touch with the skin. Metal sodium not used up don′t be thrown out without care，may add a small amount of anhydrous ethanol，make it decompose slowly.

（2）Notice while using the Bunsen flame：before turning on the switch and lighting the gas from the mouth of the tube，the tube muct be heated，otherwise thoroughly，"fire rain" will be formed even cause the fire.

（3）When detecting metal ions with flame test，the clean platinum wire with the liquid of detected ions is heated strongly in the hottest part of the flame. Pay attention to observe the characteristic flames color of ions.

4. Report Format

（1）Objectives

（2）Procedure（List in the form of table）

（3）Separation and detection of an unknown mixed solution of Na^+，K^+，NH_4^+，Mg^{2+}，Ba^{2+}，Ca^{2+}.

A mixed solution containg
Ba^{2+}，Ca^{2+}，Mg^{2+}，Na^+，K^+，NH_4^+ ⟶ Detect presence of ammonium

Add a mixed solution containing NH_4Cl，$NH_3 \cdot H_2O$，$(NH_4)_2CO_3$ until the solution precipitate completely.
Heat，centrifuge the precipitate.

$CaCO_3(s)$
$BaCO_3(s)$ ——HAc——> Ca^{2+}、Ba^{2+} ——K_2CrO_4、NH_4Ac——>

$BaCrO_4\downarrow$ ——HCl——> Ba^{2+} ——flame test——> green-yellow flame indicate Ba^{2+}

Ca^{2+} ——$(NH_4)_2C_2O_4$——> $CaC_2O_4\downarrow$ white indicate Ca^{2+}

Mg^{2+}，K^+，Na^+，NH_4^+（Ba^{2+}，Ca^{2+}）——$(NH_4)_2C_2O_4$，$(NH_4)_2SO_4$——>

$BaSO_4$，$CaSO_4$

Add HNO_3 until dryness. （eliminate NH_4^+）

K^+、Na^+

$Na_3[Co(NO_2)_6]$ HAc ——> $K_2Na[Co(NO_2)_6]$ yellow indicate K^+

Zinc uranyl acetate HAc ——> yellow-green precipitate. indicate Na^+

$NH_3 \cdot H_2O$ $(NH_4)_2HPO_4$ ——> $MgNH_4PO_4$ White indicate Mg^{2+}

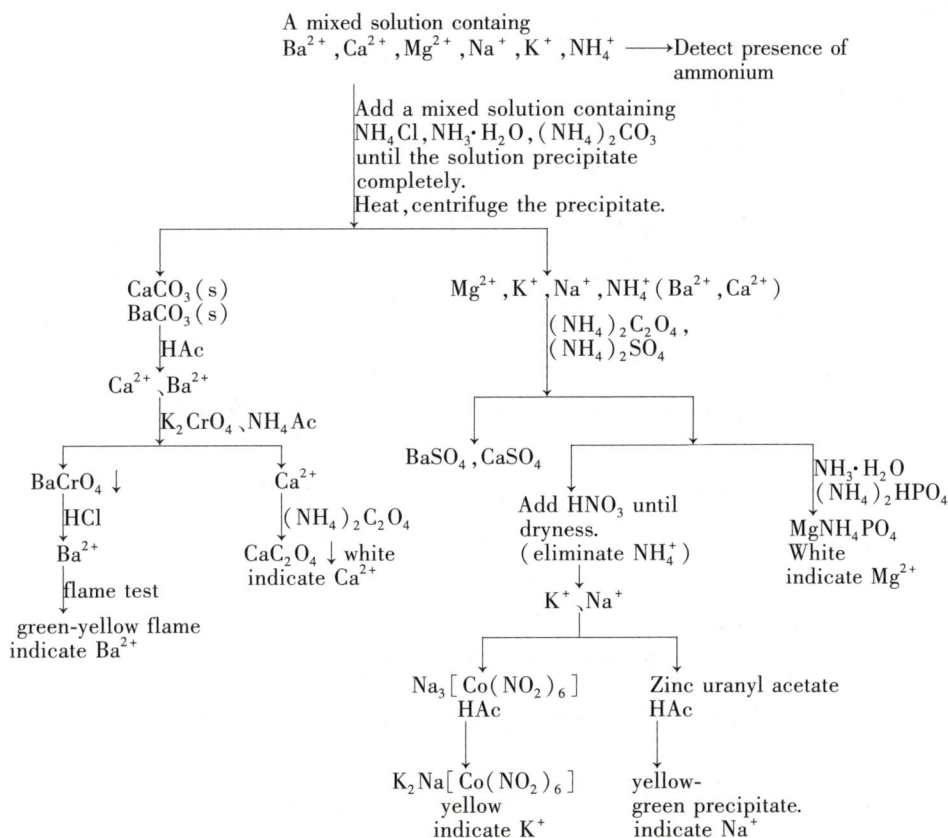

实验五　铬、锰、铁、钴、镍

PPT

目的要求 •••

1. 了解铬和锰的各种重要价态化合物的生成和性质。

2. 了解铬、锰各种价态之间的转化。

3. 了解铬、锰化合物的氧化还原性以及介质对氧化还原反应的影响。

4. 了解 +2 价和 +3 价铁、钴、镍的氢氧化物的制备和性质。

5. 了解 +2 价铁盐的还原性和 +3 价铁的氧化性。

6. 了解铁、钴、镍的硫化物和配合物的生成。

实验原理

1. 铬和锰分别为周期系ⅥB、ⅦB 族元素，它们都有可变的氧化态。铬的氧化态有 +2、+3、+6，其中氧化态 +2 的化合物不稳定，锰的氧化态有 +2、+3、+4、+5、+6、+7，其中氧化态 +3、+5 的化合物不稳定。锰的各种氧化态的化合物有不同的颜色。

氧化态	+2	+3	+4	+5	+6	+7
水合离子	Mn^{2+}	Mn^{3+}	无	MnO_3^-	MnO_4^{2-}	MnO_4^-
颜　色	浅桃红	深红		蓝	绿	紫

2. +2 价铬可用还原剂（如 Zn）将 +6 价铬或 +3 价铬还原而制得。

3. +3 价铬的氢氧化物呈两性，+3 价铬盐容易水解。在碱性溶液中，+3 价铬盐易被强氧化剂如 Na_2O_2 或 H_2O_2 氧化为黄色。

$$2CrO_2^- + 3H_2O_2 + 2OH^- \Longrightarrow 2CrO_4^{2-} + 4H_2O$$

铬酸盐和重铬酸盐在水溶液中存在着下列平衡：

$$2CrO_4^{2-} + 2H^+ \Longrightarrow Cr_2O_7^{2-} + H_2O$$

在酸性介质中，上述平衡向右移动，而在碱性介质中，该平衡向左移动。

4. 铬酸盐和重铬酸盐都是强氧化剂，易被还原为 +3 价铬（+3 价铬离子呈绿色或蓝色）。在酸性溶液中，$Cr_2O_7^{2-}$ 与 H_2O_2 作用生成蓝色过氧化铬 CrO_5。

$$Cr_2O_7^{2-} + 4H_2O_2 + 2H^+ \Longrightarrow 2CrO_5 + 5H_2O$$

这个反应常用来鉴定 $Cr_2O_7^{2-}$ 或 Cr^{3+}（先氧化为 CrO_4^{2-}）。

5. +2 价锰的氢氧化物呈白色，但是在空气中容易被氧化，逐渐变成棕色 MnO_2 的水合物 $MnO(OH)_2$。

6. +6 价锰酸盐可由 MnO_2 和强碱在氧化剂如 $KClO_3$ 的作用下，强热而制得。绿色的锰酸钾溶液在中性或微碱性时，MnO_4^{2-} 即发生歧化反应，生成紫色的高锰酸钾和棕黑色的 MnO_2 沉淀。

$$3K_2MnO_4 + 2H_2O \Longrightarrow 2KMnO_4 + MnO_2\downarrow + 4KOH$$

7. K_2MnO_4 可被强氧化剂氧化成 $KMnO_4$。

8. K_2MnO_4 和 $KMnO_4$ 都是强氧化剂，它们的还原产物随介质的不同而不同，例如 MnO_4^- 在酸性介质中，被还原为 Mn^{2+}，在中性介质中，被还原为 MnO_2，而在强碱性介质中和少量还原剂作用时，则被还原为 MnO_4^{2-}。

9. 在硝酸溶液中 Mn^{2+} 可以被 $NaBiO_3$ 氧化为紫红色的 MnO_4^-。通常利用这个反应来鉴定 Mn^{2+}。

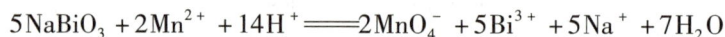

$$5NaBiO_3 + 2Mn^{2+} + 14H^+ \Longrightarrow 2MnO_4^- + 5Bi^{3+} + 5Na^+ + 7H_2O$$

10. 铁、钴、镍都是中等活泼金属。

11. Fe^{3+} 离子的外层电子层构型为 $3d^5$ 半充满稳定结构，所以它的 +3 价氧化态最稳定，其次是 Fe^{2+}，而钴、镍主要是 +2 价氧化态，按 Fe-Co-Ni 顺序，+3 价氧化态逐渐不稳定，镍的 +3 价氧化态化合物则较少见。

12. +2 价铁、钴、镍的氢氧化物及其性质如下。

	$Fe(OH)_2$	$Co(OH)_2$	$Ni(OH)_2$
颜色	白色	粉红色	苹果绿
在水中溶解情况	难溶	难溶	难溶
酸碱性	碱性	碱性	碱性
还原性	← 还原性增强		

13. +3 价铁、钴、镍的氢氧化物及其性质如下。

	$Fe(OH)_3$	$Co(OH)_3$	$Ni(OH)_3$
颜色	红棕色	褐棕色	黑色
在水中溶解情况	难溶	难溶	难溶
酸碱性	两性偏碱	两性偏碱	两性偏碱
氧化性		氧化性增强 →	

14. $Co(OH)_3$ 和 $Ni(OH)_3$ 与盐酸作用时，不能生成相应的 +3 价盐，因为它们的 +3 价盐极不稳定，而能把 Cl^- 氧化为 Cl_2。

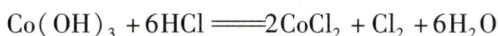

$$Co(OH)_3 + 6HCl == 2CoCl_2 + Cl_2 + 6H_2O$$

$Fe(OH)_3$ 与盐酸作用仅发生中和反应：

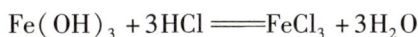

$$Fe(OH)_3 + 3HCl == FeCl_3 + 3H_2O$$

15. +2 价铁、钴、镍的硫化物难溶于水，均为黑色沉淀，易溶于稀酸，但是 CoS 和 NiS 一旦自溶液中析出，由于结构改变就不易再溶于稀酸。

16. 铁、钴、镍能生成氨合配离子：$[Co(NH_3)_6]^{2+}$、$[Ni(NH_3)_4]^{2+}$ 和 $[Ni(NH_3)_6]^{2+}$。Fe^{2+}、Fe^{3+} 只有在无水状态下可以与氨生成配合物，因为溶于水时，它们将立即分解产生 $Fe(OH)_2$、$Fe(OH)_3$ 沉淀。CN^- 与 Fe^{3+}、Co^{3+}、Fe^{2+}、Co^{2+}、Ni^{2+} 都能形成配位数为 6 或 4 的配合物。向 Fe^{3+} 溶液中加入亚铁氰化钾溶液，出现深蓝色沉淀。

$$4Fe^{3+} + 3[Fe(CN)_6]^{4-} == Fe_4[Fe(CN)_6]_3 \downarrow$$

向 Fe^{2+} 溶液中，加入铁氰化钾溶液，也出现深蓝色沉淀。

$$3Fe^{2+} + 2[Fe(CN)_6]^{3-} == Fe_3[Fe(CN)_6]_2 \downarrow$$

17. 向 Fe^{3+} 离子溶液中加入硫氰化钾溶液即出现血红色。

$$Fe^{3+} + nSCN^- \rightleftharpoons [Fe(SCN)_n]^{3-n}$$

$$n = 1, 2, 3, 4, 5, 6$$

这是 Fe^{3+} 的灵敏反应之一，可用于鉴定 Fe^{3+} 离子。Fe^{2+}、Co^{2+}、Ni^{2+} 与 SCN^- 形成的配合物有配位数 4 和 6 两类，它们在水溶液中都不稳定，但蓝色配离子 $[Co(SCN)_4]^{2-}$ 能稳定地存在于丙酮溶液中。

18. Ni^{2+} 在弱碱性条件下与丁二酮肟生成红色螯合物沉淀，这反应可用来鉴定 Ni^{2+} 离子。

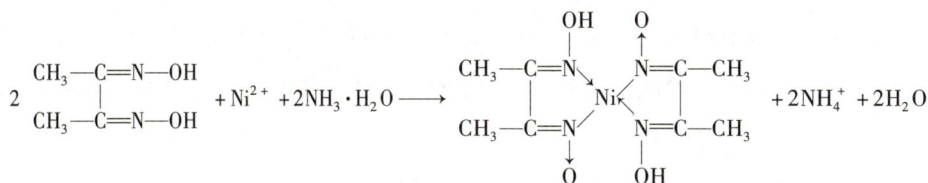

仪器和药品 •••

1. 仪器　离心机。

2. 药品

（1）**固体**　Zn 粉，KOH，$KMnO_4$，$NaBiO_3$，$FeSO_4 \cdot 7H_2O$，KSCN。

（2）**酸**　HCl 2mol/L，6mol/L，浓；HNO_3 6mol/L；H_2SO_4 2mol/L，浓。

（3）**碱**　NaOH 2mol/L，6mol/L，40%；氨水 2mol/L，6mol/L。

（4）**盐**　$CrCl_3$ 0.1mol/L；$K_2Cr_2O_7$ 0.1mol/L；Na_2SO_4 0.1mol/L；
$MnSO_4$ 0.1mol/L，0.5mol/L；NH_4Cl 1mol/L；$KMnO_4$ 0.01mol/L；Na_2CO_3 0.1mol/L；

$CoCl_2$ 0. 1mol/L；$NiSO_4$ 0. 1mol/L；$FeCl_3$ 0. 1mol/L；KI 0. 1mol/L；

$K_4[Fe(CN)_6]$ 0. 1mol/L；　$K_3[Fe(CN)_6]$ 0. 1mol/L；

丁二酮肟 1%，二苯硫腙（打萨宗），四氯化碳溶液。

3. 其他　溴水，淀粉 – KI 试纸，硫化氢饱和溶液，H_2O_2 3%，乙醚，淀粉溶液。

实验内容 •••

1. 铬

（1）+2 价铬的生成　取 10 滴 0. 1mol/L $CrCl_3$，加 10 滴 6mol/L HCl，再加少量 Zn 粉，微热至有大量气体逸出，观察溶液颜色由 Cr^{3+} 的绿色变为 Cr^{2+} 的天蓝色。吸出上部清液，加数滴浓 HNO_3，观察溶液又有什么变化？

（2）氢氧化铬的制备和性质　在盛有 10 滴 $CrCl_3$ 溶液的试管中，滴加 2mol/L NaOH 溶液，至产生氢氧化铬沉淀为止。观察沉淀的颜色。用实验证明 $Cr(OH)_3$ 呈两性，并写出反应方程式。

（3）+3 价铬的氧化和 Cr^{3+} 的鉴定　取 1 ~ 2 滴 $CrCl_3$ 溶液，加 6mol/L NaOH 过量，使生成的沉淀又复溶解，再加 3 滴 H_2O_2 溶液，加热，观察溶液颜色的变化，解释现象，并写出反应方程式，待试管冷却后，加 10 滴乙醚，然后慢慢加 6mol/L HNO_3 酸化，摇动试管，在乙醚层出现蓝色，表示有 Cr^{3+} 存在。

（4）铬酸盐和重铬酸盐的相互转变　取 5 滴 $K_2Cr_2O_7$ 溶液，加少许 2mol/L NaOH 溶液，观察颜色的变化。加 1mol/L H_2SO_4 至酸性，观察颜色的变化。解释现象，并写出反应方程式。

（5）+6 价铬的氧化性

①在 5 滴 0. 1mol/L $K_2Cr_2O_7$ 溶液中，加 5 滴 1mol/L H_2SO_4，然后滴加 0. 1mol/L Na_2SO_3，观察溶液的变化。写出反应方程式。

②在 5 滴 $K_2Cr_2O_7$ 溶液中，加 15 滴浓 HCl，加热，用润湿的淀粉 KI 试纸，检验逸出的气体。观察试纸和溶液颜色的变化。解释现象，并写出反应方程式。

2. 锰

（1）Mn^{2+} 氢氧化物的制备和性质　取 5 滴 $MnSO_4$ 溶液，加 2mol/L NaOH 溶液至沉淀完全，用吸管将沉淀连同溶液分成两份，一份迅速试验 $Mn(OH_2)$ 是否呈两性。另一份在空气中摇荡，注意沉淀颜色的变化，解释现象。

（2）Mn^{3+} 化合物的生成和性质　取 10 滴 0. 5mol/L $MnSO_4$，加 1mL 浓 H_2SO_4，用冷水冷却试管，然后再加 2 ~ 3 滴 0. 01mol/L $KMnO_4$ 溶液。观察深红色 Mn^{3+} 的生成（如果酸度不够，则易生成 MnO_2）。

$$MnO_4^- + 4Mn^{2+} + 8H^+ \Longrightarrow 5Mn^{3+} + 4H_2O$$

将所得深红色溶液用 0. 1mol/L Na_2CO_3 中和，则 Mn^{3+} 即转变为 Mn^{2+} 和 MnO_2，而有棕色沉淀产生。

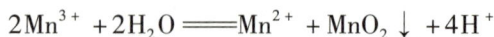

$$2Mn^{3+} + 2H_2O \Longrightarrow Mn^{2+} + MnO_2 \downarrow + 4H^+$$

（3）Mn（Ⅳ）化合物的生成　取 10 滴 0. 01mol/L $KMnO_4$ 溶液，滴加 0. 1mol/L $MnSO_4$ 溶液，观察 MnO_2 的生成。

$$2MnO_4^- + 3Mn^{2+} + 2H_2O \Longrightarrow 5MnO_2 \downarrow + 4H^+$$

（4）Mn（Ⅴ）化合物的生成　取 1mL 40% NaOH 加入试管中，加 1 滴 0. 01mol/L $KMnO_4$ 摇荡，观察 +5 价 Mn 的特征浅蓝色的出现

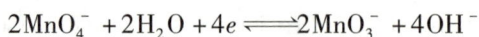

$$2MnO_4^- + 2H_2O + 4e \Longrightarrow 2MnO_3^- + 4OH^-$$

$$4OH^- - 4e \Longrightarrow 2H_2O + O_2$$
$$2MnO_4 \Longrightarrow 2MnO_3^- + O_2$$

（5）Mn（Ⅵ）化合物的生成　在 10 滴 0.01mol/L KMnO$_4$ 溶液中，加 20 滴 40% NaOH，然后加少量 MnO$_2$ 固体，加热搅动后静置片刻，离心沉降，上层清液即显 +6 价 Mn 的特征颜色。

$$2MnO_4^- + MnO_2 + 4OH^- \Longrightarrow 3MnO_4^{2-} + 2H_2O$$

取出上层绿色溶液，加 6mol/L H$_2$SO$_4$ 酸化，观察溶液颜色的变化和沉淀的析出。通过以上的实验，对 Mn 的各种价态的稳定性做出结论。

（6）MnO$_2$ 的氧化性　取少量 MnO$_2$ 固体粉末于试管中，加入 10 滴浓 HCl，微热，检验有无氯气逸出。

（7）高锰酸钾还原产物和介质的关系　在 3 支试管中各加 5 滴 0.01mol/L KMnO$_4$ 溶液，再分别加 2 滴 2mol/L H$_2$SO$_4$、水和 6mol/L NaOH，然后各加数滴 Na$_2$SO$_3$ 溶液，观察各试管中所发生的现象。写出反应方程式，并说明 KMnO$_4$ 的还原产物和介质的关系。

（8）Mn^{2+} 的鉴定　取 5 滴 0.1mol/L MnSO$_4$ 溶液于试管中，加 5 滴 6mol/L HNO$_3$，然后加少量 NaBiO$_3$ 固体，摇荡离心沉降后，上层清液呈紫色，表示有 Mn^{2+} 存在。

3. 铁、钴、镍的氢氧化物的制备和性质

（1）+2 价铁、钴、镍的氢氧化物的制备和性质

①Fe(OH)$_2$ 的制备和性质　在试管中加入 2mL 蒸馏水，加 1~2 滴 2mol/L H$_2$SO$_4$，煮沸片刻（为什么？），然后在其中溶解几粒 FeSO$_4$·7H$_2$O 晶体，同时在另一试管中煮沸 1mL 2mol/L NaOH，迅速加到 FeSO$_4$ 溶液中去（不要振荡），观察现象，静置片刻，观察沉淀颜色的变化。

②Co(OH)$_2$ 的制备和性质　取 20 滴 0.1mol/L CoCl$_2$ 溶液，加热，然后滴加 2mol/L NaOH，观察现象。静置片刻，观察沉淀颜色的变化。

③Ni(OH)$_2$ 的制备和性质　取 20 滴 0.1mol/L NiSO$_4$ 溶液，滴加 2mol/L NaOH，观察现象，并观察在空气中放置时颜色是否发生变化。

（2）+3 价铁、钴、镍的氢氧化物的制备和性质

①Fe(OH)$_3$ 的制备和性质　取 5 滴 0.1mol/L FeCl$_3$ 溶液，滴加 2mol/L NaOH，观察沉淀的颜色和形状。

②Co(OH)$_3$ 的制备和性质　取 5 滴 0.1mol/L CoCl$_2$，加入 3 滴溴水，然后加入 5 滴 2mol/L NaOH，观察沉淀的颜色。将溶液加热至沸，静置，弃去上面清液，在沉淀上滴加几滴浓 HCl，加热，用润湿的 KI - 淀粉试纸检验逸出气体，观察现象。

③Ni(OH)$_3$ 的制备和性质　取 0.1mol/L NiSO$_4$ 溶液 5 滴，同上法制备 Ni(OH)$_3$，并检验 Ni(OH)$_3$ 和浓 HCl 作用时，是否产生氯气。

4. 铁盐的氧化还原性

（1）亚铁盐的还原性　在试管中，加 10 滴新鲜配制的 FeSO$_4$ 溶液，再加 1~2 滴 2mol/L H$_2$SO$_4$，然后滴加少量 0.01mol/L KMnO$_4$ 溶液，观察 KMnO$_4$ 紫色的褪去。

（2）铁盐的氧化性　取 10 滴 0.1mol/L FeCl$_3$ 溶液，再滴加 0.1mol/L KI 溶液，观察现象，然后滴加 1 滴淀粉溶液，观察现象。

5. 铁、钴、镍的硫化物　在 3 支试管中，分别加入新鲜配制的 FeSO$_4$ 溶液、0.1mol/L CoCl$_2$ 和 0.1mol/L NiSO$_4$ 溶液各 2 滴，各约加 2 滴稀 HCl 酸化，加 H$_2$S 饱和溶液，有无沉淀产生？然后各加 2mol/L NH$_3$·H$_2$O 至碱性，观察有无沉淀产生？在各沉淀中加入稀 HCl 至酸性，再多加 2 滴观察沉淀是

否溶解？

6. 铁、钴、镍的配合物

（1）铁的配合物

① +2 价铁的配合物 取 10 滴 0.1mol/L $K_4[Fe(CN)_6]$ 溶液，滴加数滴 2mol/L NaOH 溶液，是否有 $Fe(OH)_2$ 沉淀产生，试解释之。另取 $FeCl_3$ 溶液 5 滴，滴入 2 滴 0.1mol/L $K_4[Fe(CN)_6]$ 溶液，观察现象。

② +3 价铁的配合物 取 10 滴 0.1mol/L $K_3[Fe(CN)_6]$ 溶液，滴加数滴 2mol/L NaOH 溶液，是否有 $Fe(OH)_3$ 沉淀产生？试解释之。

另取几粒 $FeSO_4$ 晶体于试管中，加 10 滴水使溶解，继续加入 2 滴 $K_3[Fe(CN)_6]$ 溶液，观察现象。

（2）钴的配合物

① +2 价钴的配合物 取 5 滴 0.5mol/L $CoCl_2$ 溶液加于试管中，加 5 滴 1mol/L NH_4Cl 溶液和过量的 6mol/L $NH_3 \cdot H_2O$，观察二氯化六氨合钴 $[Co(NH_3)_6]Cl_2$ 溶液的颜色。静置片刻，观察颜色的改变。

②取 5 滴 0.1mol/L $CoCl_2$ 溶液加于试管中，加少量固体 KSCN，再加 10 滴丙酮，由于生成配离子 $[Co(SCN)_4]^{2-}$ 溶于丙酮而呈现蓝色。

（3）镍的配合物 取 5 滴 0.1mol/L $NiSO_4$ 溶液，加 2 滴 1mol/L NH_4Cl，然后加 6mol/L 氨水，观察溶液的颜色有何变化。

取 5 滴 0.1mol/L $NiSO_4$ 溶液，加 5 滴 2mol/L 氨水，再加 1 滴 1% 二乙酰二肟（丁二酮肟），生成红色沉淀——此反应可用于鉴定 Ni^{2+} 离子。

7. 含 Fe^{2+}、Co^{2+}、Ni^{2+}、Cr^{3+}、Mn^{2+}、Zn^{2+} 的混合溶液的分离和检出

混合离子溶液 $\xrightarrow[\substack{15mol/L \text{ 氨水} \\ \text{微碱性，加热，} \\ \text{离心分离}}]{NH_4Cl \text{ 固体}}$

- $\begin{matrix} Fe(OH)_2\downarrow \\ Mn(OH)_2\downarrow \\ Cr(OH)_3\downarrow \end{matrix}$ $\xrightarrow[\substack{NaOH \\ \text{加热} \\ \text{离心分离}}]{\text{加 } H_2O_2}$

  - 溶液 (CrO_4^{2-}) $\xrightarrow{HNO_3}$ 蓝色溶液（示有 Cr^{3+}） $\xrightarrow[\text{乙醚}]{H_2O_2}$ (CrO_3)
  - 沉淀 $\begin{matrix} Fe(OH)_3 \\ MnO_2 \cdot H_2O \end{matrix}$ $\xrightarrow[\text{酸化}]{\text{浓 HCl}}$ $\begin{matrix} Fe^{3+} \\ Mn^{2+} \end{matrix}$

    - Fe^{3+} $\xrightarrow{K_4[Fe(CN)_6]}$ 蓝色↓（示有 Fe^{3+}）
    - Mn^{2+} $\xrightarrow[NaBiO_3]{HNO_3}$ 红色液（示有 Mn^{2+}）(MnO_4^-)

- 溶液
  - $[Ni(NH_3)_6]^{2+}$ $\xrightarrow{\text{丁二酮肟}}$ 红色沉淀（示有 Ni^{2+}）
  - $[Zn(NH_3)_4]^{2+}$ $\xrightarrow[\text{二苯硫宗}]{NaOH}$ 粉红色溶液（示有 Zn^{2+}）
  - $[Co(NH_3)_6]^{2+}$ $\xrightarrow{6mol/L \text{ HCl}}$ Co^{2+} $\xrightarrow[\text{丙酮}]{KSCN}$ 蓝色溶液（示有 Co^{2+}）

实 验 指 导

预习要求 •••

1. 铬、锰、铁、钴、镍属于过渡元素，试根据它们的原子结构，说明它们有哪些性质？

2. +2 价铬的稳定性要比 +3 价铬差，试根据它们的价电子构型加以说明。

3. 如何鉴别 Cr^{3+} 和 $Cr_2O_7^{2-}$？实验中为什么要加入乙醚？

4. 怎样以铬酸盐为原料制备重铬酸盐？

5. 锰元素的氧化态有哪几种？举出常见锰的化合物及其用途。

6. $KMnO_4$ 的还原产物与介质的酸碱性有何关系？

7. 怎样鉴别 Mn^{2+} 的存在？

8. 如何制备 +2 价、+3 价铁、钴、镍的氢氧化物？它们的氢氧化物有何性质？

9. 向 +3 价铁、钴、镍的氢氧化物中加入浓盐酸，各自得到的产物是什么？

10. 如何鉴别 Fe^{2+}、Fe^{3+}、Co^{2+}、Ni^{2+}？

11. 完成并配平下列化学方程式：

（1）$Cr^{3+} + Zn \longrightarrow$

（2）$Cr^{3+} + NaOH（过量）\longrightarrow$

（3）$CrO_2^- + H_2O_2 + OH^- \longrightarrow$

（4）$Cr_2O_7^{2-} + H_2O_2 + H^+ \xrightarrow{\text{乙醚}}$

（5）$Mn(OH)_2 + O_2 \longrightarrow$

（6）$MnO_4^- + SO_3^{2-} + H^+ \longrightarrow$

$\qquad MnO_4^- + SO_3^{2-} + H_2O \longrightarrow$

$\qquad MnO_4^- + SO_3^{2-} + OH^- \longrightarrow$

（7）$Mn^{2+} + NaBiO_3 + H^+ \longrightarrow$

（8）$Fe(OH)_2 + O_2 \longrightarrow$

（9）$Ni(OH)_2 + Br_2 + OH^- \longrightarrow$

（10）$MnO_4^- + Fe^{2+} + H^+ \longrightarrow$

（11）$Co(OH)_3 + HCl(Conc) \longrightarrow$

注意事项 •••

1. Mn^{3+} 存在于强酸性溶液中，实验中要保证试液足够的酸度。

2. K_2MnO_4 存在于强碱性溶液，实验中要保证试液足够的碱度。

3. $KMnO_4$ 在酸性介质中被还原成 Mn^{2+}（近无色）如得到棕色溶液，说明试液的酸度不足。

4. 在鉴别 Cr^{3+}（或 $Cr_2O_7^{2-}$）试验中，最后加 HNO_3 时要加足量，使呈酸性。

5. CoS、NiS 沉淀一旦生成由于自身结构的变化便不再溶于稀酸。

6. $[Co(NH_3)_6]Cl_2$ 为棕黄色，$[Co(NH_3)_6]Cl_3$ 为棕红色，试验中要仔细观察颜色的变化。

报告格式 •••

1. 目的

2. 原理

3. 实验内容

用表格形式表示。例：

试 剂	现 象	反应方程式	结 论
$FeCl_3 + KI$	有 I_2 生成淀粉变蓝	$2Fe^{3+} + 2I^- =\!=\!= 2Fe^{2+} + I_2$	Fe^{3+} 具有氧化性

实验后思考 •••

1. Cr（Ⅵ）作氧化剂时的介质条件是什么？在选用介质时应考虑什么问题？

2. 哪些实验说明锰发生的自身氧化还原反应？其介质条件如何？

3. $KMnO_4$ 溶液中若有 Mn^{2+} 和 MnO_2 存在，对其稳定性有何影响？

4. 实验室常用的洗液是 $K_2Cr_2O_7$ 和浓 H_2SO_4 的混合液，用久后会变绿，为什么？

5. 即使长时间通 H_2S 于 $K_2Cr_2O_7$ 或 $KMnO_4$ 酸性溶液中，也得不到硫化物沉淀，为什么？得到什么物质？写出反应式。

6. 混合液中含有 Ca^{2+}、Al^{3+}、Cr^{3+}、Mn^{2+}、Fe^{2+}、Ni^{2+}，如何分离？

7. 为什么在 $Fe(Ⅲ)$ 盐溶液中加 Na_2CO_3 放出 CO_2，在 $Fe(Ⅱ)$ 溶液中加 Na_2CO_3 不放出 CO_2？

Experiment 5　　Chromium, Manganese, Iron, Cobalt and Nickel

◆ Objectives

1. Know the properties and preparation of some compounds of chromium and manganese in their important oxidation states.

2. Know transformation of the various oxidation states of Cr and Mn.

3. Know the oxidation-reduction properties of compounds Cr and Mn. Know how the medium affects these oxidation-reduction reactions.

4. Know the preparation and properties of $Fe(Ⅱ,Ⅲ)$, $Co(Ⅱ,Ⅲ)$, $Ni(Ⅱ,Ⅲ)$ hydroxides.

5. Know the reducing property of iron($Ⅱ$) and the oxidizing property of iron($Ⅲ$).

6. Know how to prepare the sulfides and the complexes of iron, cobalt and nickel.

◆ Principles

1. Cr is an element of Group ⅥB in the periodic table and Mn is of Group ⅦB. Their oxidation states are various. Those of Mn are $+2$, $+3$, $+4$, $+5$, $+6$, $+7$. The trivalent states and the V state are unstable. The Mn compounds their oxidation states have different colors.

Oxidation state	+2	+3	+4	+5	+6	+7
Aquo ion	Mn^{2+}	Mn^{3+}	non	MnO^{3-}	MnO_4^{2-}	MnO_4^-
Color	pink	red		blue	green	dark purple

2. $Cr(Ⅱ)$ is obtained when $Cr(Ⅵ)$ or $Cr(Ⅲ)$ is reduced by reducing reagents, such as Zn.

3. Hydroxide $Cr(OH)_3$ is amphoteric. $Cr(Ⅲ)$ salts are hydrolyzed appreciably. In alkaline solution, $Cr(Ⅲ)$ can be oxidized by powerful oxidizers such as Na_2O_2 and H_2O_2 to yellow chromate.

$$2CrO_2^- + 3H_2O_2 === 2CrO_4^{2-} + 4H_2O$$

In aqueous solutions, there is an equilibrium reaction for chromate and dichromate which is shifted to the right in acid solution and to the left in alkaline solution.

$$2CrO_4^{2-} + 2H^+ === Cr_2O_7^{2-} + H_2O$$

4. Both chromates and dichromates are strong oxidizing reagents and easily reduced to $Cr(Ⅲ)$ ($Cr(Ⅲ)$ ions is green or blue.). When H_2O_2 is added to dichromate solution, a blue solution containing peroxide CrO_5 is formed. The reaction is used to identify dichromate ion and $Cr(Ⅲ)$ ion (oxidized to CrO_4^{2-} forehead).

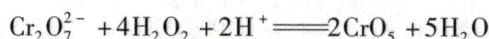

$$Cr_2O_7^{2-} + 4H_2O_2 + 2H^+ === 2CrO_5 + 5H_2O$$

5. Hydroxide $Mn(OH)_2$ is white and easily oxidized to $MnO(OH)_2$, a brown aqua compound of MnO_2.

6. When MnO_2 is fused with alkalis in the presence of oxidizing reagents like $KClO_3$, $Mn(Ⅵ)$ salt is ob-

tained. The green potassium manganate is unstable and disproportionate to purple MnO_4^- and brown precipitate MnO_2 in neutral or weakly alkaline solution.

$$3K_2MnO_4 + 2H_2O = 2KMnO_4 + MnO_2 \downarrow + 4KOH$$

7. Oxidation with Cl_2 converts K_2MnO_4 to $KMnO_4$.

8. Both K_2MnO_4 and $KMnO_4$ are strong oxidants and the reduced products vary with the media. In acids, MnO_4^- is reduced to Mn^{2+} ion. But in neutral solutions, the reduction proceeds to MnO_2, and in alkaline solutions to MnO_4^{2+} in presence of a little reducing reagent.

9. Oxidation with $NaBiO_3$ converts Mn^{2+} to dark purple MnO_4^- ion, which is usually used to identify Mn^{2+} ion.

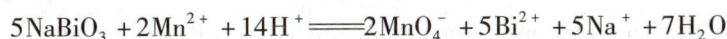

$$5NaBiO_3 + 2Mn^{2+} + 14H^+ \Longrightarrow 2MnO_4^- + 5Bi^{2+} + 5Na^+ + 7H_2O$$

10. Iron, cobalt and nickel are all metals with medium activity.

11. The trivalent state of iron is the most stable for iron because $Fe(\text{III})$ ion has the electronic configuration $3d^5$ (Single occupation of orbital). The II state of iron is the second stable. The chief oxidation state of cobalt and nickel is the II state. According to the sequence of Fe-Co-Ni, the III oxidation state gets gradually unstable. And the compounds of $Ni(\text{III})$ are rare.

12. Hydroxides $Fe(\text{II})$、$Co(\text{II})$、$Ni(\text{II})$ have the properties as follows

	$Fe(OH)_2$	$Co(OH)_2$	$Ni(OH)_2$
Color	white	pink	apple green
Solubility in water	insolable	insoluable	insoluable
Acid-base property	acieic	basic	basic
Reducing property	$\leftarrow$ getting stronger		

13. Hydroxides $Fe(\text{III})$、$Co(\text{III})$、$Ni(\text{III})$ have the following properties

	$Fe(OH)_3$	$Co(OH)_3$	$Ni(OH)_3$
Color	red-brown	dark-brown	black
Solubility in water	insolable	insoluable	insoluable
Acid-base property	amphoteric apt to basic	amphoteric apt to basic	amphoteric apt to basic
Reducing property	$\rightarrow$ getting stronger		

14. When hydrochloric acid (HCl) is added to $Co(OH)_3$ or $Ni(OH)_3$, the respective salt of the III state is not obtained because it is extraordinary unstable to oxidize Cl^- ion to Cl_2.

$$Co(OH)_3 + 6HCl \Longrightarrow 2CoCl_2 + Cl_2 + 6H_2O$$

But only takes place the neutralized reactions when hydrochloric (HCl) is added to $Fe(OH)_3$

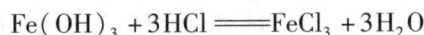

$$Fe(OH)_3 + 3HCl \Longrightarrow FeCl_3 + 3H_2O$$

15. The black sulfides of iron (II), cobalt (II). nickel(II) are aqueous insoluble but soluble in acidic solution. However, once CoS and NiS precipitate from the solution, they are not soluble in acid any more as a result of the transformation of their structures.

16. Ammonia complexes of iron, cobalt and nickel: $[Co(NH_3)_6]^{2+}$, $[Ni(NH_3)_4]^{2+}$, and $[Ni(NH_3)_6]^{2+}$ are formed easily. The analogues of $Fe(\text{II})$ and $Fe(\text{III})$ are only formed in absence of water because they are immediately decomposed to the precipitates $Fe(OH)_2$ and $Fe(OH)_3$ when they are dissolved in water. Cyanide

complexes with coordination number 4 or 6 are formed by the metal ions Fe^{3+}、Co^{3+}、Fe^{2+}、Co^{2+}、Ni^{2+} with CN ion (cyanide ion). A dark blue precipitate forms when the Fe^{3+} solution is added to the solution of potassium ferrocyanide.

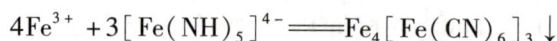

$$4Fe^{3+} + 3\left[Fe(NH)_5 \right]^{4-} =\!=\!= Fe_4 \left[Fe(CN)_6 \right]_3 \downarrow$$

The dark blue precipitate also forms when a solution of potassium ferricyanide is added to a Fe^{2+} ion solution.

17. Addition of a potassium thiocyanate solution to a Fe^{3+} ion solution gets a deep-colored red solution.

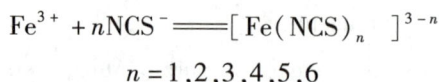

$$Fe^{3+} + n NCS^- =\!=\!= \left[Fe(NCS)_n \right]^{3-n}$$
$$n = 1,2,3,4,5,6$$

The coordination reaction is one of the sensitive reactions of Fe^{3+} ion and employed in identifying Fe^{3+}. Sulfo cyanide complexes formed by the metal ions Fe^{2+}、Co^{2+}、Ni^{2+} with SCN^- ion are four-coordinated or six-coordinated and unstable in aqueous solution. Among them the blue cobalt(Ⅱ) complex ion is stable in acetone solution.

18. When Ni^{2+} ion is combined with dimethyglyoxime in a feeble basic medium, a red chelate nickel dimethylglyoximate is precipitated. This reaction is used to identify Ni^{2+} ion.

◆ Equipment and Chemicals

1. Equipment   Centrifugal machine

2. Chemicals

(1)Solids   Zn powder, KOH, $KMnO_4$, $NaBiO_3$, $FeSO_4 \cdot 7H_2O$, $KSCN$

(2)Acids   HCl 2mol/L, 6mol/L, concentrated; HNO_3 6mol/L, H_2SO_4 2mol/L, concentrated

(3)Salts $CrCl_3$ 0.1mol/L; $K_2Cr_2O_7$ 0.1mol/L; Na_2SO_4 0.1mol/L; $MnSO_4$ 0.1mol/L, 0.5mol/L; NH_4Cl 1mol/L; $KMnO_4$ 0.01mol/L; Na_2CO_3 0.1mol/L; $CoCl_2$ 0.1mol/L; $NiSO_4$ 0.1mol/L; $FeCl_3$ 0.1mol/L; KI 0.1mol/L; $K_4 \left[Fe(CN)_4 \right]$ 0.1mol/L; $K_3 \left[Fe(NH)_6 \right]$ 0.1mol/L dimethylglyoxime 1%   dithizone solution of CCl_4

(4)Rest   Bromine water, starch iodide paper, Hydrogen sulfide saturated solution, Hydrogen peroxide 3% ether, starch solution.

◆ Procedures

1. Chromium (Cr)

(1) Preparation of Cr(Ⅱ) compounds   Add 10 drops of HCl (6mol/L) to drops of $CrCl_3$ solution (0.1mol/L) with a little Zn powder. The mixture is heated slightly until a lot of gas is given out. The color of the solution changed into sky blue from green ($Cr^{3+} \rightarrow Cr^{2+}$) is observed. Fetch the upper clear solution and add a few drops of concentrated HNO_3 to it. What happens to the solution?

(2) Preparation and property of hydroxide Cr(Ⅱ)   Add dropwise the NaOH solution (2mol/L) to a tube containing 10 drops of $CrCl_3$ solution until the precipitate $Cr(OH)_3$ forms. Observe the color of the precipitate. Test the amphoteric proprty of $Cr(OH)_3$ and write out the equation of the reaction.

(3) Oxidation and identification of Cr(Ⅲ)   NaOH solution (6mol/L) is added dropwise to a tube containing a $CrCl_3$ solution (1~2d) until the precipitate formed is redissolved. Then 3 drops of a H_2O_2 solution are added to the solution before the mixture is heated. Observe and explain the change of the color and write out the equation. When the tube is cooled down, add 10 drops of ether to it before the solution is acidified by adding slowly HNO_3(6mol/L) to it. Shake the tube, the ether layer gets blue, which suggests Cr^{3+} ion is presence.

（4）Alternation between chromate and dichromate　Some NaOH solution （2mol/L） is added to 5 drops of $K_2Cr_2O_7$. Then the mixture is acidified with H_2SO_4（1mol/L）. Explain the reason of the color changing and write out the reaction equation.

（5）Oxidation property of Cr（Ⅵ）

（a）Add a solution of H_2SO_4（1mol/L,5 d） to a solution of $K_2Cr_2O_7$（5 d） before a Na_2SO_3 solution （0.1mol/L） is dropped into the mixture. Observe the change and write out the equation.

（b）Heat the mixture of 5 drops of $K_2Cr_2O_7$ solution and 15 drops of concentrated HCl. Test the escaping gas from the mixture by a piece of wet starch iodide paper. Look at the color changes of the paper and the solution. Explain the phenomena and write out the equation.

2. Manganese（Mn）

（1）Preparation and property of hydroxide Mn（Ⅱ）　5 drops of $MnSO_4$ solution is precipitated completely by a NaOH solution （2mol/L）. Divide the mixture into two parts. One is used to test its amphoteric property. The other is shaken in the air. Notice the color of the precipitate and explain the change.

（2）Preparation and property of Mn（Ⅲ） compounds　Add 1mL of concentrated H_2SO_4 into a tube containing 10 drops of a $MnSO_4$ solution （0.5mol/L）. Cool the tube down with ice water before a $KMnO_4$ solution （0.01mol/L） is dropped into it. The dark red solution of Mn^{3+} is observed. （If the acidity is not enough, MnO_2 is usually obtained. ）

$$MnO_4^- +4Mn^{2+} +8H^+ =\!=\!=5Mn^{3+} +4H_2O$$

Neutralization of the dark red solution with Na_2CO_3 solution （0.1mol/L） converts Mn^{3+} ion into Mn^{2+} ion and MnO_2 ,a brown precipitate.

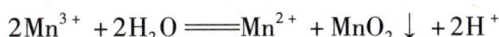

$$2Mn^{3+} +2H_2O =\!=\!=Mn^{2+} +MnO_2 \downarrow +2H^+$$

（3）Preparation of Mn（Ⅳ） compounds　Drop a solution of $MnSO_4$（0.1mol/L） into a $KMnO_4$ solution of 10 drops. Notice the showing of MnO_2

$$2MnO_4^- +3Mn^{3+} +2H_2O =\!=\!=5MnO_2 \downarrow +4H^+$$

（4）Preparation of Mn（Ⅴ） compounds　Put a NaOH solution （40%） of 1mL into a tube. Shock the tube after a drop of $KMnO_4$ solution （0.01mol/L） is added. Observe the pale blue color,the characteristic color of Mn（Ⅴ）.

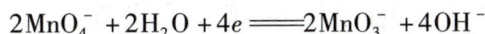

$$2MnO_4^- +2H_2O +4e =\!=\!=2MnO_3^- +4OH^-$$
$$4OH^- -4e =\!=\!=2H_2O +O_2$$
$$2MnO_4^- =\!=\!=2MnO_3^- +O_2$$

（5）Preparation of Mn（Ⅵ） compounds　Heat a mixture of 10 drops of $KMnO_4$ solution （0.01mol/L）,20 drops of 40% NaOH and a little MnO_2 solid until it is boiled. The mixture is stirred before it stays without disturbance for a while. After centrifugal separation,the clear liquid solution of the upper layer shows the character color of Mn（Ⅵ）

$$2MnO_4^- +MnO_2 +4OH^- =\!=\!=3MnO_4^{2-} +2H_2O$$

Notice what happens when the green solution of the upper layer is acidified with 6mol/L H_2SO_4. Conclude the stability of the oxidation states of Mn from these experiments.

（6）Oxidizing property of MnO_2　Heat slightly a tube containing a little MnO_2 powder and 10 drops of concentrated HCl,and test whether Cl_2 is produced.

（7）Relationship between the medium and the reduced product of $KMnO_4$ Take out three tubes. Add 5 drops of a 0.01mol/L $KMnO_4$ solution into each tube. Then add 2 drops of 2mol/L H_2SO_4, 2 drops of water, 2 drops of NaOH into the tubes respectively. Observe what happens in the tubes when a few drops of Na_2SO_3 solution are added into each tube. Write out the reaction equations and discuss the relationship between the medium and reduced products.

（8）Identification of Mn（Ⅱ） 5 drops of $MnSO_4$ solution （0.1mol/L） and 5 drops of HNO_3 （6mol/L） are added into a tube. Then add a little $NaBiO_3$ solid into the mixture after the tube is shaken. The upper clear purple solution separated by a centrifugal machine shows the existence of Mn^{2+} ion.

3. Preparations and properties of Hydroxides of iron, cobalt and nickel.

（1）Preparations and properties of hydroxides of iron（Ⅱ）, cobalt（Ⅱ）, nickel（Ⅱ）

（a）Preparation and property of Iron（Ⅱ）hydroxide 2mL of distilled water acidified with 2mol/L H_2SO_4 （1~2d） is boiled for a while （why?）before a few grains of $FeSO_4 \cdot 7H_2O$ crystal are dissolved in it. Then 1mL of NaOH （2mol/L） boiled is quickly added to the $FeSO_4$ solution. （Don't shake the tube）. Notice the color change. Let it stay undisturbed for a few minutes.

（b）Preparation and property of $Co(OH)_2$ Drop 2mol/L NaOH to a $CoCl_2$ solution （0.1mol/L）of 20 drops that has been heated. Let the mixture stay undisturbed for a while. Observe that the precipitate changes color.

（c）Preparation and property of $Ni(OH)_2$ Drop a 2mol/L NaOH solution to 20 drops of $Ni(OH)_2$ （0.1mol/L） which has been heated. Take the mixture in the air. Observe whether the precipitate changes color.

（2）Preparation and properties of Hydroxides of Fe（Ⅲ）, Co（Ⅲ）, Ni（Ⅲ）

（a）Preparation and property of $Fe(OH)_3$ Drop a 2mol/L NaOH solution to a tube containing 5 drops of a $FeCl_3$ solution （0.1mol/L）. Observe the color and the shape of the precipitate formed.

（b）Preparation and property of $Co(OH)_3$ Drops of bromine water are added to 5 drops of a 0.1mol/L $CoCl_2$ solution before 5 drops of a NaOH solution （2mol/L） is added. A black solid is precipitated from the solution. Boil the mixture before it stays without disturbance. Drop concentrated HCl to the precipitate after the upper clear solution is thrown away. Test the giving-off gas with a piece of wet starch iodide paper when the mixture is heated.

（c）Preparation and property of $Ni(OH)_3$ Fetch 5 drops of $NiSO_4$ solution （0.1mol/L） to prepare $Ni(OH)_3$ precipitate according to the preparation of $Co(OH)_3$. And certify whether the redox reaction of $Ni(OH)_3$ with concentrated HCl forms Cl_2 gas.

4. Properties of iron salts

（1）Reducing the property of ferrous salt. Drop a little $KMnO_4$ solution （0.01mol/L） into a tube containing 10 drops of a fresh $FeSO_4$ solution acidified by 1~2 drop of H_2SO_4（2mol/L）. Observe the purple color of $KMnO_4$ vanishing.

（2）Oxidizing property of ferric salt Drop a KI solution （0.1mol/L） to 10 drops of 0.1mol/L $FeCl_3$ solution. What happens? Then add a drop of starch solution to the mixture and observe what happens again.

5. Sulfides of iron, colalt and nickel
Take three tubes. Add two drops of a fresh $FeSO_4$ solution, two drops of a $CoCl_2$ solution （0.1mol/L）, two drops of a $NiSO_4$ solution （0.1mol/L） to the tubes respectively. Then add 2 drops of dilute HCl and a saturated H_2S solution to each tube. What happens? Then all the mix-

tures are basic with NH_3H_2O (2mol/L). What happens again? Add dilute HCl to the precipitates and acidify them. Are the precipitates dissolved? If not, add 2 drops more of HCl solution again.

6. Complexes of iron, cobalt and nikel

(1) Iron (II) complexes　Add a few drops of NaOH solution (2mol/L) to $K_4[Fe(CN)_6]$ solution (0.1mol/L, 10 drops). Explain why there is no precipitate from the solution. Fetch another 6 drops of $FeCl_3$ solution. Add 2 drops of $K_4[Fe(CN)_6]$ solution to it. Observe what happens.

(2) Iron (III) complexes　Add a few drops of 2mol/L NaOH to $K_3[Fe(CN)_6]$ (10 drops, 0.1mol/L). Notice whether $Fe(OH)_3$ is formed. And explain the reason. Put a few grains of $FeSO_4$ crystal into a tube. And dissolve the $FeSO_4$ with water. Then add 2 drops of a $K_3[Fe(CN)_4]$ solution. Notice the reaction that takes place in the tube.

(3) Ammonia complexes of cobalt　Add 5 drops of NH_4Cl solution (1mol/L) and excess of 6mol/L NH_3H_2O to a tube containing 0.5mol/L $CoCl_2$ solution (5 drops). Observer the forming of a yellowish brown solution of $[Co(NH_3)_6]Cl_6$. Wait for a few minutes. The complex solution changes slowly to reddish brown. Why?

(4) Sulfo cyanic cobalt(II) complexes　Add a few grains of KSCN crystal to a tube containing 5 drops of a 0.1mol/L $CoCl_2$ solution before acetone is dropped into the tube. The acetone layer is blue owing to the forming of the complex ion $[Co(NCS)_4]^{2-}$

(5) Complexes of Nickel　Five drops of $NiSO_4$ solution is mixed with 2 drops of NH_4Cl (0.1mol/L) before NH_3H_2O(0.1mol/L) is added. Notice the color changes.

7. Separation and identification of $Fe^{2+}, Co^{2+}, Ni^{2+}, Cr^{3+}, Zn^{2+}$ from their mixture solution

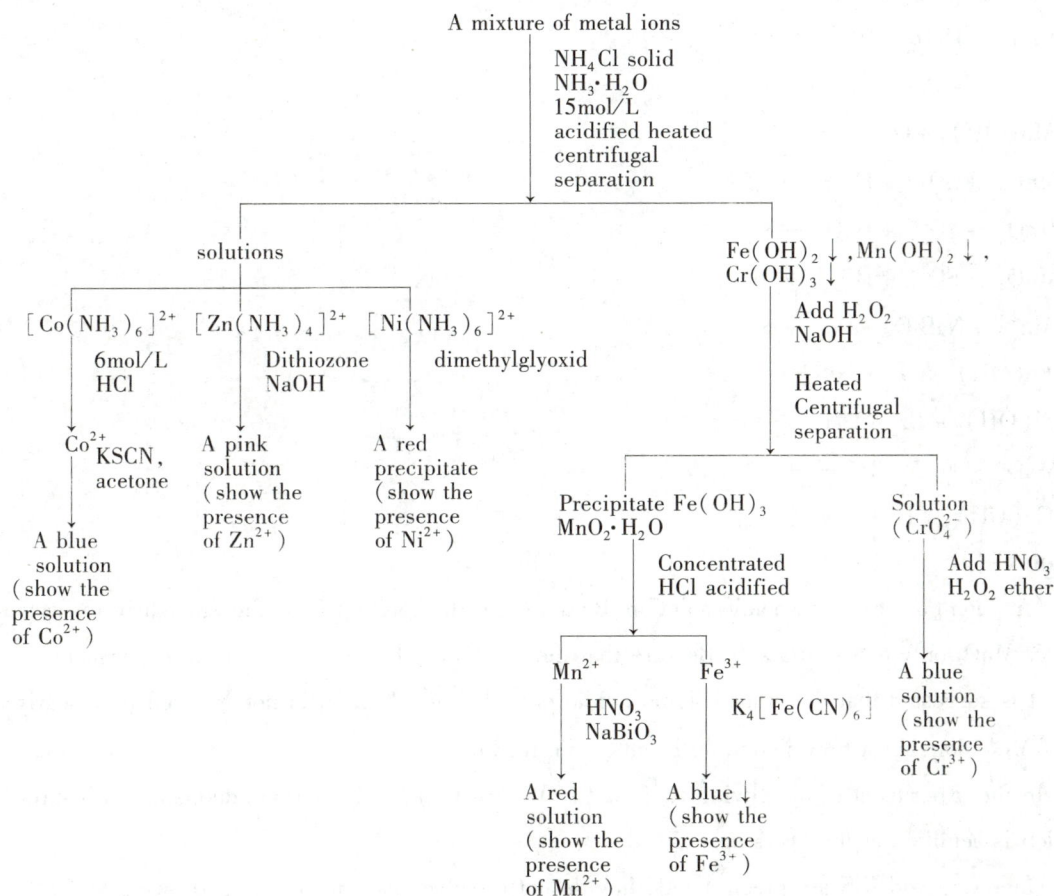

◆ Instructions

1. Requirements

（1）Iron, manganese, chromium, cobalt and nickel are all transition metal elements. Discuss their properties on the basis of their atomic structure.

（2）Explain the reason why $Cr(\mathrm{II})$ is more unstable than $Cr(\mathrm{III})$ according to their electronic configurations.

（3）How do you identify Cr^{3+} and $Cr_2O_7^{2-}$? Why must ether be added in the experiment?

（4）How is dichromate prepared from chromate?

（5）What are the oxidation states of manganese? Give names of some common managanese compounds and their uses.

（6）Discuss that the oxidant of $KMnO_4$ is pH dependent.

（7）How is Mn^{2+} ion identified?

（8）How do you prepare the hydroxides of iron, cobalt, nickel in their divalent and trivalent states? Discuss their properties.

（9）What is obtained when concentrated HCl is added to the hydroxides of $Fe(\mathrm{III})$, $Co(\mathrm{III})$ and $Ni(\mathrm{III})$ respectively?

（10）How do you identify the metal ions of Fe^{2+}, Fe^{3+}, Co^{2+}, Ni^{2+}?

（11）Finish and equilibrate the following reaction equations:

（a）$Cr^{3+} + Zn \longrightarrow$

（b）$Cr^{3+} + NaOH(excess) \longrightarrow$

（c）$CrO_2^- + H_2O_2 + OH^- \longrightarrow$

（d）$Cr_2O_7^{2-} + H_2O_2 + H^+ \longrightarrow$

（e）$Mn(OH)_2 + O_2 \longrightarrow$

（f）$MnO_4^- + SO_3^{2-} + H^+ \longrightarrow$

$MnO_4^- + SO_3^{2-} + H_2O \longrightarrow$

$MnO_4^- + SO_3^{2-} + H^+ \longrightarrow$

（g）$Mn^{2+} + NaBiO_3 + H^+ \longrightarrow$

（h）$Fe(OH)_2 + O_2 \longrightarrow$

（i）$Ni(OH)_2 + Br_2 + OH^- \longrightarrow$

（j）$MnO_4^- + Fe^{2+} + H^+ \longrightarrow$

（k）$Co(OH)_3 + HCl(Conc) \longrightarrow$

2. Notes

（1）Mn^{3+} ion only exists in acidic solution. Be sure that the solution is acidic enough in the experiment.

（2）K_2MnO_4 only exists in alkali. Be sure that the solution is basic enough in the experiment.

（3）It is suggested that the reagent is no acidic enough when $KMnO_4$ is not reduced to a nearly colorless solution of Mn^{2+} ion, but a brown solution in an acidic medium.

（4）In the experiment of identifying Cr^{3+} or $Cr_2O_7^{2-}$, HNO_3 must be added adequately to insure that reaction solution is acidified at the last step.

（5）Once CoS and NiS are precipitated, they do not dissolve again in dilute acid owing to the transformation of their structures.

(6) $[Co(NH_3)_6]Cl_2$ is yellowish brown and $[Co(NH_3)_6]Cl_3$ is reddish brown. Observe attentively the color change.

3. Report format

(1) Objectives

(2) Principles

(3) Content

reagent	phenomena	reaction equation	conclusions
$FeCl_3 + KI$	turn to blue	$2Fe^{3+} + 2I^- = 2Fe^{2+} + I_2$	Fe^{3+} is oxidative

4. Questions

(1) Which medium is chosen when $Cr(VI)$ is used as an oxidant?

What is taken into account when a medium is chosen?

(2) What experiments show that disproportionate of manganese takes place? What are the media?

(3) Discuss the stability of a $KMnO_4$ solution in the presence of Mn^{2+} and MnO_2.

(4) The common washing solution in the lab, a mixture of $K_2Cr_2O_7$ and concentrated H_2SO_4, gets green when it is used for a long time. Why?

(5) The sulfide precipitate is not formed when piping H_2S gas into $K_2Cr_2O_7$ or $KMnO_4$ of acidic solution even for a long time. Why? What is formed? Write out the reaction equation.

(6) How do you separate the metal ions Cr^{2+}、Al^{3+}、Cr^{3+}、Mn^{2+}、Fe^{2+} and Ni^{2+} from their mixture?

(7) CO_2 is given off when Na_2CO_3 is added to $Fe(III)$ solution. Conversely, CO_2 is not given off when Na_2CO_3 is added to a $Fe(II)$ solution. Why?

Words

hydrolysis	水解	reducing property	还原性
oxidation state	氧化态	oxidizing property	氧化性
oxidation-reduction properties	氧化还原性质	neutralized reaction	中和反应
		equilibrium reaction	平衡反应
coordination number	配位数	peroxide	过氧化物
coordination reaction	配位反应	hydroxides	氢氧化物
chelate	螯合物	ferroferricyanide	氰铁酸亚铁
ferric chloride	$FeCl_3$	potassium ferricyanide	$K_3[Fe(CN)_6]$
ferric sulfate	$Fe(SO_4)_3$	potassium sulfocyanide	KSCN
ferrous sulfate	$FeSO_4$	ferric potassium ferrocyanide	
ferrous sulfide	FeS		亚铁氰化铁钾,钾盐铁蓝
Ferric thiocyanide	$Fe(SCN)_3$	chromium	铬
cobaltous hydroxide	$Co(OH)_2$	iron	铁
cobaltous chloride	$CoCl_2$	nickel	镍

cobalt(Ⅱ) hexaa-ammine ion	$\left[Co(NH_3)_6\right]^{2+}$	cobalt(Ⅲ) hexaammine ion	$\left[Co(NH_3)_6\right]^{3+}$
cobaltous sulfide	CoS	hydrogen peroxide	H_2O_2
cobaltic hydroxide	$Co(OH)_3$	nickel ammine ion	
alkaline solution	碱性溶液		$\left[Ni(NH_3)_6\right]^{2+},\left[Ni(NH_3)_4\right]^{2+}$
concentrated hydrochl-oric acid	浓 HCl	nickel dimethylglyoxime	丁二酮肟镍
dilute hydrochloric acid	稀 HCl	nickel chloride	$NiCl_2$
sulphuric acid	H_2SO_4	nickel sulfate,nickelous sulfate	$NiSO_4$
nitric acid	HNO_3	nickel nitrite	$Ni(NO_3)_2$
cyanide	CN^-	nickel hydroxide	$Ni(OH)_2$
ammonia complexes	氨配合物	nickelic hydroxide	$Ni(OH)_3$
potassium manganate	K_2MnO_4	potassium hydroxide	KOH
potassium permanganate	$KMnO_4$	sodium bismuthate	$NaBiO_3$
manganese dioxide	NiS,MnO_2	sodium sulfate	Na_2SO_4
chromic oxide	Cr_2O_3	ammonium chloride	NH_4Cl
chromic chloride,chromium chloride	$CrCl_3$	sodium carbonate	Na_2CO_3
		dimethylglyoxime	丁二酮肟
		dithizone	二苯硫腙
chromium hydrate,chromium hydroxide	$Cr(OH)_3$	bromine water	Br_2 水
		starch iodide paper	碘淀粉试纸
chromium peroxide	$Cr(O_2)_2O$	hydrogen sulfide saturated solution	硫化氢饱和溶液
potassium chromate	K_2CrO_4		
potassium dichromate	$K_2Cr_2O_7$	ether	乙酸乙酯
dichromide	$H_2Cr_2O_7$	starch solution	淀粉溶液
chromide	H_2CrO_4	manganese	锰
potassium ferrocyanide	$K_4\left[Fe(CN)_6\right]$	cobalt	钴

实验六　铜、银、锌、镉、汞

目的要求 •••

1. 了解铜、银、锌、镉和汞的氢氧化物的性质。
2. 了解铜、银、锌、镉和汞的配合物的形成和性质。
3. 了解铜、银、锌、镉和汞的离子的分离和鉴定。

实验原理 •••

在周期系ⅠB与ⅡB族元素中，重要的元素是铜、银、锌、镉、汞。铜的化合物与锌的化合物性质相似，银、镉、汞三元素化合物的性质相近似。

1. 铜、锌

（1）氧化物和氢氧化物 CuO（黑色），Cu_2O（红色），ZnO（白色）

$Cu(OH)_2$、$Zn(OH)_2$ 是两性氢氧化物，尤以 $Zn(OH)_2$ 的两性更突出，溶于强碱生成配合物。

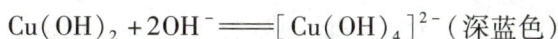

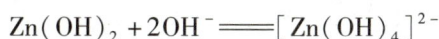

$$Cu(OH)_2 + 2OH^- =\!=\!= [Cu(OH)_4]^{2-}（深蓝色）$$

$$Zn(OH)_2 + 2OH^- =\!=\!= [Zn(OH)_4]^{2-}$$

它们受热脱水，分别生成 CuO 和 ZnO。

（2）Cu^{2+} 和 Cu^+ 的转化 铜的元素电位图是

$$Cu^{2+} \xrightarrow{0.157} Cu^+ \xrightarrow{0.52} Cu$$

因为 $E^{\ominus}_{Cu^+/Cu} > E^{\ominus}_{Cu^{2+}/Cu^+}$，可见 Cu^+ 不稳定，容易按下式发生歧化反应：

$$2Cu^+ \rightleftharpoons Cu^{2+} + Cu \quad K = 10^{6.08}$$

因为 Cu^{2+} 为弱氧化剂，当有 Cu 或其他还原剂存在的条件下，在能生成难溶的亚铜盐时才能被还原。例如将 $CuCl_2$ 溶液与铜屑（和 $NaCl$ 混合后）加热可生成白色的氯化亚铜 $CuCl$ 沉淀。

又如 Cu^{2+} 溶液中加入 KI 时，Cu^{2+} 被 I^- 还原得白色 CuI 沉淀。

$$2Cu^{2+} + 4I^- \rightleftharpoons 2CuI \downarrow + I_2$$

（3）配合物 Cu^{2+} 可与 Cl^-、NH_3 形成稳定程度不同的配离子，Cu^{2+} 大多以 dsp^2 杂化轨道成键，形成平面正方形的内轨型配合物。

Zn^{2+} 的配离子几乎是以 sp^3 杂化轨道成键，形成四面体外轨型配合物。

Cu^+ 能与卤离子（除 F^- 外）、CN^-、SCN^- 等离子形成 $[CuX_2]^-$ 型配离子，但需在过量的配位剂存在时，上述配离子才稳定。这些离子用水稀释时，将形成 CuX 沉淀，例如：

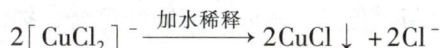

$$2[CuCl_2]^- \xrightarrow{加水稀释} 2CuCl \downarrow + 2Cl^-$$

2. 银、镉、汞 Ag^+、Cd^{2+}、Hg^{2+} 离子都是无色的，由它们组成的化合物一般也是无色的，但 Ag^+、Cd^{2+}、Hg^{2+} 离子都是具有 18 电子外壳，离子半径大，极化力强，变形性较大，它们能与易变形的负离子发生较强的极化作用，以至它们形成的化合物往往有很深的颜色和较低的溶解度，例如：

Ag_2S 黑色（难溶）　　HgS 黑色（难溶）　　CdS 黄色（难溶）

AgI 黄色（难溶）　　HgI_2 红色（极微溶）　　CdI_2 黄绿（可溶）

Ag_2O 棕色（难溶）　　HgO 红色或黄色（极难溶）　　CdO 棕灰（难溶）

（1）氧化物和氢氧化物 Ag_2O、HgO、Hg_2O（不稳定立即分解为 HgO 和 Hg 故为黑色）和 CdO 都难溶于水和碱，而易溶于 HNO_3（HgO，CdO 也可溶于盐酸）。

$AgOH$、$Hg(OH)_2$、$Hg_2(OH)_2$、$Cd(OH)_2$ 都是碱性占优势的，特别是 $AgOH$ 接近于强碱性。$AgOH$、$Hg(OH)_2$、$Hg_2(OH)_2$ 很不稳定，从溶液析出沉淀后，立即分解为它们的氧化物。

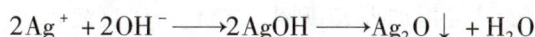

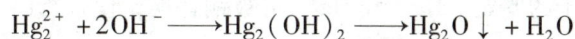

$$2Ag^+ + 2OH^- \longrightarrow 2AgOH \longrightarrow Ag_2O \downarrow + H_2O$$

$$Hg^{2+} + 2OH^- \longrightarrow Hg(OH)_2 \longrightarrow HgO \downarrow + H_2O$$

$$Hg_2^{2+} + 2OH^- \longrightarrow Hg_2(OH)_2 \longrightarrow Hg_2O \downarrow + H_2O$$

（2）Hg^{2+} 与 Hg_2^{2+} 的转化 Hg_2^{2+} 盐在一定条件下也可发生歧化反应。例如 Hg_2Cl_2 与 NH_3 反应，先生成氨基氯化亚汞 Hg_2NH_2Cl 白色沉淀，Hg_2NH_2Cl 进一步歧化为氨基氯化汞 $HgNH_2Cl$ 白色沉淀和黑色 Hg。

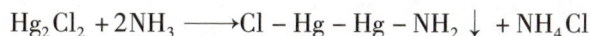

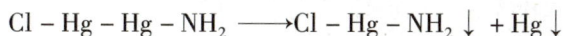

$$Hg_2Cl_2 + 2NH_3 \longrightarrow Cl-Hg-Hg-NH_2 \downarrow + NH_4Cl$$

$$Cl-Hg-Hg-NH_2 \longrightarrow Cl-Hg-NH_2 \downarrow + Hg \downarrow$$

由于有 Hg 析出，故显黑色，这一反应可以用来鉴定 Hg_2^{2+} 离子。

Hg_2I_2（黄绿色）在过量的 KI 溶液中也会发生歧化反应，生成$[HgI_4]^{2-}$和 Hg。

（3）配合物　Ag^+和Hg^{2+}都可形成配位数 2（sp 杂化轨道成键）的直线型和配位数为 4（sp^3 杂化轨道成键）的四面体型配合物。Cd^{2+}主要形成配位数为 4（sp^3 杂化轨道成键）的四面体型配合物。

Ag^+能与Cl^-、Br^-、SCN^-、I^-、CN^-形成配位数为 2 或 4 的配合物，Hg^{2+}也能与它们形成$[HgX_4]^{2-}$型的稳定配合物。

难溶于水的 AgCl，AgBr，AgSCN，AgI 和 AgCN 都能溶于具有相同离子或溶于与Ag^+配位能力更强的盐溶液中，汞盐也有这种相似性质。例如：

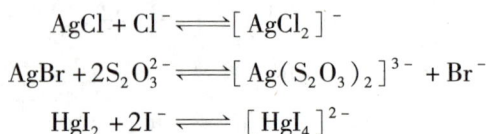

$$AgCl + Cl^- \rightleftharpoons [AgCl_2]^-$$

$$AgBr + 2S_2O_3^{2-} \rightleftharpoons [Ag(S_2O_3)_2]^{3-} + Br^-$$

$$HgI_2 + 2I^- \rightleftharpoons [HgI_4]^{2-}$$

Ag^+和Cd^{2+}可与过量氨水作用，分别生成配合物离子$[Ag(NH_3)_2]^+$和$[Cd(NH_3)_4]^{2+}$，但Hg^{2+}与过量氨水作用时，在无大量NH_4^+存在条件下，并不生成配合物而生成氨基氯化汞$HgNH_2Cl$沉淀。

Ag^+、Cd^{2+}、Hg^{2+}形成配合物时，随配位体浓度不同，可形成一系列中间型配合物，例如Hg^{2+}与Cl^-存在下述平衡：

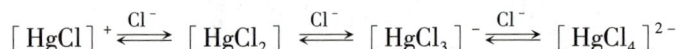

$$[HgCl]^+ \xrightarrow{Cl^-} [HgCl_2] \xrightarrow{Cl^-} [HgCl_3]^- \xrightarrow{Cl^-} [HgCl_4]^{2-}$$

3. 离子的鉴定

（1）Cu^{2+}离子的鉴定　Cu^{2+}离子与黄血盐$K_4[Fe(CN)_6]$生成红褐色$Cu_2[Fe(CN)_6]$沉淀。通常用于鉴定Cu^{2+}的方法是借浓氨水与其作用，得到深蓝色的$[Cu(NH_3)_4]^{2+}$配离子。

（2）Zn^{2+}离子的鉴定　Zn^{2+}可与无色二苯硫腙生成粉红色螯合物：

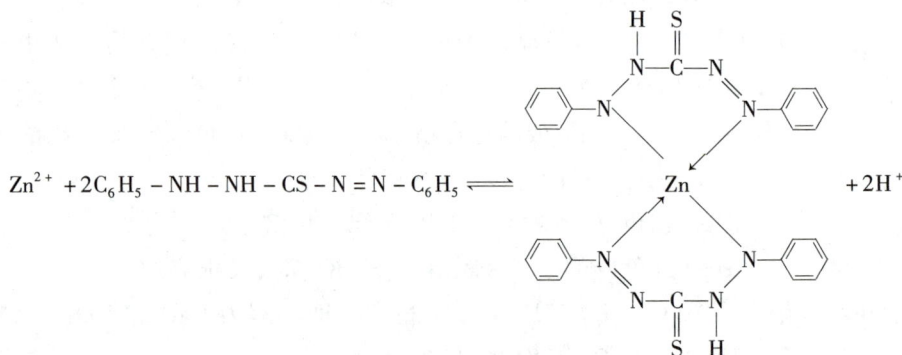

$$Zn^{2+} + 2C_6H_5-NH-NH-CS-N=N-C_6H_5 \rightleftharpoons \qquad\qquad +2H^+$$

（3）Ag^+离子的鉴定　Ag^+可被葡萄糖还原。

Ag^+与Cl^-可生成白色 AgCl 沉淀，此沉淀易溶于氨水生成$[Ag(NH_3)_2]^+$，利用此反应可与其他阳离子氯化物沉淀分离，得到的溶液再加入 KI 则生成黄色 AgI 沉淀。

（4）Cd^{2+}离子的鉴定　Cd^{2+}离子可生成黄色 CdS 沉淀。

（5）Hg^{2+}与Hg_2^{2+}离子的鉴定　Hg^{2+}和Hg_2^{2+}可被铜置换，使汞析出，在铜的表面上呈现白色光亮的斑点。

仪器和药品 •••

1. 仪器　离心机。

2. 药品

（1）固体　Cu 屑。

（2）酸　HCl 2mol/L，浓。

（3）碱 NaOH 2mol/L，6mol/L；$NH_3 \cdot H_2O$ 2mol/L，6mol/L。

（4）盐 $CuSO_4$ 0.1mol/L；$ZnSO_4$ 0.1mol/L；KI 0.1mol/L，饱和；KSCN 饱和；

$K_4[Fe(CN)_6]$ 10%；$CuCl_2$ 1mol/L；$AgNO_3$ 0.1mol/L；$Cd(NO_3)_2$ 0.1mol/L；

$Hg(NO_3)_2$ 0.1mol/L；$Hg_2(NO_3)_2$ 0.1mol/L；$HgCl_2$ 0.1mol/L；

KBr 0.1mol/L；$Na_2S_2O_3$ 0.1mol/L。

3. 其他 淀粉溶液，二苯硫腙（打萨宗），四氯化碳溶液，葡萄糖溶液 10%。

实验内容 •••

1. 铜和锌

（1）铜和锌的氢氧化物

①在 3 支试管中，各取 6 滴 0.1mol/L $CuSO_4$ 溶液，并分别加 2mol/L NaOH 溶液，观察沉淀的生成。其中两支试管分别加入 2mol/L 盐酸和过量的 6mol/L NaOH 溶液，将另 1 支试管加热，观察各试管中产生的现象。

②在两试管中各取 5 滴 0.1mol/L $ZnSO_4$ 溶液，并分别加入 2mol/L NaOH 溶液观察沉淀的生成，然后分别加入 2mol/L 盐酸和过量 6mol/L NaOH 溶液，观察各试管中产生的现象。

（2）铜和锌的配合物 分别于两试管中加入 10 滴 0.1mol/L $CuSO_4$ 和 0.1mol/L $ZnSO_4$，加入 6mol/L $NH_3 \cdot H_2O$ 制取它们的配离子，并分别加入 1mol/L NaOH 溶液，观察有无沉淀重新产生。

（3）+2 价铜的氧化性和 +1 价铜的配合物

①取两支离心管，分别加入 5 滴 0.1mol/L $CuSO_4$，20 滴 0.1mol/L KI 溶液，离心沉降，分离，于清液中检查是否有 I_2 存在。把沉淀用蒸馏水洗涤二次，观察沉淀的颜色。

取一份沉淀，加入饱和 KI 溶液至沉淀刚好溶解。将此溶液加蒸馏水稀释，观察又有沉淀产生。

取另一份沉淀，加入饱和 KSCN 溶液至沉淀刚好溶解，然后再用水稀释，观察沉淀又析出，试解释之。

②取 10 滴 1mol/L $CuCl_2$ 溶液于一小试管中，加入 3～4 滴浓 HCl，再加入少许铜屑，加热至沸，待溶液呈泥黄色，停止加热。取出少量这种溶液并用水稀释，观察是否有白色沉淀产生。解释现象，写出反应方程式。

（4）Cu^{2+} 和 Zn^{2+} 的鉴定

①Cu^{2+} 的鉴定 取 2 滴 0.1mol/L $CuSO_4$ 溶液，于点滴板上，加 2 滴 10% $K_4[Fe(CN)_6]$ 溶液，$Cu_2[Fe(CN)_6]$ 红褐色沉淀生成，表示 Cu^{2+} 存在。

②Zn^{2+} 的鉴定 取 2 滴 0.1mol/L $ZnSO_4$ 溶液，加 6mol/L NaOH 至溶解，再加 10 滴二苯硫腙，水溶液呈粉红色示有 Zn^{2+} 存在。

2. 银、镉、汞

（1）银、镉、汞的氧化物和氢氧化物的生成和性质 在 4 支试管中，分别加入 5 滴 0.1mol/L $AgNO_3$、0.1mol/L $Cd(NO_3)_2$、0.1mol/L $Hg(NO_3)_2$ 和 0.1mol/L $Hg_2(NO_3)_2$，再加数滴 2mol/L NaOH，观察有无沉淀产生和沉淀颜色。说明每支试管中的沉淀是氧化物还是氢氧化物。

（2）和氨水的反应

①取 5 滴 0.1mol/L $AgNO_3$ 逐滴加入 2mol/L 氨水，观察沉淀的生成。继续滴加 2mol/L 氨水，观察沉淀的溶解，然后加入 2 滴 2mol/L NaOH 溶液观察有无沉淀产生。解释原因，并写出反应方程式。

②取 5 滴 0.1mol/L $Cd(NO_3)_2$ 溶液，用 6mol/L 氨水按照（1）的方法操作，解释观察到的现象，写出反应方程式。

③取 5 滴 0.1mol/L HgCl₂ 溶液，滴加 2mol/L 氨水，观察沉淀的生成，再加入过量的 2mol/L 氨水，沉淀是否溶解，写出反应方程式。

④取 5 滴 0.1mol/L Hg₂(NO₃)₂ 溶液，加数滴 6mol/L 氨水，观察沉淀的生成，再加入过量的 6mol/L 氨水，沉淀是否溶解？写出反应方程式。

根据上面实验比较银、镉、汞的盐类与氨水的反应有什么不同？

（3）银和汞的其他配合物

①取 5 滴 0.1mol/L AgNO₃，加 10 滴 0.1mol/L KBr 溶液，观察沉淀的颜色，离心分离，弃去清液，在沉淀中逐滴加 0.1mol/L Na₂S₂O₃ 溶液，搅拌，观察沉淀是否溶解，解释现象，写出反应方程式。

②取 5 滴 0.1mol/L Hg(NO₃)₂，加 10 滴 0.1mol/L KI 溶液，观察沉淀的颜色，再加过量 KI 溶液，观察沉淀是否溶解，解释现象，写出反应方程式。

③取 5 滴 0.1mol/L Hg₂(NO₃)₂ 溶液，加 10 滴 0.1mol/L KI 溶液，观察沉淀的颜色，再加入过量 KI 溶液，观察沉淀的变化，解释现象，写出反应方程式。

（4）银镜的制备　在 1 支洁净的试管中，加 2mL 0.1mol/L AgNO₃ 溶液，滴加 2mol/L 氨水至起初生成的沉淀刚好溶解，然后加 3 滴 10% 葡萄糖溶液，将试管于水浴中加热，试管内壁即有一层光亮的金属银生成（如何洗下管壁上的 Ag）。

实 验 指 导

预习要求

1. $Cu(OH)_2$ 和 $Zn(OH)_2$ 具有哪些性质？

2. 将 KI 加到 $CuSO_4$ 溶液中，是否会得到 CuI_2 沉淀？CuI 沉淀为什么可溶于浓 KI 中，也可溶于浓 KCNS 溶液中？CuI 是否溶于浓盐酸中？

3. 在银盐、镉盐和汞盐溶液中，加入 KOH 溶液，是否均可得到相应的氢氧化物？

4. 在银、镉、汞盐溶液中，加入氨水能否得到相应的配合物？

5. $Hg(NO_3)_2$ 和 $Hg_2(NO_3)_2$ 溶液中，各加少量或过量 KI 溶液，将分别产生什么？

6. $ZnSO_4$ 和 $CdSO_4$ 溶液中，各加少量或过量氨水，将分别产生什么？

7. Zn^{2+} 和 Cd^{2+} 的氢氧化物是否都溶于酸和碱？将 NaOH 加到汞盐溶液中，会产生什么？

8. 完成下列反应

（1）$Cu^{2+} + I^- \longrightarrow$

（2）$CuI + I^-$（过量）$\longrightarrow$ (　　) $\xrightarrow{H_2O}$

（3）$CuI + SCN^-$（过量）$\longrightarrow$ (　　) $\xrightarrow{H_2O}$

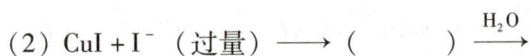

（4）$AgNO_3 + NH_3 \cdot H_2O \longrightarrow$ (　　) $\xrightarrow{NH_3 \cdot H_2O（过量）}$

$Cd(NO_3)_2$

$HgCl_2$

$Hg_2(NO_3)_2$

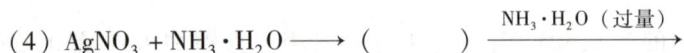

（5）$Ag^+ + OH^- \longrightarrow$

Cd^{2+}

Hg^{2+}

Hg_2^{2+}

(6) $AgBr + S_2O_3^{2-} \longrightarrow$

(7) $Hg^{2+} + I^- \longrightarrow ($ 　　 $) \xrightarrow{I^-（过量）}$

(8) $Hg_2^{2+} + I^- \longrightarrow ($ 　　 $) \xrightarrow{I^-（过量）}$

注意事项

1. $HgCl_2$ 为毒品，使用时要注意安全。

2. $CuCl_2$ 溶液与铜屑在浓盐酸存在下加热，时间须稍长些，否则有时现象不甚明显。

报告格式

1. 目的

2. 实验内容

试　样	加入试剂	现　象	反应方程式
结论			

实验后思考

1. 怎样分离 Fe^{3+}、Cu^{2+}、Zn^{2+} 混合离子？

2. 怎样分离 Ag^+、Cd^{2+}、Hg^{2+} 混合离子？

3. 试设计一分析流程图，分离和检出 Ag^+、Zn^{2+}、Cu^{2+} 混合溶液。

4. 汞盐和亚汞盐的性质有何不同？通过实验你可得到几种区别它们的方法？

5. 根据实验结果填写下表。

氢氧化物 / 性质	$Cu(OH)_2$	$AgOH$	$Zn(OH)_2$	$Cd(OH)_2$	$Hg(OH)_2$
酸碱性					
稳定性					
颜色					
形成配合物能力					

Experiment 6　Copper, Silver, Zinc Cadmium and Mercury

◆ Objectives

1. Understand the properties of the hydroxides copper, silver, zinc, cadmium and mercury.

2. Understand the preparation and properties of the complexes of copper, silver, zinc, cadmium and mercury.

3. Understand the isolation and identifying of the ions copper, silver, zinc, cadmium and mercury.

◆ Principles

Copper, silver, zinc, cadmium and mercury are important elements of group ⅠB and group ⅡB in the period table. We will discuss their properties in two groups with respect to the properties of copper compounds familiar to of zinc compounds and of silver compounds familiar to of cadmium and mercury compounds.

1. Copper and zinc

（1）Oxides and hydroxides

CuO（black）,Cu_2O（red）,ZnO（white）.

$Cu(OH)_2$ and $Zn(OH)_2$ are both amphoteric,particularly $Zn(OH)_2$. They are not only soluble in acids, but also in concentrated aqueous alkalis,in which hydroxo complexes are formed.

$$Cu(OH)_2 + 2OH^- \!=\!\!=\!\! [Cu(OH)_4]^{2-}$$
$$Zn(OH)_2 + 2OH^- \!=\!\!=\!\! [Zn(OH)_4]^{2-}$$

The hydroxides are readily to dehydrate to the oxides CuO and ZnO when heated.

（2）Inversion of Cu^{2+} and Cu^+.

In aqueous media,as may be seen from the standard potentials,copper（Ⅰ）is unstable by a small margin with respect to copper（Ⅱ）and metal：

$$Cu^{2+} \xrightarrow{+0.157} Cu^+ \xrightarrow{+0.52} Cu$$

The disproportionate is formulated as：

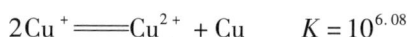

$$2Cu^+ \!=\!\!=\!\! Cu^{2+} + Cu \qquad K = 10^{6.08}$$

A copper（Ⅰ）salt protected by its insolubility can be formed when Cu^{2+},a weak oxidizing reagent,is reduced by copper or other reducing reagent. For example,cuprous chloride,a white precipitate,is obtained when cupric chloride is boiled with copper powder（mixed with sodium chloride）; cuprous iodide is also obtained when a Cu^{2+} ions solution is added to an iodide ions solution.

$$2Cu^{2+} + 4I^- \!=\!\!=\!\! 2CuI(s) + I_2$$

（3）Complexes The complex ions with various stabilities can be formed with copper（Ⅱ）ions reacting with chloride ions or ammonia. Most of copper（Ⅱ）complexes are inner-orbital complexes with square dsp^2 hybrid orbital.

Whilst,zinc complexes almost are out-orbital complexes with tetrahedral sp^3 hybrid orbital. With halide ions or cyanide ions CN^- or thiocyanide ions SCN^-,copper（Ⅱ）ions form the copper（Ⅰ）complexes formula as $[CuX_2]^-$. The complexes are stable in their corresponding concentrated ligand solutions; if dilution the complexes decompose and CuX is precipitated.

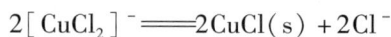

$$2[CuCl_2]^- \!=\!\!=\!\! 2CuCl(s) + 2Cl^-$$

2. Silver,cadmium and mercury Ag^+ ion,Cd^{2+} ion and Hg^{2+} ion are all colorless. Usually,the compounds of them are also colorless. They have long ionic radii and strong polarizing power and are readily deformed as result of their 18 electronic shields. Therefore,many of compounds they form are found in dark colors and sparingly soluble in water.

（1）Oxides and hydroxides Ag_2O,HgO,Hg_2O（which is more unstable with respect to the rapid disproportionation to the black products HgO and Hg. ）and CdO are all insoluble in water and alkalis,but soluble in nitric acid. HgO and CdO can dissolve in chloric acid.

$AgOH$,$Hg(OH)_2$,$Hg_2(OH)_2$ and $Cd(OH)_2$ are all amphoteric aptly to basic. And $AgOH$ is almost a strong base. $AgOH$,$Hg(OH)_2$,$Hg_2(OH)_2$ are so unstable that they decompose to the oxides as soon as they precipitate from the solutions.

$$2Ag^+ + 2OH^- \longrightarrow 2AgOH \longrightarrow Ag_2O \downarrow + H_2O$$
$$Hg^{2+} + 2OH^- \longrightarrow Hg(OH)_2 \longrightarrow HgO \downarrow + H_2O$$
$$Hg_2^{2+} + 2OH^- \longrightarrow Hg_2(OH)_2 \longrightarrow Hg_2O \downarrow + H_2O$$

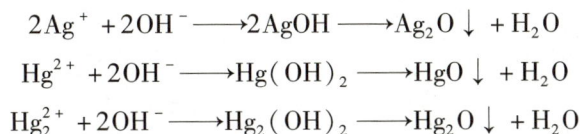

（2）Inversion of Hg^{2+} and Hg_2^{2+}　The general method for preparation of Hg（Ⅰ）compounds is by the action of metallic mercury on Hg（Ⅱ）compounds. Form the measured values of E^{θ} for the half – reactions.

$$Hg^{2+} \underset{}{\overset{+0.92}{\rule{1.5cm}{0.4pt}}} Hg_2^{2+} \overset{0.79}{\rule{1.5cm}{0.4pt}} Hg$$

It can be seen that Hg_2^{2+} in aqueous solution is only just stable respect to the disproportionate.

$$Hg_2^{2+} \Longleftrightarrow Hg^{2+} + Hg(s)$$

Reagents, which form insoluble Hg^{2+} salts or stable Hg^{2+} complexes upset the equilibrium and decompose Hg_2^{2+} salts. Ammonia and iodide ions, for example, do so with formation of Hg and $HgNH_2Cl$ and $[HgI_4]^{2-}$ respectively. The reactions can be formulated as follows:

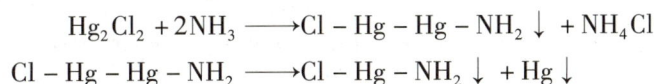

$$Hg_2Cl_2 + 2NH_3 \longrightarrow Cl-Hg-Hg-NH_2 \downarrow + NH_4Cl$$
$$Cl-Hg-Hg-NH_2 \longrightarrow Cl-Hg-NH_2 \downarrow + Hg \downarrow$$

（3）Complexes　Ag^+ and Hg^{2+} complexes are formed with chloride, bromine, iodide, cyanide and thiocyanide ions in aqueous solution. Ag^+ complexes are liner and *sp* orbital hybridized with coordination number two; Hg^{2+} complexes are tetrahedral with coordination number four and sp^3 hybrid orbital. Cd^{2+} forms tetrahedral sp^3 hybrid orbital complexes.

With respect to the formation of the complexes, the insoluble salts, such as AgCl, AgBr, AgSCN, AgI and AgCN, are soluble in the solutions of certain ligands. And Hg（Ⅱ）salts do so.

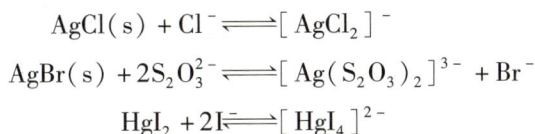

$$AgCl(s) + Cl^- \Longleftrightarrow [AgCl_2]^-$$
$$AgBr(s) + 2S_2O_3^{2-} \Longleftrightarrow [Ag(S_2O_3)_2]^{3-} + Br^-$$
$$HgI_2 + 2I^- \Longleftrightarrow [HgI_4]^{2-}$$

Ag^+ or Cd^{2+} solution reacts with an excess ammonia solution to form ammonia complexes $[Ag(NH_3)_2]^+$ or $[Cd(NH_3)_4]^{2+}$. Hg^{2+} ions do not so, but convert to a precipitate $HgNH_2Cl$.

Many of the complexes of silver（Ⅰ）, cadmium（Ⅱ）and mercury（Ⅱ）with a certain ligand do not contain a unique form but a wide range of forms. The chloride complexes of Hg（Ⅱ）, for example, exist in aqueous solution formulated by the equilibriums:

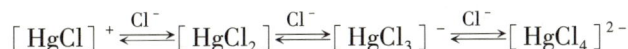

$$[HgCl]^+ \underset{}{\overset{Cl^-}{\Longleftrightarrow}} [HgCl_2] \overset{Cl^-}{\Longleftrightarrow} [HgCl_3]^- \overset{Cl^-}{\Longleftrightarrow} [HgCl_4]^{2-}$$

3. Identifying copper（Ⅱ）, silver（Ⅰ）, zinc（Ⅱ）, cadmium（Ⅱ）, mercury（Ⅰ）and mercury（Ⅱ）ions

（1）Identifying copper（Ⅱ）ions　The reaction of copper（Ⅱ）ions with yellow prussiate of potash $K_4[Fe(CN)_6]$ forming a reddish brown precipitate is sensitive, but usually we identify copper（Ⅱ）ions applying the reaction of copper（Ⅱ）with aqueous ammonia forming a dark blue solution.

（2）Identifying zinc（Ⅱ）ions　Zn^{2+} ions can react with colorless solution of dithizone to give a pink chelate.

$$Zn^{2+} + 2C_6H_5-NH-NH-CS-N=N-C_6H_5 \rightleftharpoons$$

(structure of zinc dithizone complex) $+ 2H^+$

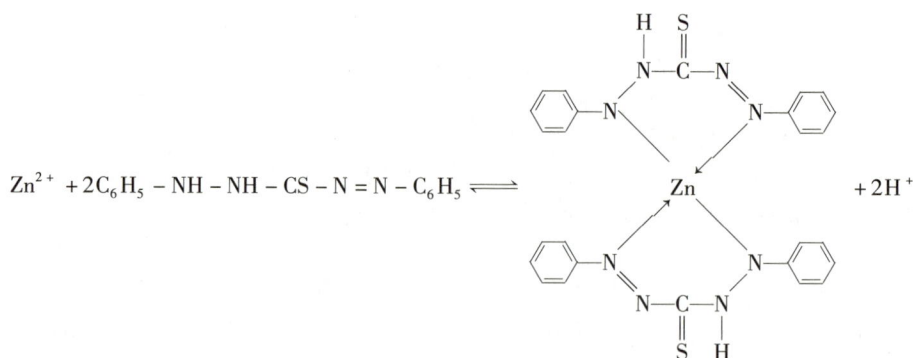

（3）Identifying silver(Ⅰ) ions　Adding Ag^+ ions to chloride ions gives AgCl, a white precipitate, which dissolves readily in an aqueous ammonia solution forming ammonia complexes. Then isolate the liquid from the mixture. When KI is added to the complexes solution, silver(Ⅰ) is precipitated again in the form of AgI, which is yellow.

（4）Identifying cadmium(Ⅱ) ions　A solution of Cd^{2+} ions gives a characteristic yellow precipitate of formula CdS on treatment with sulfide ions.

（5）Identifying mercury(Ⅰ) and mercury(Ⅱ) ions　On treatment with metal copper, Hg^{2+} and Hg_2^{2+} ions are reduced to metallic mercury, which form an alloy with copper known as amalgam. Hence, a brightly white spot comes out on the surface of copper.

◆ Equipment and Chemicals

1. Equipment　Centrifugal machine

2. Chemicals

（1）Solid　Copper powder

（2）Acids　HCl 2mol/L, concentrated

（3）Bases　NaOH 2mol/L, 6mol/L;　$NH_3 \cdot H_2O$ 2mol/L, 6mol/L

（4）Salts　$CuSO_4$, 0.1mol/L; $ZnSO_4$, 0.1mol/L; KI 0.1mol/L, saturated; KSCN, saturated; $K_4[Fe(CN)_6]$, 10%; $CuCl_2$, 1mol/L; $AgNO_3$, 0.1mol/L; $Cd(NO_3)_2$, 0.1mol/L; $Hg(NO_3)_2$, 0.1mol/L; $Hg_2(NO_3)_2$, 0.1mol/L; KBr, 0.1mol/L; $Na_2S_2O_3$, 0.1mol/L

3. Others　Starch solution; dithizone solution of CCl_4; glucose solution, 10%

◆ Procedures

Copper and zinc

1. Hydroxides

（1）Take three tubes and add six drops of $CuSO_4$ solution and a little of NaOH solution (2mol/L) into each one. The blue solid is precipitated. Treat the precipitate with HCl solution (2mol/L) in the first test tube and with of NaOH solution (6mol/L) in the second tube, and heat the precipitate in the third tube. Compare the three and find out what is different.

（2）Take two tubes and add five drops of $ZnSO_4$ solution and a little of NaOH solution (2mol/L) into each one. The white solid is precipitated. Treat the precipitate with HCl solution (2mol/L) in the first test tube and with excess NaOH solution (6mol/L) in the second tube. Explain the phenomenon.

2. Complexes　Add 10 drops of $CuSO_4$ solution and $ZnSO_4$ solution into two tube respectively, and then add $NH_3 \cdot H_2O$ (6mol/L) in excess. The ammonia complexes are obtained. Add 1mol/L NaOH solution into those complexes. Observe if the complexes are decomposed to give precipitates.

3. The oxidizing copper(II) and the copper(I) complexes

(1) Take two centrifugal tubes and add five drops of $CuSO_4$ solution (0.1mol/L) and 20 drops of KI solution (0.1mol/L) into each one. The white solid with formula CuI is precipitated. Isolate the precipitate from the mixture with a centrifugal machine. Test if there is I_2 in the clear upper solution. And observe the color of the precipitates washed with distilled water twice.

Into one precipitate, add saturated KI solution until the solid dissolves. It forms iodide complexes of copper (I), which are diluted with distilled water to generate the precipitate CuI again.

Into the other precipitate, add saturated KSCN solution until the solid dissolves. It forms the thiocyanide complexes of copper(I); on dilution decompose to the precipitate CuI and CuSCN again.

(2) Add 10 drops of $CuCl_2$ solution and 3~4 drops of concentrated HCl and a little of copper powder into a test tube, which is boiled until the solution becomes dirt yellow. Take a little of the solution and dilute it with distilled water. Notice if some of white solid shows.

4. Identifying Cu^{2+} and Zn^{2+} ions

(1) Identifying Cu^{2+} ions   Take 2 drops of 0.1mol/L $CuSO_4$ solution to a spot-plate, and add 2 drops of 10% $K_4[Fe(CN)_6]$ solution to it. You can find a reddish brown precipitate showing.

(2) Identifying Zn^{2+} ions   Take 2 drops of 0.1mol/L $ZnSO_4$ solution. Add 6mol/L NaOH solution to it until the solid dissolves. Sequentially, add 10 drops of dithizone solution. Shock the mixture and let it rest for while. A pink solution is formed in the aqueous phase.

Silver, cadmium, mercury

1. Oxides and hydroxides   Add five drops of 0.1mol/L $AgNO_3$ solution, 0.1mol/L $Cd(NO_3)_2$ solution, 0.1mol/L $Hg(NO_3)_2$ solution and 0.1mol/L $Hg_2(NO_3)_2$ solution into each of four tubes respectively. Sequentially, add some drops of NaOH solution (2mol/L). Tell what is precipitated in each tube.

2. Reactions with aqueous ammonia.

(1) Take 5 drops of 0.1mol/L $AgNO_3$ solution. Drop 2mol/L ammonia into it and observe the precipitate forms, then dissolve it in excess ammonia. Treat the solution obtained with 2 drops of 2mol/L NaOH solution and observe if any of precipitate is given again.

(2) Take 5 drops of 0.1mol/L $Cd(NO_3)_2$ solution. Treat it with 6mol/L ammonia according to the method as above.

(3) Take 5 drops of 0.1mol/L $HgCl_2$ solution. Add 2mol/L ammonia to get a precipitate at first. Then, to which, add extra ammonia solution of 6mol/L.

(4) Take 5 drops of 0.1mol/L $Hg_2(NO_3)_2$ solution. Add 6mol/L ammonia and observe the precipitate forms. Sequentially, add extra ammonia solution of 6mol/L to find out if the precipitate dissolves.

What can be concluded from the experiment? What is different between the reactions of silver, cadmium and mercury salts with aqueous ammonia?

3. Complexes

(1) Combining 5 drops of 0.1mol/L $AgNO_3$ solution and 10 drops of 0.1mol/L KBr solution gives a pale yellow precipitate, then which is isolated with a centrifugal machine. Treat the solid with adding 0.1mol/L $Na_2S_2O_3$ solution drop by drop. Stir the mixture until the solid is dissolved.

（2）Combine 5 drops of 0. 1mol/L $Hg(NO_3)_2$ solution and 10 drops of 0. 1mol/L KI solution. Observe the color of the precipitate. Then add extra 0. 1mol/L KI solution continually. Observe if the solid is dissolved.

（3）Take 5 drops of 0. 1mol/L $Hg_2(NO_3)_3$ solution and treat it with the method as above. Observe the reaction of Hg_2^{2+} ions with I^-.

4. Reaction of silver mirror   Add 0. 1mol/L $AgNO_3$ solution 2mL into a clean test tube. Drop a 2mol/L aqueous ammonia solution into it until the precipitate just dissolves. Next, add three drops of 10% glucose solution into the mixture. Heat the tube with water bath. In a while, a shining mirror of silver is formed on the inner surface of the tube. How to clean the tube with a silver mirror?

◆ Instructions

1. Requirements

（1）Which properties do hydroxides of copper(Ⅱ) and zinc(Ⅱ) have?

（2）Is CuI_2 formed when KI is added into $CuSO_4$ solution? Why is CuI soluble not only in concentrated KI solution but also in concentrated KSCN solution? Is CuI stable in concentrated HCl?

（3）Can you obtain the corresponding hydroxides on reactions of silver, cadmium and mercury salts solutions with KOH solution?

（4）Can you obtain the corresponding ammonia complexes on reactions of silver, cadmium and mercury salts solutions with aqueous ammonia?

（5）What different compounds are generated when a little of or extra KI solution is added into $Hg_2(NO_3)_2$ and $Hg(NO_3)_2$ solution respectively?

（6）What different compounds are generated when a little of or extra aqueous ammonia solution is added into $ZnSO_4$ and $CdSO_4$ solution respectively?

（7）Are both the hydroxides of Zn(Ⅱ) and Cd(Ⅱ) soluble either in an acid or in an alkaline? What is formed when mercury salts react with NaOH?

2. Notes

（1）Be careful when using $HgCl_2$, which is toxic.

（2）The period of boiling the mixture of $CuCl_2$ solution with copper powder should be a little longer; otherwise, the phenomenon is not obvious sometimes.

3. Report format

（1）Objectives

（2）Procedure

Chemicals	Chemicals added	Phenomenon	Equations of reactions
Conclusion			

4. Questions

（1）How do you separate Fe^{2+}, Cu^{2+} and Zn^{2+} ions from their mixture?

（2）How do you separate Ag^+, Cd^{2+} and Hg^{2+} ions from their mixture?

（3）Design a flow chart for separating and detecting Ag^+, Zn^{2+} and Cu^{2+} ions from their mixture.

（4）What is the difference of the properties of mercury（Ⅰ）and mercury（Ⅱ）salts? How many methods do you know to identify them?

（5）Fill the following table according to the result of experiment：

Hydroxide / Property	Cu(OH)$_2$	AgOH	Zn(OH)$_2$	Cd(OH)$_2$	Hg(OH)$_2$
Acidity and basicity					
Stability					
Color					
Capability of forming complexes					

书网融合……

习题1   习题2

第五章 无机化合物综合性制备实验

实验一 五水合硫酸铜的制备

实验目的

1. 学习以废铜和工业硫酸为主要原料制备 $CuSO_4 \cdot 5H_2O$ 的原理和方法。
2. 掌握并巩固无机化合物制备过程中灼烧、水浴加热、减压过滤和结晶等基本操作。

实验原理

$CuSO_4 \cdot 5H_2O$ 易溶于水，难溶于无水乙醇。加热时失水。

$CuSO_4 \cdot 5H_2O$ 的生产方法有多种，如电解液法、氧化铜法。本实验选择以废铜和工业硫酸为主要原料制备 $CuSO_4 \cdot 5H_2O$ 的方法，先将铜粉灼烧成氧化铜，然后再将氧化铜溶于适当浓度的硫酸中。反应如下：

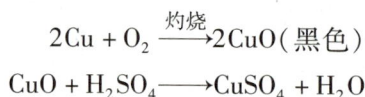

$$2Cu + O_2 \xrightarrow{\text{灼烧}} 2CuO(\text{黑色})$$
$$CuO + H_2SO_4 \longrightarrow CuSO_4 + H_2O$$

由于废铜及工业硫酸不纯，制得的溶液中除生成硫酸铜外，还含有其他一些可溶性或不溶性的杂质。不溶性杂质在过滤时可除去，可溶性杂质 Fe^{2+} 和 Fe^{3+}，一般需用氧化剂（如 H_2O_2）将 Fe^{2+} 氧化为 Fe^{3+}，然后调节 pH，并控制至 3 左右（注意不要使溶液的 pH ≥ 4，若 pH 过大，会析出碱式硫酸铜的沉淀），再加热煮沸，使 Fe^{3+} 水解成为 $Fe(OH)_3$ 沉淀而除去。反应如下：

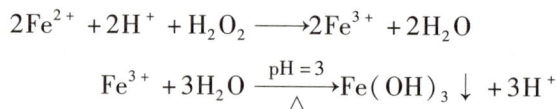

$$2Fe^{2+} + 2H^+ + H_2O_2 \longrightarrow 2Fe^{3+} + 2H_2O$$
$$Fe^{3+} + 3H_2O \xrightarrow[\triangle]{pH=3} Fe(OH)_3 \downarrow + 3H^+$$

将除去杂质的 $CuSO_4$ 溶液进行蒸发，冷却结晶，减压过滤后得到蓝色 $CuSO_4 \cdot 5H_2O$。

仪器和药品

1. 仪器 托盘天平、煤气灯、瓷坩埚、坩埚钳、泥三角、铁架台、布氏漏斗、吸滤瓶、烧杯、点滴板、玻璃棒、量筒、蒸发皿、滤纸、剪刀。

2. 药品 Cu 粉、H_2SO_4（3mol/L）、H_2O_2（3%）、$K_3[Fe(CN)_6]$（0.1mol/L）、$CuCO_3$（C.P.）、pH 试纸。

实验步骤

1. 氧化铜的制备 把洗净的瓷坩埚经充分灼烧干燥并冷却后，在托盘天平上称取 3.0g 废 Cu 粉放入其内。将坩埚置于泥三角上，用煤气灯氧化焰小火微热，使坩埚均匀受热，待 Cu 粉干燥后，加大火焰用高温灼烧，并不断搅拌，搅拌时必须用坩埚钳夹住坩埚，以免打翻坩埚或使坩埚从泥三角上掉落。灼烧至 Cu 粉完全转化为黑色 CuO（约 20min），停止加热并冷却至室温。

2. 粗 $CuSO_4$ 溶液的制备 将冷却后的 CuO 倒入 100mL 小烧杯中，加入 18mL 3mol/L H_2SO_4（工业纯），微热使之溶解。

3. CuSO₄溶液的精制　在粗 $CuSO_4$ 溶液中，滴加 2mL 3% H_2O_2，将溶液加热，检验溶液中是否还存在 Fe^{2+}（如何检验）。当 Fe^{2+} 完全氧化后，慢慢加入 $CuCO_3$ 粉末，同时不断搅拌直到溶液 pH = 3，在此过程中，要不断地用 pH 试纸测试溶液的 pH，控制溶液 pH = 3，再加热至沸（为什么？）趁热减压过滤，将滤液转移至洁净的烧杯中。

4. CuSO₄·5H₂O 晶体的制备　在精制后的 $CuSO_4$ 溶液中，滴加 3mol/L H_2SO_4 酸化，调节溶液至 pH = 1 后，转移至洁净的蒸发皿中，水浴加热蒸发至液面出现晶膜时停止。在室温下冷却至晶体析出。然后减压过滤，晶体用滤纸吸干后，称重。计算产率。

实　验　指　导

预习要求 ●●●

1. 设计硫酸铜制备的流程图。
2. 在制备硫酸铜的过程中，如何除可溶性杂性 Fe^{2+} 和 Fe^{3+}？

基本操作 ●●●

掌握灼烧、水浴加热、减压过滤和结晶等基本操作。

注意事项 ●●●

1. 难溶性杂质可用过滤的方法除去。
2. 可溶性杂质 Fe^{2+} 和 Fe^{3+}，需用氧化剂将 Fe^{2+} 氧化为 Fe^{3+}，然后调节 pH，再加热煮沸，使 Fe^{3+} 变成 $Fe(OH)_3$ 沉淀除去。

报告格式 ●●●

1. 目的
2. 原理（反应式表示）
3. 制备流程图
4. 产量及产率的计算

思考题 ●●●

（1）在粗 $CuSO_4$ 溶液中 Fe^{2+} 杂质为什么要氧化为 Fe^{3+} 后再除去？为什么要调节溶液的 pH = 3？pH 太大或太小有何影响？

（2）为什么要在精制后的 $CuSO_4$ 溶液中调节 pH = 1 使溶液呈强酸性？

（3）蒸发、结晶制备 $CuSO_4·5H_2O$ 时，为什么刚出现晶膜即停止加热而不能将溶液蒸干？

（4）如何清洗坩埚中的残余物 Cu 和 CuO 等？

Experiment 1　Synthesis of Copper Sulfate Pentahydrate

◆ Objectives

1. To learn the principles and procedures for the synthesis of copper sulfate pentahydrate using copper waste and industrial sulfuric acid.

2. To further enhance the basic inorganic synthesis skills such as combustion, heating with water bath, vacuum filtration and crystallization.

◆ Principles

$CuSO_4 \cdot 5H_2O$ is readily soluble in water and insoluble in absolute alcohol. It dehydrates when heated.

There are many methods to synthesize $CuSO_4 \cdot 5H_2O$, including electrolysis and oxidation. In this experiment, copper waste and industrial sulfuric acid are used as starting materials to synthesize $CuSO_4 \cdot 5H_2O$. Copper powder is first combusted to form copper oxide, and then copper oxide is dissolved in sulfuric acid of proper concentration. The chemical equations are as follows:

$$2Cu + O_2 \xrightarrow{\text{Combustion}} 2CuO(\text{black})$$
$$CuO + H_2SO_4 \longrightarrow CuSO_4 + H_2O$$

Because of the impurities in copper waste and industrial sulfuric acid, there are a number of other soluble and insoluble impurities formed in the reaction solution, in addition to the formation of copper sulfate. The insoluble impurities can be removed by filtration. To remove soluble impurities such as Fe^{2+} and Fe^{3+}, typically oxidizing agents (such as H_2O_2) are first used to oxidize Fe^{2+} to Fe^{3+}. Then the pH value is adjusted to 3 (the pH of solution should not be more than 4, otherwise, alkaline copper sulfate may precipitate out). After the solution is heated to boil, Fe^{3+} hydrolyzes to form $Fe(OH)_3$ precipitate and then it is removed. The chemical equations are as follows:

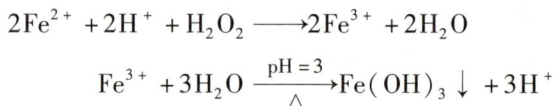

$$2Fe^{2+} + 2H^+ + H_2O_2 \longrightarrow 2Fe^{3+} + 2H_2O$$
$$Fe^{3+} + 3H_2O \xrightarrow[\triangle]{pH=3} Fe(OH)_3 \downarrow + 3H^+$$

The purified $CuSO_4$ solution is evaporated, cooled to crystallize and vacuum filtered to obtain blue $CuSO_4 \cdot 5H_2O$.

◆ Equipment and Chemicals

1. Equipment　Weighing balance, Bunsen burner, porcelain crucible, crucible tongs, triangle, Buchner funnel, filter flask, beaker, TLC plates, glass rod, graduated cylinder, evaporation dish, filter paper and scissors.

2. Chemicals　Copper powder, H_2SO_4(3mol/L), H_2O_2(3%), $K_3[Fe(CN)_6]$(0.1mol/L), $CuCO_3$(C. P.), pH litmus paper.

◆ Procedures

1. Synthesis of copper oxide　The porcelain crucible is cleaned, thoroughly dried by combustion and then cooled to room temperature. 3.0g of copper powder is weighed and placed in the crucible. The crucible is placed on the triangle, mildly and evenly heated with the Bunsen burner. When the Cu powder is dry, the flame is raised to combustion and the crucible is continually stirred. During stirring, the crucible must be held with the tongs to prevent it from falling off the triangle. After the Cu powder is completely converted to black CuO by combustion, stop heating and cool to room temperature.

2. Preparation of crude $CuSO_4$ solution　The cooled CuO is transferred into a 100mL beaker and mixed with 18mL 3mol/L H_2SO_4(industrial purity). The mixture is mildly heated to dissolve the powder.

3. Purification of $CuSO_4$ solution　2mL 3% H_2O_2 is added drop wise into the crude $CuSO_4$ solution. The mixture is heated and detected for any Fe^{2+} present (how to detect?). After Fe^{2+} is completely oxidized, $CuCO_3$ powder is added slowly and the solution is continually stirred until the pH reaches 3. During this process, the pH of the solution needs to be frequently tested and controlled at pH = 3, and then the mixture is heated to boil （why?) and hot filtered under vacuum. The filtrate is transferred into a clean beaker.

4. Preparation of CuSO$_4$·5H$_2$O crystals　3mol/L H$_2$SO$_4$ is added drop wise to the purified CuSO$_4$ solution to adjust the pH of the solution to 1. The mixture is transferred to the clean evaporation dish and heated with the water bath until crystal membrane appears on the liquid surface. The mixture is cooled to room temperature and crystallization occurs. The crystals are collected by vacuum filtration and dried with filter paper. Then they are weighed and the yield is calculated.

◆ Questions

1. Why is it necessary to oxidize Fe^{2+} impurities to Fe^{3+} before it is removed from the crude CuSO$_4$ solution? Why does the pH of the solution need to be adjusted to 3? What kind of effect will it have if the pH is too high or too low?

2. Why is it necessary to adjust the pH of purified CuSO$_4$ solution to 1, which is highly acidic?

3. During the evaporation and crystallization of CuSO$_4$·5H$_2$O, why is it necessary to stop heating as soon as the crystal membrane starts to appear instead of heating the solution until dryness?

4. How to remove the residual Cu and CuO from the crucible?

实验二　四碘化锡的制备

PPT

实验目的 ●●●

1. 学习利用非水溶剂（石油醚）制备四碘化锡。
2. 掌握回流、水浴加热等基本操作。

实验原理 ●●●

四碘化锡是橙红色的针状晶体，密度 4.48g/cm^3，熔点 144.5℃，沸点 364.5℃，180℃时就有较高的蒸气压。溶于水、较易水解，在热水中可被分解。易溶于二硫化碳、四氯化碳、苯和热的石油醚等有机溶剂中，在冷的石油醚中溶解度较小。在乙醇等溶液中，四碘化锡与碱金属的碘化物作用生成 M$_2$(SnI$_6$)黑色晶状化合物。

因四碘化锡容易水解，不宜在水溶液中制备，一般采用干法合成，即在加热条件下锡与碘蒸气进行反应或利用非水溶剂的合成方法进行制备。可被选用的非水溶剂有四氯化碳、冰醋酸等，本实验采用低沸点的石油醚为溶剂（沸点 60~90℃），它是含碳较少的烷烃（主要为戊烷、己烷）混合物，属于惰性溶剂。制备的简要过程是将碘溶解在石油醚中，在加热的条件下与锡片反应而制得：

$$Sn + 2I_2 \longrightarrow SnI_4$$

仪器和药品 ●●●

1. 仪器　台秤、圆底烧瓶（30mL）、回流冷凝器、烧杯（250mL、30mL）、温度计、恒温水浴及铁架台等加热器具。

2. 药品　石油醚（60~90℃）、锡箔、碘、0.1mol/L AgNO$_3$、0.1mol/L Pb(NO$_3$)、饱和 KI、丙酮等。

实验步骤

1. 四碘化锡的制备（图 5 - 1）

用托盘天平，在已洗净、烘干的 30mL 圆底烧瓶中，称取约 0.5g 晶体碘，再称取约 0.2g 锡箔，并将其剪成碎片后加入烧瓶中，再加入 10mL 石油醚。装配好回流冷凝器等反应装置，在水浴上加热使反应混合物沸腾。控制水浴温度在 85 ~ 95℃之间，调节冷凝水的流量，使含碘石油醚的冷凝液不高于冷凝器的中间部位，保持回流状态直到反应完全为止（至冷凝下来的石油醚液滴由紫色变为无色）。停止加热撤掉水浴，待不沸腾后取下冷凝管，趁热用倾泻法将溶液倒入 50mL 洁净、干燥的小烧杯内，使未反应的锡片留在烧瓶内。烧瓶内壁及剩余锡箔上沾有的 SnI_4 晶体，可用 1 ~ 2mL 热石油醚洗涤，洗涤液合并到上述小烧杯内，将其放到冰浴中冷却、结晶。结晶完成后用倾泻法把上层清液（母液石油醚）沿玻璃棒小心倒入回收瓶，最后将盛有结晶的小烧杯置于水浴上干燥，称量产品。计算产率。

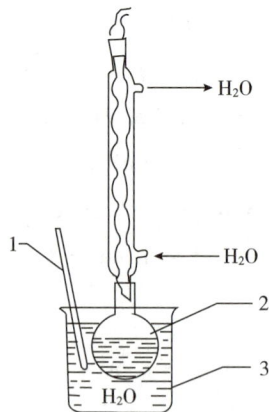

图 5 - 1　SnI_4 制备装置图

1. 温度计；2. 圆底烧瓶；
3. 烧杯

2. 性质实验

（1）取少量 SnI_4 加入 2mL 蒸馏水，用 pH 试纸检验溶液的酸碱性。

（2）将上述溶液分装于两支试管中，其中一支加入几滴 0.1mol/L $AgNO_3$，另一支加入几滴 0.1mol/L $Pb(NO_3)_2$，观察现象并写出离子反应方程式。

（3）取少量 SnI_4 加入丙酮溶解，分装于两支试管中。其中一支加入几滴水，另一支加入等量的饱和 KI 溶液，观察现象并进行解释。

实　验　指　导

预习要求

1. 设计四碘化锡制备的流程图。
2. 为什么只能在非水溶剂中才能制备四碘化锡。

基本操作

掌握回流、水浴加热等基本操作。

注意事项

1. 反应过程中应控制水浴温度在 85 ~ 95℃之间。
2. 反应完成后，停止加热，撤掉水浴，待不沸腾后取下冷凝管。
3. 母液石油醚倒入回收瓶。

报告格式

1. 目的
2. 原理（反应式表示）
3. 制备流程图
4. 产量、产率计量
5. 性质实验

1. 如沸腾过于剧烈、冷凝效果又差，碘与石油醚蒸气从冷凝器逸出或回流液滴仍有颜色就停止反应，会对实验结果有何影响？

2. 本实验以哪一种原料过量为好？为什么？

3. 实验中不用水浴加热而直接用小火加热可否？为什么？

Experiment 2　Synthesis of Tin Tetraiodide

◆ Objectives

1. To learn the inorganic synthesis of tin tetraiodide using a non-aqueous solvent (petroleum ethers).

2. To practice basic operations such as refluxing, heating with water bath, etc.

◆ Principles of Experiment

Tin tetraiodide is a red-orange color, needle-shaped crystal with a density of 4. 48g/cm^3, m. p. of 144. 5℃, b. p. of 364. 5℃ and substantial vapor pressure at 180℃. It is soluble in water and hydrolyzes relatively easily. It decomposes in hot water. It is readily soluble in organic solvents such as carbon disulfide, carbon tetrachloride, benzene and hot petroleum ethers, and its solubility in cold petroleum ethers is relatively low. In solvents such as ethanol, tin tetraiodide reacts with iodides of alkali metals to form black crystalline $M_2[SnI_6]$ compounds.

Because tin tetraiodide readily hydrolyzes, it is not suitable to be synthesized in aqueous solutions. It is usually dry synthesized, that is, synthesized from the reaction between tin and iodine vapor or with non-aqueous solvents at elevated temperatures. Non-aqueous solvents can be selected from carbon tetrachloride, glacial acetic acid, etc. In this experiment low-boiling petroleum ethers (b. p. 60~90℃) is used as solvent. It is a mixture of low hydrocarbons (mostly pentanes and hexanes) and is an inert solvent. The synthesis consists of dissolving iodine in petroleum ethers, followed by its reaction with tin metal at elevated temperatures:

$$Sn + 2I_2 \longrightarrow SnI_4$$

◆ Equipment and Chemicals

1. Equipment　Weighing balance, round bottom flask (30mL), reflux condenser, beakers (250mL and 30mL), thermometer, constant temperature bath and other heating equipment including ring-stand.

2. Chemicals　Petroleum ethers (b. p. 60~90℃), tin foil, iodine, 0. 1mol/L AgNO$_3$, 0. 1mol/L Pb(NO$_3$)$_2$, saturated KI solution, acetone, etc.

◆ Procedures

1. Synthesis of tin tetraiodide　0. 5g of crystalline iodine is weighed with a balance and placed in a clean, dry 30mL round bottom flask. Then 0. 2g of tin foils are weighed, cut into small pieces and placed in the same flask followed by the addition of 10mL petroleum ethers. After the reaction apparatus including the reflux condenser is set up, the flask is heated with the water bath until the reaction mixture boils. Temperature of the water bath is controlled to be between 85~95℃, and the flow rate of cooling water is adjusted to keep the condensed liquid to be below the middle section of the condenser. Reflux is maintained until the reaction is completed (the color of condensed petroleum ethers solution changes from purple to colorless). The heating is then stopped with the removal of water bath. The condenser is removed when the solution no longer boils. The solution is decanted hot into a clean, dry 30mL beaker, leaving the unreacted tin foils in the reaction flask. Residual SnI_4 crystals left on the inside wall of the flask or on the tin foils should be collected with $1 \sim 2$mL of hot petroleum ethers and then combined with the solutions in the above-mentioned beaker. The beaker is cooled in an ice bath to start crystallization. When crystallization is complete, the top clear solution (petroleum ethers mother liquor) is carefully decanted into the recycling flask with a glass rod. The small flask containing the crystals is dried with the water bath. The product is weighed and the yield is then calculated.

2. Property tests

(1) Small amount of SnI_4 is mixed with 2mL of distilled water. The pH litmus paper is used to test the acidity/causticity of the solution.

(2) The above solution is divided into two test tubes. Several drops of 0.1mol/L $AgNO_3$ is added into one test tube and 0.1mol/L $Pb(NO_3)_2$ into the other test tube. Observe the change and write down the ionic reaction equations.

(3) Small amount of SnI_4 is dissolved with acetone, and the solution is divided into two test tubes. Several drops of water added into one the tubes and equal amount of saturated KI solution into the other. Observe and explain.

◆ Questions

1. What effect does it have on the experimental results if the reaction is stopped when the solution boils too violently, or condensation is insufficient, or iodine and petroleum ethers evaporate from the condenser, or the condensed liquid still has color?

2. Which starting material should be in excess amount in this experiment? Why?

3. Is it right to heat with small flames instead of water bath in this experiment? Why?

实验三　硫酸亚铁铵的制备 📱微课1

PPT

目的要求

1. 了解复盐的制备方法。
2. 掌握水浴加热和减压过滤等操作。

实验原理

铁屑易与稀硫酸反应，生成硫酸亚铁：

$$Fe + H_2SO_4 \!=\!\!=\!\! FeSO_4 + H_2 \uparrow$$

硫酸亚铁与等摩尔的硫酸铵在水溶液中相互作用，生成溶解度较小、浅蓝色的硫酸亚铁铵$FeSO_4 \cdot (NH_4)_2SO_4 \cdot 6H_2O$

$$FeSO_4 + (NH_4)_2SO_4 + 6H_2O \!=\!\!=\!\! FeSO_4 \cdot (NH_4)_2SO_4 \cdot 6H_2O$$

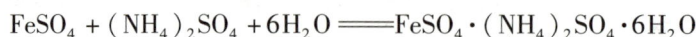

一般亚铁盐在空气中都易被氧化，但形成复盐后却比较稳定，不易被氧化。

仪器和药品 ●●●

1. 仪器 台秤，恒温水浴，抽滤水泵，蒸发皿，锥形瓶，10mL 量筒，25mL 比色管，石棉网。

2. 药品

（1）固体 碎铁屑、硫酸铵。

（2）酸 H_2SO_4 3mol/L；HCl 3mol/L。

（3）盐 Na_2CO_3 10%；KSCN 0.1mol/L。

实验内容 ●●●

1. 制备步骤

（1）铁屑的净化（去油污） 在台称上称取 2g 铁屑放于锥形瓶中，然后加入 20mL 10% Na_2CO_3 溶液，在电炉上微微加热约 10 分钟，用倾泻法去碱液，用蒸馏水将铁屑冲洗干净。

（2）硫酸亚铁的制备 往盛有铁屑的锥形瓶中加入 15mL 3mol/L H_2SO_4，在水浴上加热，使铁屑和硫酸反应至不再有气泡冒出为止。趁热抽滤，用 5mL 热蒸馏水洗涤残渣。滤液转至蒸发皿中。将锥形瓶中的和滤纸上的未反应铁屑用滤纸吸干后称重。从反应的铁屑的量求算出生成的 $FeSO_4$ 理论产量。

（3）硫酸亚铁铵的制备 根据上面计算出来的 $FeSO_4$ 的理论产量，按照 $FeSO_4$ 比 $(NH_4)_2SO_4$ 为 1：0.75 的质量比，称取固体硫酸铵，加到硫酸亚铁溶液中，水浴上蒸发浓缩至表面出现晶体膜为止，自然冷却后，减压抽滤，便得到硫酸亚铁铵晶体。干燥后称重，计算产率。

2. 产品纯度检验

Fe^{3+} 的限量检查 称 2g 产品于 50mL 比色管中，加入 30mL 不含氧的蒸馏水使之溶解。加 4mL 3mol/L HCl 和 2mL 0.1mol/L KSCN，继续加不含氧的蒸馏水至 50mL 刻度。摇匀，所呈现的红色与标准试样比较，检验产品级别。

标准试样的制备如下。

取含有下列数量 Fe^{3+} 的溶液 30mL：

Ⅰ级试剂：0.10mg

Ⅱ级试剂：0.20mg

Ⅲ级试剂：0.40mg

然后与产品同样处理（标准试样均由实验室准备）。

实　验　指　导

预习要求 ●●●

1. 如何除去废铁屑表面的油污？

2. 制备硫酸亚铁时，怎样鉴别反应已进行完全？

3. $FeSO_4$ 溶液在空气中很容易被氧化，在制备硫酸亚铁铵的过程中，怎样防止 Fe^{2+} 氧化成 Fe^{3+}？

4. 怎样计算硫酸亚铁铵的产率？是根据铁的用量还是硫酸铵的用量？

5. 如何制备不含氧的蒸馏水？为什么配制硫酸亚铁铵试液时要不含氧的蒸馏水？

基本操作 ●●●

1. 掌握减压过滤的操作。

2. 练习倾泻法洗涤的方法。

3. 了解产品限度分析方法。

注意事项 •••

1. 由于铁屑含有杂质砷，本实验在合成过程中，有剧毒气体 AsH_3 放出，它能刺激和麻痹神经系统。故实验需在通风橱中进行。

2. 在 $FeSO_4$ 溶液中加入固体 $(NH_4)_2SO_4$ 后，必须充分搅动，至 $(NH_4)_2SO_4$ 完全溶解后，才能进行蒸发浓缩。

3. 加热浓缩时间不宜过长。浓缩到一定体积后，需在室温放置一段时间，以待结晶析出、长大。

报告格式 •••

1. 目的

2. 原理

3. 实验步骤

（1）用流程图表示制备过程

（2）计算 $FeSO_4 \cdot 7H_2O$ 及硫酸亚铁铵的理论产量

（3）产率：$\dfrac{实际产量}{理论产量} \times 100\%$

（4）纯度检查

实验后思考 •••

1. 本实验反应中，是铁过量，还是 H_2SO_4 过量？为什么要这样选择？

2. 反应结束以后，为什么要趁热抽滤？为什么需用热水洗涤残渣？

3. 为什么本实验在蒸发浓缩时，溶液应控制在酸性（$pH = 2 \sim 3$）？

Experiment 3   Preparation of Ferrous Ammonium Sulphate Hexahydrate(FAS)

◆ Objectives

1. To learn how to prepare double salts.

2. To master the operation of heating in the water bath and filter by suction.

◆ Principles

The ferrous ion is converted into ferric ion easily when it is open to the air, but the ferrous ion in FAS cannot be easily oxidized.

Ferrous sulfate（$FeSO_4$）, which can be obtained by reacting iron powder with diluted sulfuric acid（H_2SO_4）, reacts with ammonium sulfate（$(NH_4)_2SO_4$）in equi molar ratio in aqueous solution. Ferrous ammonium sulphate hexahydrate（$FeSO_4 \cdot (NH_4)_2SO_4 \cdot 6H_2O$）with less solubility crystallizes from the solution as pale blue monoclinic crystal.

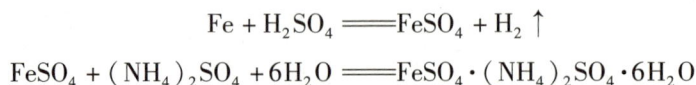

$$Fe + H_2SO_4 \Longrightarrow FeSO_4 + H_2 \uparrow$$

$$FeSO_4 + (NH_4)_2SO_4 + 6H_2O \Longrightarrow FeSO_4 \cdot (NH_4)_2SO_4 \cdot 6H_2O$$

◆ Equipment and chemicals

1. Equipment Platform balance, graduated cylinder (10mL), Erlenmeyer flask, filter flask, Buchner funnel, evaporating dish, thermostatic water bath, comparison tube (50mL).

2. Chemicals Iron powder, ammonia sulfate ($(NH_4)_2SO_4$); H_2SO_4 3mol/L,

HCl 3mol/L; Na_2CO_3 10%, KSCN 0.1mol/L.

◆ Procedures

1. Procedures

(1) 2g iron powder and 20mL 10% Na_2CO_3 are put into an Erlenmeyer flask which is heated over an electric cooker for about 10 minutes. Remove the solution by tilt-pour process; wash the iron powder with distilled water.

(2) Add 15mL 3mol/L H_2SO_4 into the Erlenmeyer flask, which is heated in the water bath at $60 \sim 70\,^{\circ}\text{C}$ until the reaction is completed. Filter by suction while hot; wash the filter residue with 5mL warm water. The filtrate is put into a clean evaporating dish. On the basis of the remains of the iron powder, calculate the ferrous sulfate's theoretical yield.

(3) For each 0.1g ferric sulfate, 0.075g ammonia sulfate is added as solid reagent into the evaporating dish. Stir the mixture to dissolve the ammonia sulfate. Then the evaporating dish is heated in the boiling water bath until a layer of tiny crystals is observed. Cool the concentrate and filter by suction. Dry the pale blue crystal and weigh it. Calculate the percentage yield of the product.

2. Purity examination of the product Dissolve 2g ferrous ammonium sulphate hexahydrate with 30mL oxygen-free distilled water in a 50mL comparison tube. Add 3mol/L HCl 4mL and 0.1mol/L KSCN 2mL. Fill to the mark with oxygen-free distilled water and mix the solution. Compare the color with that of the series of standard samples to determine the purity grade of the product.

Preparation of the standard samples:

Add 30mL solution containing ferric ion (the content of Fe^{3+} in the various solutions shown below) into comparison tubes. The following procedure is the same as that of the product.

Grade Ⅰ: 0.10mg

Grade Ⅱ: 0.20mg

Grade Ⅲ: 0.40mg

◆ Instructions

1. Requirements

(1) How to remove the oil from the surface of iron powder?

(2) How to determine whether the reaction to produce ferric sulfate is completed?

(3) The solution of ferrous sulfate is easy to be oxidized. How to avoid the oxidation when preparing the ferrous ammonium sulphate hexahydrate?

(4) How to calculate the yield of the product?

(5) How to get oxygen-free distilled water? Why should the water dissolving the ferrous ammonium sulphate hexahydrate be oxygen-free distilled water?

2. Operation

(1) Master the operation of filter by suction

（2）Practice washing solid by tilt-pour process

（3）Understand the analytical method of impurity limitation

3. Notes

（1）The iron powder having low As content may give out poisonous gas AsH_3, so the reaction for ferrous sulfate should be carried out in the fume cupboard.

（2）After ammonium sulfate is added to the evaporating dish, the mixture must be stirred thoroughly until the ammonium sulfate dissolves.

（3）The time for evaporation should not be too long and the concentrate should be kept under room temperature for a while to produce the crystal of $FeSO_4 \cdot (NH_4)_2SO_4 \cdot 6H_2O$.

4. Report format

（1）Objectives

（2）Principle

（3）Procedures

（a）Illustrate the procedure with a flow chart

（b）Calculate the theoretic yield of $FeSO_4 \cdot 7H_2O$ and FAS

（c）Percentage yield

（d）Purity examination

5. Questions

（1）Between the iron powder and the sulfuric acid in the secondary step, which one should be present in considerable excess?

（2）Why should we filter the solution while it is still hot when the reaction for ferric sulfate is complete? Why do we need warm water to wash the filter residue?

（3）Why should the pH of the solution be kept at 2～3 during evaporation?

Words

sulfate	硫酸盐	solution	溶液
ferrous	亚铁的	aqueous solution	水溶液
ferric	三价铁的	graduated cylinder	量筒
ammonium	铵根	flask	烧瓶
hexa-	（前缀）六	Erlenmeyer flask	锥形瓶
ferrous ammonium sulphate（FAS）	硫酸亚铁铵	evaporating dish	蒸发皿
		comparison tube	比色管
ferrous ammonium sulphate hexahydrate	六水硫酸亚铁铵	Buchner funnel	布氏漏斗
		fume cupboard	通风橱
ferrous sulfate	硫酸亚铁	concentrate	浓缩液
ammonium sulfate	硫酸铵	flow chart	流程图
dilute	稀的,稀释的	analysis	分析
sulfuric acid	硫酸	oxidize	氧化

double salt	复盐	oxidation	氧化（作用）
ion	离子	water bath	水浴
crystal	晶体	filter	过滤器，过滤
monoclinic crystal	单斜晶	filtrate	过滤，滤液
crystallize	结晶	filter flask	吸滤瓶
yield	产量	tilt-pour process	倾泻法
actual yield	实际产量	filter by suction	抽滤

实验四　葡萄糖酸锌 $Zn(C_6H_{11}O_7)_2 \cdot 3H_2O$ 的制备 📱微课2

PPT

目的要求 ••••

1. 了解葡萄糖酸锌（治疗人体缺锌药物）的制备方法。
2. 学会锌盐的含量测定。

实验原理 ••••

葡萄糖酸钙与等摩尔的硫酸锌反应式：

$$Ca(C_6H_{11}O_7)_2 + ZnSO_4 + 3H_2O \longrightarrow Zn(C_6O_{11}O_7)_2 \cdot 3H_2O + CaSO_4 \downarrow$$

仪器和药品 ••••

1. 仪器　台天平，恒温水浴，抽滤装置，酸式滴定管，电炉，蒸发皿，烧杯，量筒等。

2. 药品　葡萄糖酸钙；$ZnSO_4 \cdot 7H_2O$；95% 乙醇；$NH_3 - NH_4Cl$ 缓冲溶液；0.1mol/L EDTA – 2Na 标准溶液；铬黑 T 指示剂。

实验内容 ••••

1. 制备步骤　量取 80mL 蒸馏水置烧杯中，加热至 80～90℃，加入 13.4g $ZnSO_4 \cdot 7H_2O$ 使完全溶解，将烧杯放在 90℃的恒温水浴中，再逐渐加入葡萄糖酸钙 20.0g，并不断搅拌。在 90℃水浴上静置保温 20min。趁热抽滤（用两层滤纸），滤液移至蒸发皿中（滤渣为 $CaSO_4$，弃去），将滤液在沸水浴上浓缩至黏稠状（体积约为 20mL，如浓缩液有沉淀系 $CaSO_4$，需过滤掉）。滤液冷至室温，加 20mL 95% 乙醇（降低葡萄糖酸锌的溶解度），并不断搅拌，此时有大量的胶状葡萄糖酸锌析出，充分搅拌后，用倾泻法去除乙醇液。于胶状沉淀上，再加 20mL 95% 乙醇，充分搅拌后，沉淀慢慢转变成晶体状，抽滤至干，即得粗品（母液回收）。

重结晶。粗品加水 20mL，加热（90℃）至溶解，趁热抽滤，滤液冷至室温，加 20mL 95% 乙醇，充分搅拌，结晶析出后，抽滤至干，即得精品，在 50℃烘干。

2. 含量测定　准确称取 0.8g 葡萄糖酸锌，溶于 20mL 水中（可微热），加 10mL NH_3-NH_4Cl 缓冲溶液，加铬黑 T 指示剂 4 滴，用 0.1mol/L EDTA-2Na 标准溶液滴定至溶液呈蓝色。样品中锌的含量计算如下：

$$锌的含量\% = \frac{C_{\text{EDTA-2Na}} \cdot V_{\text{EDTA-2Na}} \times 65}{W_S \times 1000} \times 100\%$$

式中，$C_{\text{EDTA-2Na}}$ 为浓度，mol/L；$V_{\text{EDTA-2Na}}$ 为体积，mL；W_S 为样品的质量，g。

实　验　指　导

预习要求 •••

1. 设计葡萄糖酸锌制备的流程图。
2. 为什么葡萄糖酸钙和硫酸锌的反应需保持在 90℃ 的恒温水浴中？
3. 葡萄糖酸锌可以用哪几种方法进行结晶？
4. 如何测定葡萄糖酸锌中锌的含量？

基本操作 •••

1. 掌握恒温水浴的使用操作。
2. 学习用乙醇为溶剂进行重结晶的方法。

注意事项 •••

1. 反应需在 90℃ 恒温水浴中进行。这是由于：温度太高，葡萄糖酸锌会分解；温度太低，则葡萄糖酸锌的溶解度降低。

2. 用酒精为溶剂进行重结晶时，开始有大量胶状葡萄糖酸锌析出，不易搅拌，可用竹棒代替玻棒进行搅拌。

3. 滤液需在沸水浴中浓缩。

4. 用铬黑 T 做指示剂，以 EDTA-2Na 标准溶液测定葡萄糖酸锌时，注意观察终点颜色的变化。

报告格式 •••

1. 目的
2. 原理（反应式表示）
3. 制备流程图
4. 产量，产率计算
5. 含量测定、数据记录和结果处理

		（一）	（二）
记录部分	W_s 葡萄糖酸锌，g		
	$C_{EDTA-2Na}$，mol/L		
	$V_{EDTA-2Na}$，mL		
	锌的摩尔质量，g/mol		
计算	含锌% $= \dfrac{C_{EDTA-2Na} \cdot V_{EDTA-2Na} \times 65}{W_s \times 1000} \times 100\%$		

实验后思考 •••

查阅有关资料，了解微量元素锌与人体健康的关系。

Experiment 4　Preparation and Content Assay of Zinc Gluconate

◆ Objectives

1. To learn how to prepare zinc gluconate.

2. To learn how to determine the content of zinc salt.

◆ Principles

Calcium gluconate reacts with equal molar zinc sulfate:

$$Ca(C_6H_{11}O_7)_2 + ZnSO_4 + 3H_2O \longrightarrow Zn(C_6H_{11}O_7)_2 \cdot 3H_2O + CaSO_4 \downarrow$$

◆ Equipment and chemicals

1. Equipment　Platform balance, graduated cylinder, beaker, evaporating dish, electric cooker, acid-type burette, thermostatic water bath.

2. Chemicals　Calcium gluconate, heptahydrate zinc sulfate, 95% ethanol, ammonia-ammonium chloride buffer solution, 0.1mol/L disodium ethylenediaminetetraacetate standard solution, eriochrome black solution.

◆ Procedures

1. Preparation procedure of zinc gluconate　Add 80mL distilled water to a beaker. Heat it to 80 ~ 90℃, then add 13.4g $ZnSO_4 \cdot 7H_2O$ and dissolve them completely. Put the beaker in water bath and keep it at 90℃. Then 20g calcium gluconate is added gradually and stirred constantly. After 30 minutes, filter it quickly by vacuum filter (double filter paper). The filtrate is transferred to an evaporating dish (filter cake discarded). It is concentrated and becomes sticky(the volume of approximately 20mL). Then the filtrate is cooled to room temperature. 20mL of 95% ethanol is added (to reduce the solubility of zinc gluconate)to the filtrate. Then stir continually. A lot of colloidal zinc gluconate was separated out. After stirring completely, the precipitate slowly changes into crystal. Then by filtering, the crude product is obtained. It is dissolved by adding 20mL water at 90℃, filter warmly, then the filtrate is cooled to room temperature. Add 20mL 95% ethanol, stir strongly, and the crystal is separated out. After filtering and baking at 50℃, pure product is obtained.

2. Content assay　Weigh 0.8g zinc gluconate accurately, dissolve it in 20mL of water (sometimes need to be heated). Add 10mL of ammonia-ammonium chloride buffer solution and 4 drops of eriochrome black solution. Titrate the solution with 0.1mol/L disodium ethylenediaminetetraacetate (EDTA) standard solution, until the solution is deep blue in color. The zinc concentration in the sample can be calculated as follows:

$$Zn\% = \frac{C_{EDTA \cdot 2Na} \cdot V_{EDTA \cdot 2Na} \times 65}{W_S \times 1000} \times 100\%$$

$(C \cdot V)_{EDTA \cdot 2Na}$ is the concentration of EDTA·2Na(C, mol/L) and volume (V, mL); W_S is the weight of the sample(g).

◆ Instructions

1. Requirements

(1) Design flow chart of the preparation of zinc gluconate.

(2) Why must we keep the temperature at 90℃ when zinc sulfate reacts with calcium gluconate?

（3）What methods can be used to recrystallize zinc gluconate?

（4）How can zinc content of zinc gluconate be determined?

2. Operation

（1）Master the operation of thermostatic water bath.

（2）Learn the recrystallization method with ethanol as a solvent.

3. Notes

（1）The reaction must be undertaken at constant temperature water bath boiler of 90℃. Too high temperature may lead to the decomposition of zinc gluconate, while too low temperature may cause the solubility of zinc gluconate decreasing.

（2）When recrystallized with ethanol as solvent, small amount of zinc gluconate may appear at the beginning. Chopstick, often replaces glass rod for the solution is very sticky and hard to stir.

（3）The filtrate is concentrated in boiling water bath.

（4）When titrating with 0.1mol/L disodium EDTA standard solution, observe the color change carefully.

4. Report format

（1）Objectives

（2）Principles（in reaction formula）

（3）Flow chart of preparation

（4）Yield, percentage yield

（5）Record data and result dealing

		1	2
Record part	W(zinc gluconate), g		
	$C_{\text{EDTA}\cdot 2\text{Na}}$, mol/L		
	$V_{\text{EDTA}\cdot 2\text{Na}}$, mL		
	molar weight of zinc, g/mol		
Calculation	$\text{Zn\%}=\dfrac{C_{\text{EDTA}\cdot 2\text{Na}}\cdot V_{\text{EDTA}\cdot 2\text{Na}}\times 65}{W_{\text{S}}\times 1000}\times 100\%$		

5. Questions Refer to other materials and understand the important role of zinc playing in human body.

Words

zinc	锌	thermostatic water bath	恒温水浴
zinc gluconate	葡萄糖酸锌	calcium gluconate	葡萄酸钙
platform balance	台秤	heptahydrate zinc sulfate	七水硫酸锌
buffer solution	缓冲液	measuring cylinder	量筒
ethylenediaminetetraacetate	EDTA 溶液	beaker	烧杯
eriochrome black solution	铬黑 T 溶液	evaporating dish	蒸发皿
electric cooker	电炉		

实验五　药用氯化钠的制备、性质及杂质限度检查 微课3

PPT

目的要求 •••

1. 掌握药用氯化钠的制备原理和方法。

2. 初步了解药品的鉴别及检查方法。

实验原理 •••

1. 食盐是能溶于水的固态物质。对于其中所含杂质可用以下方法除去。

（1）机械杂质如泥沙可采取过滤法除去。

（2）一些能溶解的杂质可根据其性质借助于化学方法除去。如加入 $BaCl_2$ 溶液可使 SO_4^{2-} 生成 $BaSO_4$ 沉淀，加入 Na_2CO_3 溶液可使 Ca^{2+}、Mg^{2+}、Fe^{3+}、Ba^{2+} 等离子生成难溶物沉淀，先后滤去。

（3）少量可溶性杂质如 Br^-、I^-、K^+ 等离子，可根据溶解度不同，在重结晶时，使其残留在母液中弃去。

2. 鉴别试验是被检药品组成离子的特征试验（这里指的是氯化钠的组成离子 Na^+、Cl^-）。

3. 钡盐、硫酸盐、钾盐、钙镁盐的限度检查，是根据沉淀反应原理，样品管和标准管在相同条件下进行比浊试验，样品管不得比标准管更浑浊。

4. 重金属系指 Pb、Bi、Cu、Hg、Sb、Sn、Co、Zn 等金属的离子，它们在一定条件下能与 H_2S 或 Na_2S 作用而显色。《中国药典》规定反应在弱酸性条件下进行，用稀醋酸调节。实验证明，在 $pH = 3$ 时，PbS 沉淀最完全，反应式为：

$$Pb^{2+} + S^{2-} \longrightarrow PbS \downarrow$$

重金属的检查，是在相同条件下进行比色试验。

仪器和药品 •••

1. 仪器　电炉，蒸发皿，布氏漏斗，铂丝棒，烧杯 250mL 两个，台天平，奈氏比色管。

2. 药品

（1）固体　NaCl（粗盐）。

（2）酸　浓 HCl，1/50mol/L；H_2SO_4 6mol/L。

（3）碱　NaOH 2mol/L，1/50mol/L；Na_2CO_3 饱和溶液。

（4）盐　$BaCl_2$ 25%；KI 10%；钙盐溶液；$AgNO_3$ 0.25mol/L；KBr 10%；镁盐溶液；醋酸铀酰锌溶液；$(NH_4)_2C_2O_4$ 试液；Na_2HPO_4 试液；NH_4Cl 试液。

3. 其他　淀粉 - 碘化钾试纸；溴麝香草酚蓝试液。

实验内容 •••

1. 食盐精制

（1）操作步骤以流程图表示如下。

粗食盐 50g

$$\begin{bmatrix} 主要成分 & 氯化钠（NaCl） \\ 杂质成分 & 有机物、泥沙以及 Ca^{2+}、Mg^{2+}、Fe^{3+}、SO_4^{2-}、Br^-、I^-等 \end{bmatrix}$$

↓小火炒至有机物炭化

↓移烧杯中，加适量水加热至溶解

↓趁热倾泻法过滤①

沉淀甲　　　　滤液甲

弃去（炭化　　↓加热近沸

物、泥沙）　　↓在搅拌下逐滴加入 25% BaCl₂ 试液至沉完全②

↓煮沸、稍冷、减压抽滤

沉淀乙　　　　滤液乙

弃去（BaSO₄）　↓加热

↓加饱和 Na₂CO₃ 使沉淀完全，并加

NaOH 溶液，使 pH 为 10～11

↓煮沸、放冷、减压抽滤，滤液转至蒸发皿中

沉淀丙弃去 [CaCO₃，　　滤液丙

BaCO₃，Fe(OH)₃，　　↓用 HCl 中和至 pH 3～4

Mg(OH)₂·3MgCO₃]　　↓加热蒸发至稠厚糊状，有大量 NaCl 析出时

↓趁热减压抽滤

母液（氯化钠的饱和溶液　　粗品 NaCl

含有 Br⁻，I⁻，K⁺ 等离子）　↓重结晶

精制 NaCl

↓烘干

母液　　成品 NaCl

母液合并，生←
产上可套用

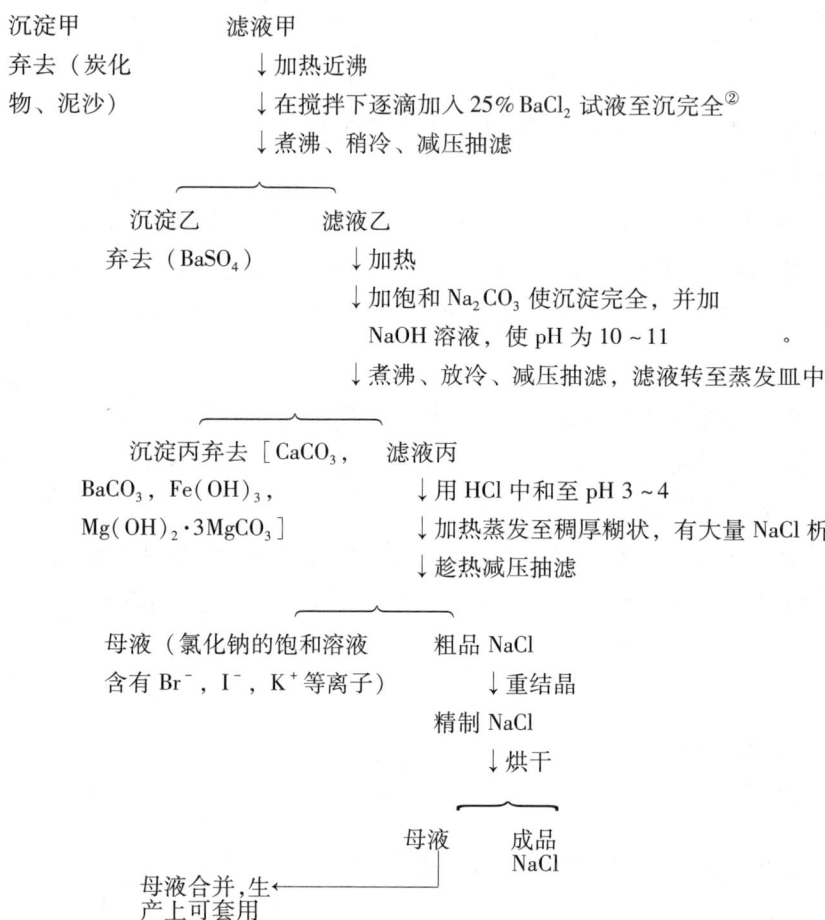

① 倾泻法过滤，多在沉淀易于下沉的条件下使用。是指尽量将沉淀保留于烧杯中，待溶液澄清后，只将澄清液除去。

② 检查沉淀是否完全，可吸滤少量上层清液置试管中，加 1～2 滴沉淀剂，无浑浊生成即表示已沉淀完全。

2. 鉴别反应　氯化钠溶液（1∶10）的制备（供下列鉴别应用）。

（1）钠盐

①焰色反应　取铂丝，用浓盐酸润湿后在无色火焰中灼烧，至火焰不显颜色为止（表示铂丝已洁净）。后蘸取氯化钠溶液，置无色火焰中灼烧，火焰出现持久鲜黄色。

②沉淀反应　取氯化钠溶液 1～2 滴，用 3mol/L HAc 3 滴使酸化，加醋酸铀酰锌试液 10 滴，用玻璃棒摩擦试管内壁，即渐渐析出醋酸铀酰锌钠淡黄绿色沉淀。反应如下：

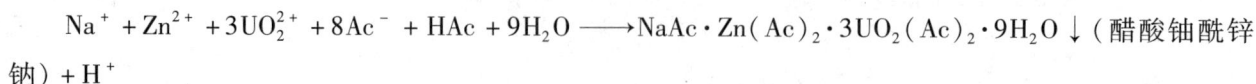

$$Na^+ + Zn^{2+} + 3UO_2^{2+} + 8Ac^- + HAc + 9H_2O \longrightarrow NaAc \cdot Zn(Ac)_2 \cdot 3UO_2(Ac)_2 \cdot 9H_2O \downarrow （醋酸铀酰锌$$

钠）$+ H^+$

（2）氯化物　生成氯化银沉淀。取氯化钠溶液 1～2 滴，加 0.25mol/L 硝酸银试液 2 滴，即发生白

色凝乳状沉淀。滴加 6mol/L 氨试液，沉淀溶解，再加 6mol/L HNO_3 至酸性，又有白色沉淀生成。

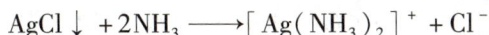

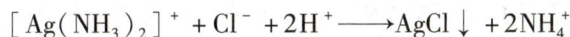

$$Cl^- + Ag^+ \longrightarrow AgCl\downarrow（白）$$

$$AgCl\downarrow + 2NH_3 \longrightarrow [Ag(NH_3)_2]^+ + Cl^-$$

$$[Ag(NH_3)_2]^+ + Cl^- + 2H^+ \longrightarrow AgCl\downarrow + 2NH_4^+$$

3. 检查　成品氯化钠需进行以下各项质量检查试验。

（1）溶液的澄清度　取本品 5g，加水至 25mL，应溶解成无色澄明的溶液。

（2）酸碱度　取本品 5g，加新鲜蒸馏水 50mL 溶解后，加溴麝香草酚蓝指示液 2 滴。如溶液呈黄色，滴加 1/50mol/L NaOH 溶液使其变蓝色，所消耗的 1/50mol/L NaOH 溶液不得超过 0.1mL。如显蓝色或绿色，滴加 1/50mol/L HCl 溶液使其变黄色，所消耗的 1/50mol/L HCl 溶液不得超过 0.2mL。

氯化钠为强酸强碱所生成的盐，在水溶液中应呈中性。但在制备过程中，可能夹杂少量酸或碱，所以药典把其水溶液的 pH 限制在很小范围内。溴麝香草酚蓝指示液的变色范围是 pH 6.0～7.6，由黄色到蓝色。

（3）碘化物与溴化物　取本品 1g，加蒸馏水 3mL 溶解后，加三氯甲烷 1mL，并注意逐滴加入用等量蒸馏水稀释的氯试液，随滴随振摇，三氯甲烷层不得显紫堇色、黄色或橙色。

对照试验　分别取 10% 的碘化物、溴化物溶液各 1mL，分置于两试管内，各加三氯甲烷 1mL，同上法逐滴加入用等量蒸馏水稀释的氯试液，随滴随振摇。氯试液氧化 I^- 离子释出碘，使三氯甲烷层显紫红色，氯试液氧化 Br^- 离子释出溴，使三氯甲烷层显黄色或橙黄色。

$$2Br^- + Cl_2 \longrightarrow Br_2 + 2Cl^-$$

$$2I^- + Cl_2 \longrightarrow I_2 + 2Cl^-$$

（4）钡盐　取本品 4g，用蒸馏水 20mL 溶解、过滤，滤液分为两等份。一份中加稀硫酸 2mL，另一份中加蒸馏水 2mL，静置 2h，两份溶液应同样澄明。

（5）钙盐与镁盐　取本品 4g，加蒸馏水 20mL 溶解后，加氨试液 2mL，摇匀，分为两等份。一份加草酸铵试液 1mL，另一份加磷酸氢二钠试液 1mL，氯化铵试液数滴，5 分钟内均不得发生浑浊。

对照试验

①取钙盐溶液 1mL，加氨试液至微碱性，加草酸铵试液 1mL，溶液有白色结晶析出。反应式为：

$$Ca^{2+} + C_2O_4^{2-} \longrightarrow CaC_2O_4\downarrow 白色$$

②另取镁盐溶液 1mL，加入氨水和氯化铵数滴，再逐滴加磷酸氢二钠溶液，有白色沉淀析出，反应式为：

$$Mg^{2+} + HPO_4^{2-} + NH_4^+ + OH^- \longrightarrow MgNH_4PO_4\downarrow 白色 + H_2O$$

（6）硫酸盐　本品含硫酸盐，依下法检查，如发生浑浊，与标准硫酸钾 1mL 配成的对照标准溶液比较，不得更浓（0.002%）。

取 50mL 奈氏比色管两支，甲管中加标准硫酸钾溶液 1mL，加蒸馏水稀释至约 35mL 后，加 1mol/L 盐酸 5mL，氯化钡试液 5mL，再加适量蒸馏水至刻度，摇匀。

取本品 5g 置乙管中，加适量的蒸馏水溶解，至约 35mL，加 1mol/L 盐酸 5mL，溶液应澄明，如不澄明可用滤纸过滤，加氯化钡试液 5mL，用蒸馏水稀释至刻度，摇匀。

甲乙两管放置 10 分钟后，置比色管架上，在光线明亮处双眼自上而下透视。比较两管的浑浊度，乙管发生的浑浊度不得大于甲管。

标准硫酸钾溶液的制备　精密称取硫酸钾 0.1813g，置 1000mL 的容量瓶中，加适量的蒸馏水使溶解并稀释至刻度，摇匀即得（每 1mL 相当于 0.1mg 的 SO_4^{2-}）。

（7）铁盐　取本品 5g，置于 50mL 奈氏比色管中，加蒸馏水 35mL 溶解，加稀盐酸 5mL，过硫酸铵约 30mg，再加硫氰酸铵试液 5mL，补加适量的蒸馏水至刻度，摇匀，如显色与标准铁溶液 1.5mL 用同法处理后制得的标准管的颜色比较不得更深（0.0003%），反应式为：

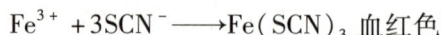

$$Fe^{3+} + 3SCN^- \longrightarrow Fe(SCN)_3 \text{ 血红色}$$

标准铁溶液的制备　精密称取未风化的硫酸铁铵 0.8630g，置于 1000mL 的容量瓶中，加蒸馏水溶解后，加稀盐酸 2mL，用蒸馏水稀释至刻度，摇匀。精密量取溶液 10mL，置于 100mL 容量瓶中，加稀盐酸 0.5mL，用蒸馏水稀释至刻度，摇匀即得（每 1mL 相当于 0.01mg 的 Fe）。

（8）钾盐　取本品 5g，置奈氏比色管中。加蒸馏水 20mL 溶解，加 3mol/L 醋酸 2 滴（使 pH 为 5~6），加 0.1mol/L 四苯硼钠溶液 2mL，加水至刻度，如显浑浊，与标准硫酸钾 0.5mL 制成的对照标准溶液比较，不得更浓（0.01%）。反应式为：

$$K^+ + B(C_6H_5)_4^- \longrightarrow KB(C_6H_5)_4 \downarrow \text{ 白色}$$

标准硫酸钾溶液的制备　精密称取 105℃ 干燥至恒重的硫酸钾 2.2280g，置 1000mL 容量瓶中，加适量蒸馏水溶解，加水至刻度，摇匀即得（1mL 相当于 1mg 的 K）。

四苯硼钠溶液的配制　取四苯硼钠 [NaB(C_6H_5)_4] 1.5g，置乳钵中加水 10mL 研磨后，再加水 40mL 研匀，用质密的滤纸过滤即得。

（9）重金属　取 50mL 比色管两支，于第一管中加标准铅溶液 1mL（每 1mL = 0.01mg Pb），加稀醋酸 2mL，再加水稀释至 25mL，于第二管中加样品 5g，蒸馏水 23mL 溶解后，加稀醋酸 2mL。再于两管中分别加硫化氢试液 10mL，摇匀，在暗处放置 10min，然后进行比色。如果第二管所显颜色比第一管为浅，说明重金属不超过规定限度。

附：上述标准铅溶液 1mL 是根据具体情况计算出来的，计算方法如下。

因《中国药典》中规定氯化钠的重金属检查项目："取样品 5g，加蒸馏水 23mL 溶解后，加醋酸 2mL 依法检查，含重金属不得超过 2ppm"，并没有直接给出标准铅溶液的取用量，需要按下述次序自行计算。

①先根据供试品的取用量和重金属的限量，求得重金属含量（mg），题给条件：NaCl 的取用量为 5g，其中含重金属的限量为 2ppm。

则允许重金属（Pb）的含量为：

$$5 \times 1000 \times 0.000002 = 0.01 \text{mg}（Pb）$$

②再计算标准铅溶液的取用量。

因标准铅溶液每 1.0mL 含 Pb 0.01mg，需 0.01mg Pb 作为标准进行对照。应取标准铅溶液的体积为：0.01mg/（0.01mg/1mL）= 1mL。

实　验　指　导

预习要求 ●●●

1. 海盐除含有 NaCl 外，还含哪些杂质？如何除去？

2. 食盐精制过程中，加试剂的次序，为什么必须先加 $BaCl_2$，再加 Na_2CO_3，最后加盐酸？是否可改变加入的次序？

3. 粗食盐中所含 K^+、Br^-、I^- 等离子是怎样除去的？

4. 在加入沉淀剂将 SO_4^{2-}、Ca^{2+}、Mg^{2+} 等离子变成沉淀以除去它们时，加热与不加热对分离操作各有何影响？如何检查这些离子是否沉淀完全？

5. 如何除去过量的沉淀剂 $BaCl_2$、Na_2CO_3 和 NaOH？

6. 在调 pH 过程中，若加入的盐酸过量，怎么处理？为何要调成弱酸性？

7. 在浓缩过程中，能否把溶液蒸至近干？为什么？

8. 什么是重结晶？根据你所得粗品 NaCl 晶体的量计算应加入多少水量，使之溶解为宜？

9. 在检查产品纯度时，能否用自来水溶解食盐，为什么？

基本操作 •••

通过本实验训练，要求掌握：

（1）沉淀产生与检查沉淀完全的操作；

（2）正确进行减压抽滤；

（3）蒸发浓缩、结晶；

（4）重结晶。

注意事项 •••

1. 将粗食盐加水（自来水）至全部溶解（其量根据食盐溶解度计算）为限，用水量不能过多，以免给以后蒸发浓缩带来困难。

2. 减压抽滤时，要注意防止回吸。

3. 在加沉淀剂过程中，溶液煮沸时间不宜过长，以免水分蒸发而使 NaCl 晶体析出。若发现液面有晶体析出时，可适当补充些蒸馏水。

4. 加热浓缩时，当大量 NaCl 晶体析出，要不断搅拌（玻璃棒尽量平放在溶液中）以破坏表层薄膜，防止 NaCl 晶体外溅。

5. 浓缩时不可蒸发至干，要保留少量水分，以使 Br^-、I^-、K^+ 等离子随母液去掉，并在抽滤时用玻璃瓶盖尽量将晶体压干。

6. 正确使用奈氏比色管，注意平行条件，用水稀释至刻度后再摇匀。

报告格式 •••

1. 目的

2. 原理

3. 实验步骤（用流程图表示）

4. NaCl 精品产量并计算产率

5. 写出 Na^+、Cl^- 离子的鉴别反应式，并用图表列出杂质限度检查结果

实验后思考 •••

根据实验体会，总结本实验中提高 NaCl 精品的产率和质量的关键。

Experiment 5　Preparation of Medicinal Sodium Chloride and Examination of Impurities′ Limitation

◆ Objectives

1. To master principle and method of how to prepare medicinal sodium chloride.

2. To learn how to identify and check medicines.

◆ Principles

1. Sodium chloride is soluble in aqueous solution, so the impurities in sodium chloride can be removed with

the processes showed below.

（1）The insoluble impurities are removed by filtration.

（2）Some soluble impurities can be removed by precipitation basing on their chemical properties. For example, sulfate can be separated as $BaSO_4$ by solution of $BaCl_2$; Ca^{2+}, Mg^{2+}, Ba^{2+}, Fe^{3+} etc. can be removed as insoluble precipitates.

（3）Some low content soluble impurities, such as Br^-, I^-, K^+, having different solubility in sodium chloride, can be removed by recrystallization. They will be retained in the mother liquid and moved away.

2. The limit tests of barium, sulfate, potassium, calcium and magnesium are carried out in comparison tubes by addition of corresponding precipitate agents. Under same conditions, any opalescence produced in the test solutions should not be pronounced than that of the reference solutions.

3. Heavy metals (comprising the ions of Pb, Bi, Cu, Hg, Sb, Sn, Co, Zn and other metals) can be colored by sulfide ion under the specified test conditions. The test of heavy metallic impurities is carried out by comparing the color of solutions with the corresponding reference solutions under same conditions.

◆ Equipment and Chemicals

1. Equipment Electric cooker, evaporating dish, Buchner funnel, platinum wire, beakers (250mL), platform balance, color-comparison tubes (25mL, 50mL)

2. Chemicals

（1）Solid raw sodium chloride

（2）Acid concentrated hydrochloric acid, dilute hydrochloric acid (1/50mol/L), sulfuric acid (6mol/L)

（3）Alkali sodium hydroxide solution (2mol/L; 1/50mol/L), saturated solution of sodium carbonate

（4）Salt $BaCl_2$(25%), KI(10%), Ca^{2+} TS, $AgNO_3$(0.25mol/L), KBr(10%), Mg^{2+} TS, zinc uranylacetate solution, $(NH_4)_2C_2O_4$ TS, Na_2HPO_4 TS, NH_4Cl TS, ammonia TS

3. Others KI-starch test paper, bromothymol blue TS

◆ Procedures

1. Purification of NaCl

The procedure is illustrated as the following flow chart.

Raw NaCl(impurities: organic compounds, sludge, sand, Ca^{2+}, Mg^{2+}, Fe^{3+}, SO_4^{2-}, Br^-, I^-, etc.)

↓ heat moderately to carbonize the organic impurities

↓ dissolve the solid in a beaker while heating

↓ filter with suction by tilt-pour process while hot

filter residue   filtrate

(sludge, sand   ↓ heat to keep the solution on the simmer

carbonaceous deposit)   ↓ add 25% $BaCl_2$ solution drop by drop and stir

                                  ↓ boil and cool down, filter by suction

        filter residue   filtrate

        discard ($BaSO_4$)   ↓ heat

                                  ↓ add saturated solution of Na_2CO_3, then add

                                     solution of NaOH to adjust the pH to 10～11

                                  ↓ boil, and cool down, filter by suction

```
filter residue                          filtrate
discard[ CaCO₃ , BaCO₃ ,               ↓ neutralize with HCl until the pH 3 ~ 4
   Fe( OH)₃ , Mg( OH)₂ , 3MgCO₃ ] ↓    ↓ evaporate the solution to be thick paste , a lot of
                                          crystal precipitate
                                        ↓ vacuum filter while hot

mother liquid ( saturated              crude NaCl
solution of NaCl containing
   Br⁻ , I⁻ , K⁺ )                      ↓ recrystallize
                                        refined NaCl
                                        ↓ dry

mother liquid( recycled )              product
```

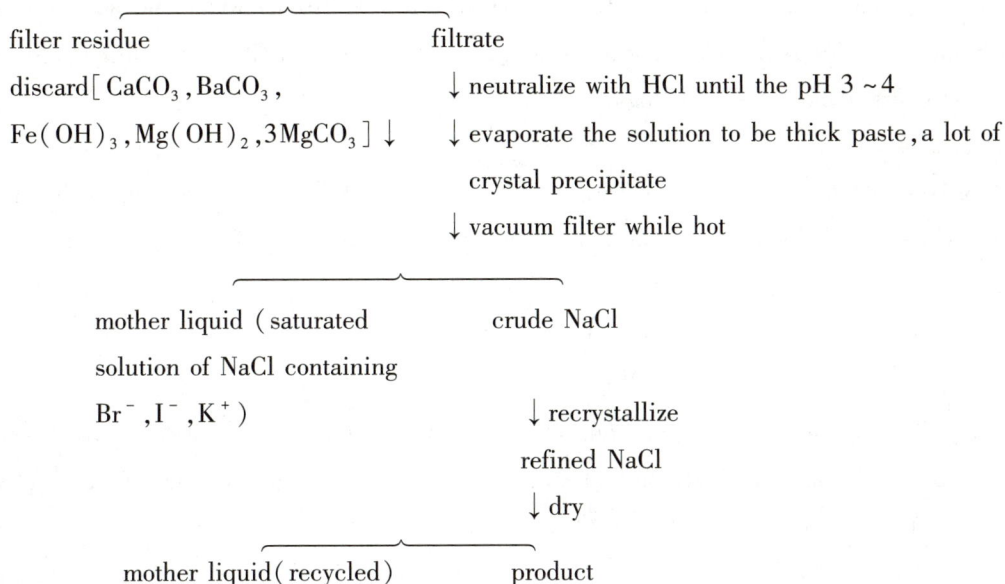

2. Identification test

An aqueous solution (1 in 10) of the product is prepared for the identification test.

(1) Sodium

(a) Dry test(flame color)   Prepare a platinum wire by burning it on a non-luminous flame after moistening it with concentrated hydrochloric acid until the flame is colorless. Moisten the test solution on the platinum wire ; it imparts an intense yellow color to a non-luminous flame.

(b) Precipitation   Add two drops of solution in a tube , acidified with three drops of 3mol/L acetic acid , and add 10 drops of zinc uranylacetate TS , if necessary , rub the inside wall of the test tube with a glass rod. A yellow precipitate is formed.

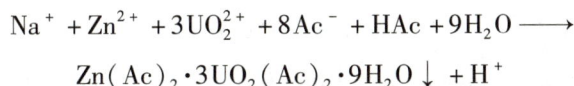

$$Na^+ + Zn^{2+} + 3UO_2^{2+} + 8Ac^- + HAc + 9H_2O \longrightarrow$$
$$Zn(Ac)_2 \cdot 3UO_2(Ac)_2 \cdot 9H_2O \downarrow + H^+$$

(2) Chloride   2 drops of the tested solution yield a white、curdy precipitate with 2 drops of 0. 25 mol/L silver nitrate TS. With addition of 6mol/L ammonia TS , the precipitate dissolve. White precipitate is formed again by acidifing the solution with 6mol/L nitric acid :

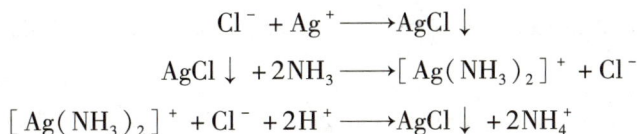

$$Cl^- + Ag^+ \longrightarrow AgCl \downarrow$$
$$AgCl \downarrow + 2NH_3 \longrightarrow [Ag(NH_3)_2]^+ + Cl^-$$
$$[Ag(NH_3)_2]^+ + Cl^- + 2H^+ \longrightarrow AgCl \downarrow + 2NH_4^+$$

3. Limit test of impurity

The product should be checked with the limit tests described below.

(1) Appearance of solution   Dissolve 5g in 25mL of distilled water. The solution should be clear and colorless.

(2) Acidity or alkalinity   Dissolve 5g in carbon dioxide-free water , and dilute with the same solvent to 50mL. Add 2 drops of bromothymol blue TS : Not more than 0. 1mL of 0. 02mol/L hydrochloric acid (HCl) or 0. 02mol/L sodium hydroxide (NaOH) is required to change the color of the solution.

The content of free acid or alkali , if there is any , in medicinal NaCl should not be more than the limit. And the pH range of bromothymol blue is 6. 0 ~ 7. 6 , yellow to blue.

（3）Iodide（I^-）and bromide（Br^-）　Dissolve 1g in 3mL of distilled water and add 1mL of chloroform. Cautiously introduce, dropwise, with constant shaking, dilute chlorine（Cl_2）TS（lin2）: the chloroform does not acquire a violet, yellow, or orange color.

Control test:

Add 1mL of 10% iodide TS and bromide TS in two tubes respectively, the following step is similar to that of the test solution. The color of chloroform in one tube（containing the iodide TS）is violet while the other tube is yellow or orange.

$$2Br^- + Cl_2 \longrightarrow Br_2 + 2Cl^-$$
$$2I^- + Cl_2 \longrightarrow I_2 + 2Cl^-$$

（4）Barium（Ba^{2+}）　Dissolve 4.0g in 20mL of distilled water, filter if necessary, and divide the solution into two equal portions. To one portion add 2mL of diluted sulfuric acid, and to the other add 2mL of water: the solution should be equally clear after standing for 2 hours.

（5）Calcium（Ca^{2+}）and magnesium（Mg^{2+}）　Dissolve 4.0g in 20mL of distilled water, add 2mL of ammonia TS and divide the mixture into two equal portions. Treat one portion with 1mL of ammonium oxalate TS and the other portion with 1mL of dibasic sodium phosphate TS and a few drops of ammonium chloride TS: no opalescence is produced within 5 minutes.

Control test.

Calcium – Pipet 1mL of calcium TS to a tube, alkalied with ammonia TS, add ammonium oxalate TS: while crystal is precipitated.

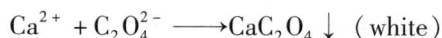

$$Ca^{2+} + C_2O_4^{2-} \longrightarrow CaC_2O_4 \downarrow （white）$$

Magnesium – Pipet 1mL of magnesium TS to a tube, add a few drops of ammonia TS and ammonium chloride, drop in dibasic sodium phosphate（Na_2HPO_4）TS: white precipitate is separated from the solution.

$$Mg^{2+} + HPO_4^{2-} + NH_4^+ + OH^- \longrightarrow MgNH_4PO_4 \downarrow （white） + H_2O$$

（6）Sulfate（SO_4^{2-}）　Reference preparation – Into a 50mL color-comparison tube pipet 1mL of standard potassium sulfate solution, and dilute with water to about 35mL. Add 5mL of 1mol/L hydrochloric acid and 5mL of barium chloride TS. Dilute with water to volume and mix.

Test preparation-Into a 50mL color-comparison tube place 5g of the product, and dissolve the solid with about 35mL of distilled water, filter if necessary. Add 5mL of barium chloride TS, dilute with water to volume and mix.

Procedure – Allow the two tubes to stand for 10 minutes, and view downward. Any opalescence produced in the latter tube is not more pronounced than that of the standard tube.

Standard potassium sulfate Dissolve 0.1813g of potassium chloride with distilled water in 1000mL volumetric flask, dilute with water to volume and mix. This solution contains the equivalent of 0.1mg of sulfate per ml.

（7）Ferric salt（Fe^{3+}）　Dissolve 5g with 35mL of distilled water in a 50mL color-comparison tube, add 5mL of dilute hydrochloric acid and approximate 30mg of ammonium persulfate, add 3mL of 30% ammonium thiocyanate（NH_4SCN）solution and sufficient water to produce 50mL, mix well. Any color produced is not more intense than that of a reference solution using 1.5mL of standard iron solution（0.0003%）.

$$Fe^{3+} + 3SCN^- \longrightarrow Fe（SCN）_3$$

Standard iron solution – Dissolve 863.0mg of ferric ammonium sulfate in water, add 2mL of dilute hydrochloric acid, and dilute with water to 1000mL. Pipet 10mL of this solution into a 100mL volumetric flask, add 0.5mL of dilute hydrochloric acid, dilute with water to volume, and mix. This solution contains the equivalent of 0.01mg of iron per ml.

（8）Sylvite（K^+）　Dissolve 5g of product with distilled water in a color-comparison tube, and adjust with 2 drops of acetic acid to a pH between 5 and 6. Add 2mL of 0.1mol/L sodium tetraphenylboron solution and dilute with water to 50mL, mix. Any opalescence produced is not more pronounced than that of a reference solution using 0.5mL of standard potassium sulfate solution（0.02%）.

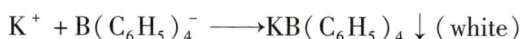

$$K^+ + B(C_6H_5)_4^- \longrightarrow KB(C_6H_5)_4 \downarrow (white)$$

Standard potassium sulfate solution Dissolve 2.2280g of potassium sulfate, previously dried at 105℃ and weighed accurately, with water in 1000mL volumetric flask, dilute to volume, and mix. This solution contains the equivalent of 1mg of potassium per ml.

Sodium tetraphenylboron solution Triturate 1.5g of sodium tetraphenylboron with 10mL of water, then add 40mL of water triturate again and filter.

（9）Heavy metals　Reference preparation – Into a 50mL color-comparison tube pipet 1mL of standard lead solution（20 μg of Pb）, add 2mL of dilute acetic acid and dilute with water to 25mL.

Test preparation – Dissolve 5g of the product with 23mL of distilled water in another 50mL color-comparison tube, add 2mL of dilute acetic acid.

Procedure – To each of the two tubes add 10mL of hydrogen sulfide TS, dilute to 50mL and mix. The tubes are allowed to stand for 10 minutes in dark place. If the color of the solution from the standard preparation is not darker than that of the solution from the standard preparation, it pronounces that the content of heavy metals is not more than the limits.

◆ Instructions

1. Requirements

（1）What impurities are contained in raw solid of sodium chloride? And how to remove these impurities?

（2）Why, during the purification of NaCl, should the agents be added sequentially: $BaCl_2$, Na_2CO_3, and HCl? Can we change the order of agents?

（3）How to remove the impurities of K^+, Br^-, I^- and other ions?

（4）To remove SO_4^{2+}, Ca^{2+}, Mg^{2+} etc as precipitate by addition corresponding precipitate agents, what influence does heating or not heating the solution have on the result? How to determine whether these ions are removed entirely?

（5）How to remove the excess precipitate agents: $BaCl_2$, Na_2CO_3 and NaOH?

（6）During the adjustment of pH of the solution with HCl, what can we deal with the excess HCl? Why should we adjust the solution to be weak acidic? Can we adjust the solution to be weak alkaline?

（7）Can we evaporate the condensed solution to dryness? Why?

（8）What is recrystallization? How much water should be appropriate to dissolve the product during recrystallization?

（9）Can tap water be used to dissolve the resulting product when we check the impurity limitation? Why?

2. Operation

（1）Precipitate impurity ions and check whether the ions precipitate completely

（2）Vacuum filter（filter with suction）

（3）Evaporate solution and crystallize

（4）Recrystallize

3. Notes

（1）Water used to dissolve the raw material should be proportional to the material. Too much water will prolong the following evaporation time.

（2）When we remove the impurities by addition precipitate agents, the time of boiling should not be too long. Otherwise, some of NaCl will separate from the hot solution. If some NaCl crystal, some few distilled water should be added to the solution.

（3）During evaporation, the crystal membrane of NaCl on the surface of the condensed solution should be ruptured with a parallel lying glass rod, otherwise, the crystal will splatter everywhere.

（4）Some soluble impurities, such as Br^-, I^- and K^+, will be removed away together with the mother liquid. So the solution should not be evaporated to dryness. Furthermore, the resulting crystal should be pressed with a glass stopper during filtration.

（5）Use the comparison tubes correctly. And compare the sample tubes with the corresponding standard tubes under same conditions.

4. Report format

（1）Objectives

（2）Principle

（3）Procedure（illustrated with a flow chart）

（4）The yield of refined NaCl and its percentage yield

（5）Write out the reaction equations of identification of Na^+, Cl^-. And illustrate the results of impurity limit tests with a table.

5. Questions Summarize the results of the tests and give conclusion on how to get a higher yield.

Words

medicinal	药用的	magnesium	镁
sodium chloride	氯化钠	comparison-tube	比色管
aqueous	水的	opalescence	乳白光
soluble	可溶解的	pronounced	显著的
insoluble	不可溶解的	concentrated	浓缩的
filtration	过滤	diluted	稀释的
filter	过滤	hydrochloric acid	盐酸
precipitate	沉淀	sulfuric acid	硫酸
precipitation	沉淀反应	acidic acid	醋酸
sulfate	硫酸盐	ammonia	氨

solubility	溶解性	ammonium	铵根（离子）
recrystallization	重结晶	hydroxide	氢氧化物
TS	（缩略语）试液	saturated	饱和的
zinc uranylacetate	醋酸铀酰锌	non-luminous flame	无色的火焰
bromothymol blue	溴麝香草酚蓝	acidity	酸度
sludge	淤泥	alkalinity	碱度
simmer	（使某物）保持	iodide	碘化物
	在接近沸点	bromide	溴化物
vacuum filter（filter with suction）	抽滤	chloroform	三氯甲烷
		agitation	搅动
tilt-pour process	倾泻法	triturate	粉碎
carbonate	碳酸盐	ammonium oxalate	草酸铵
carbonaceous deposit	碳质沉积	volumetric flask	容量瓶
curdy	凝乳状的	sodium tetraphenylboron	四苯硼钠
raw	未加工的	acetic acid	醋酸
platinum wire	铂丝	evaporate	蒸发
mother liquid	母液	control test	对照实验
barium	钡	pipet	移取
potassium	钾	membrane	膜
calcium	钙	glass stopper	玻璃塞

实验六　高锰酸钾的制备

PPT

实验目的 ●●●

1. 了解碱熔法分解矿石以及制备高锰酸钾的原理和方法。
2. 掌握浸取、减压过滤、蒸发结晶、重结晶等基本操作。
3. 学习气体钢瓶的使用操作或启普发生器的使用操作。

实验原理 ●●●

高锰酸钾的制备方法有多种，方法之一是以软锰矿（主要成分为 MnO_2）为原料制备高锰酸钾。制备过程一般分为两步，首先氧化制备锰酸钾，然后再将锰酸钾转化为高锰酸钾。根据 Mn 的电势图可知：

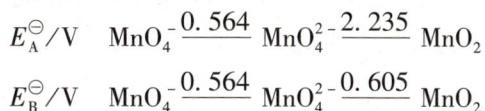

$$E_A^\ominus/V \quad MnO_4^- \frac{0.564}{} MnO_4^{2-} \frac{2.235}{} MnO_2$$

$$E_B^\ominus/V \quad MnO_4^- \frac{0.564}{} MnO_4^{2-} \frac{0.605}{} MnO_2$$

MnO_4^{2-} 不稳定，在酸性介质中极易发生歧化反应，而在碱性介质歧化反应趋势小，并且反应速度也慢，所以只适宜存在于碱性介质之中；因此将矿石转化为锰酸盐首选碱熔的方法。即将软锰矿在较强氧化剂氯酸钾存在下与碱共熔，先氧化为锰酸钾：

$$MnO_2 + KClO_3 + 6KOH \xrightarrow{熔融} 3K_2MnO_4 + KCl + 3H_2O$$

然后再将锰酸钾转化为高锰酸钾，一般可利用歧化反应或氧化的方法。如利用歧化反应，可加酸或通 CO_2 气体，使反应顺利进行，如 CO_2 法：

$$3MnO_4^{2-} + 2CO_2 \longrightarrow 2MnO_4^- + MnO_2 + 2CO_3^{2-}$$

反应后，过滤除去 MnO_2，再蒸发浓缩即可析出高锰酸钾的晶体。此方法操作简便，基本无污染，但锰酸钾的转化率仅为 2/3，其余 1/3 则转变为 MnO_2。

通过重结晶可获得精制的高锰酸钾（溶解度为：60℃ 22.1g/100g 水；20℃ 6.34g/100g 水；0℃ 2.83g/100g 水）。

实验要求 •••

1. 设计高锰酸钾的制备方法。明确所用仪器，并用流程图表示制备过程。
2. 方案经教师修改后，完成制备实验，并将产品重结晶。
3. 计算产率。
4. 完成实验报告（实验原理、实验过程、结果与讨论）。

Experiment 6   Synthesis of Potassium Permanganate

◆ Objectives

1. To understand the decomposition of ore using alkali fusion method, and the principles and procedures to synthesize potassium permanganate.

2. To practice basic operations such as extraction, vacuum filtration, evaporation crystallization and recrystallization.

3. To learn to use gas cylinders and reaction apparatus.

◆ Principles

There are multiple methods to synthesize potassium permanganate, and one of them is by using pyrolusite (with MnO_2 as the main component). The synthesis consists of two steps: first, synthesis of potassium manganate by oxidation reactions, followed by conversion of potassium manganate to potassium permanganate. From the electric potential chart, it is known that:

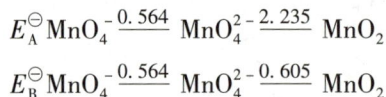

$$E_A^\ominus \; MnO_4^- \xrightarrow{0.564} MnO_4^{2-} \xrightarrow{2.235} MnO_2$$
$$E_B^\ominus \; MnO_4^- \xrightarrow{0.564} MnO_4^{2-} \xrightarrow{0.605} MnO_2$$

MnO_4^{2-} is unstable and readily disproportionates in acidic medium. It is less likely to disproportionate in alkaline medium, and the reaction is much slower. Therefore, it is only suitable to be stored in alkaline medium. Alkaline fusion is the most desired method to convert ore to manganates. That is, in the presence of a strong oxidizing agent such as potassium chlorate, pyrolusite is first fused with alkaline to form potassium manganate:

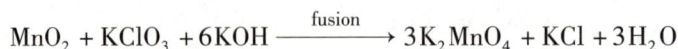

$$MnO_2 + KClO_3 + 6KOH \xrightarrow{fusion} 3K_2MnO_4 + KCl + 3H_2O$$

Then, potassium manganate is converted to potassium permanganate with disproportionation reaction or oxidation reaction. If disproportionation reaction is used, acid or CO_2 gas can be added to promote the reaction. For the CO_2 method:

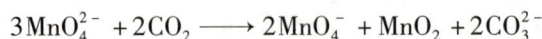

$$3MnO_4^{2-} + 2CO_2 \longrightarrow 2MnO_4^- + MnO_2 + 2CO_3^{2-}$$

After the reaction, MnO_2 is removed by filtration and potassium permanganate crystals are collected through evaporation concentration. This method is simple and essentially free of contamination. However, the conversion of potassium manganate is only around 2/3 with the rest 1/3 converted to MnO_2.

High purity potassium permanganate is obtained through recrystallization (solubility: 60℃, 22.1g/100g of water; 20℃, 6.34g/100g of water; 0℃, 2.83g/100g of water).

◆ Requirements

(1) Use above method to design the synthesis of potassium permanganate. Identify experimental equipment and demonstrate the synthesis procedures with a flow chart.

(2) After the experimental design being approved by the instructor, carry out the experiment and recrystallize the product.

(3) Calculate the yield.

(4) Complete the experiment report (Principles, Procedures, and Results & Discussion).

实验七　三草酸合铁(Ⅲ)酸钾的制备、组成测定及表征 📱微课4

PPT

实验目的

1. 掌握配合物的制备、定性、化学分析等基本操作。
2. 通过设计性实验的基本训练，增强分析问题与解决问题的能力。

实验原理

三草酸合铁（Ⅲ）酸钾易溶于水（溶解度0℃，4.7g/100g；100℃，117.7g/100g），难溶于乙醇。110℃下可失去全部结晶水，230℃时分解。此配合物对光敏感，受光照射分解变为黄色：

$$2K_3[Fe(C_2O_4)_3] \xrightarrow{光} 3K_2C_2O_4 + 2FeC_2O_4 + 2CO_2$$

因其具有光敏性，所以常用来作为化学光量计。另外它也是一些有机反应良好的催化剂。其合成工艺路线有多种，方法之一是首先由硫酸亚铁铵与草酸反应制备草酸亚铁：

$$(NH_4)_2Fe(SO_4)_2 \cdot 6H_2O + H_2C_2O_4 \longrightarrow FeC_2O_4 \cdot 2H_2O \downarrow + (NH_4)_2SO_4 + H_2SO_4 + 4H_2O$$

然后在过量草酸根存在下，用过氧化氢氧化草酸亚铁即可得到三草酸合铁（Ⅲ）酸钾，同时有氢氧化铁生成：

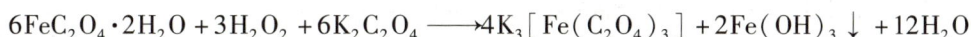

$$6FeC_2O_4 \cdot 2H_2O + 3H_2O_2 + 6K_2C_2O_4 \longrightarrow 4K_3[Fe(C_2O_4)_3] + 2Fe(OH)_3 \downarrow + 12H_2O$$

加入适量草酸可使Fe(OH)₃转化为三草酸合铁（Ⅲ）酸钾配合物：

$$2Fe(OH)_3 + 3H_2C_2O_4 + 3K_2C_2O_4 \longrightarrow 2K_3[Fe(C_2O_4)_3] \cdot 3H_2O$$

再加入乙醇，放置即可析出产物的结晶。其后几步总反应式为：

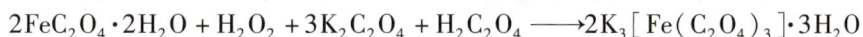

$$2FeC_2O_4 \cdot 2H_2O + H_2O_2 + 3K_2C_2O_4 + H_2C_2O_4 \longrightarrow 2K_3[Fe(C_2O_4)_3] \cdot 3H_2O$$

实验要求

1. 用上述方法设计三草酸合铁（Ⅲ）酸钾的制备方法。明确所用仪器，并用流程图表示制备过程。

2. 方案经教师修改后，完成制备实验，并定性检定 K^+、Fe^{3+}、$C_2O_4^{2-}$。

3. 计算产率。

4. 完成实验报告（实验原理、实验过程、结果与讨论）。

Experiment 7   Synthesis, Composition Analysis and Characterization of Potassium Trioxalatoferrate (Ⅲ)

◆ Objectives

1. To learn basic operations such as synthesis, characterization and chemical analysis of coordination compounds.

2. To develop analytical and problem solving skills through the practice of experimental design.

◆ Principles

Potassium trioxalatoferrate (Ⅲ) is readily soluble in water (solubility: $0℃, 4.7g/100g$; $100℃, 117.7g/100g$) and insoluble in ethanol. It completely dehydrates at $110℃$ and decomposes at $230℃$. This coordination compound is sensitive to light and decomposes to yellow color when exposed to light:

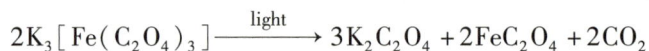

$$2K_3[Fe(C_2O_4)_3] \xrightarrow{\text{light}} 3K_2C_2O_4 + 2FeC_2O_4 + 2CO_2$$

Because of its sensitivity to light, it is commonly used as a chemical photometric agent. Additionally, it is an effective catalyst for some organic reactions. There are many routes for its synthesis, one of which is the preparation of ferrous oxalate through the reaction between ammonium ferrous sulfate and oxalate acid:

$$(NH_4)_2Fe(SO_4)_2 \cdot 6H_2O + H_2C_2O_4 \longrightarrow FeC_2O_4 \cdot 2H_2O \downarrow + (NH_4)_2SO_4 + H_2SO_4 + 4H_2O$$

Then, with the presence of excess amount of oxalate anions, oxidation of ferrous oxalate by hydrogen peroxide produces potassium trioxalatoferrate and ferric hydroxide:

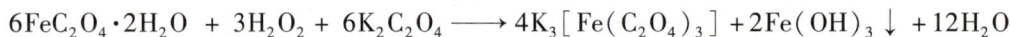

$$6FeC_2O_4 \cdot 2H_2O + 3H_2O_2 + 6K_2C_2O_4 \longrightarrow 4K_3[Fe(C_2O_4)_3] + 2Fe(OH)_3 \downarrow + 12H_2O$$

$Fe(OH)_3$ can be converted to trioxalatoferrate coordination compound with proper amount of oxalate acid:

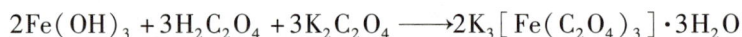

$$2Fe(OH)_3 + 3H_2C_2O_4 + 3K_2C_2O_4 \longrightarrow 2K_3[Fe(C_2O_4)_3] \cdot 3H_2O$$

The product crystallizes after the addition of ethanol. The overall chemical equation for the final few steps of reaction is:

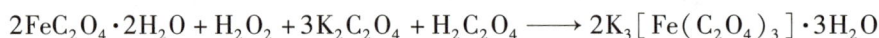

$$2FeC_2O_4 \cdot 2H_2O + H_2O_2 + 3K_2C_2O_4 + H_2C_2O_4 \longrightarrow 2K_3[Fe(C_2O_4)_3] \cdot 3H_2O$$

◆ Requirements

1. Design the experiment to synthesize potassium trioxalatoferrate (Ⅲ) using above reactions. Identify experimental equipment and demonstrate the synthesis procedures with a flow chart.

2. After the experimental design being approved by the instructor, carry out the experiment and qualitatively identify K^+, Fe^{3+} and $C_2O_4^{2-}$.

3. Calculate the yield.

4. Complete the experiment report (Principles, Procedures, and Results & Discussion).

书网融合……

微课1   微课2   微课3   微课4   习题

第六章　化学分析实验

实验一　称量练习和溶液的配制

PPT

实验目的 •••

1. 掌握直接称量法、减重称量法和固定重量称量法；掌握溶液配制的方法。
2. 了解测量仪器的精密度和有效数字的意义。
3. 学习电子分析天平的使用方法；学习如何正确规范记录实验原始数据。

实验原理 •••

1. 天平是一种重量计量的仪器，可以承载称量规定大小的物体。由于不同实验对质量准确度的要求不同，所以称量时会使用不同精密度的天平。在基础化学实验中常用的天平有托盘天平和分析天平。托盘天平一般可精确称量至0.1g，通常用作粗略的称量。分析天平的质量称量结果要比托盘天平准确和精密，普通分析天平可以精确称量至0.0001g。电子分析天平以其操作简便、称量准确可靠等优点在实际中应用广泛，是许多实验室必备的实验仪器。当被称量物放在电子天平的秤盘上后，几乎立即就能显示出质量数值，它还可以与打印机、计算机、记录仪等联用。

2. 实验中通常对不同溶液的浓度精密度有不同要求，例如，调节酸度用的硫酸溶液与定量分析中用的硫酸标准溶液在浓度精度要求上是不同的，在配制不同浓度精密度要求的溶液时要采用不同方法进行配制。配制浓度准确度要求不高的溶液时，用托盘天平、量筒和烧杯等低准确度的仪器配制就可以满足实验的要求。而配制浓度准确度要求高的溶液时，实验时所称取的质量和体积都要求精确，这就要用到分析天平、容量瓶、移液管等高准确度的仪器。

仪器和药品 •••

1. **仪器**　电子分析天平、托盘天平、称量瓶、烧杯（50mL，100mL）、容量瓶（100mL）。
2. **药品**　NaCl（摩尔质量58.45g/mol）固体。

实验内容 •••

1. 分析天平称量练习

（1）称量准备　称量前应先检查天平放置是否水平，天平箱内是否清洁等。

（2）直接称量　将需要称量的物品，如洁净干燥的小烧杯、称量瓶、坩埚等，直接置于天平秤盘上称量读数。

（3）固定重量称量　固定重量称量法适用于称量不宜吸水、在空气中稳定的样品。先将空的称量容器（如称量纸、称量瓶或小烧杯等）放在天平秤盘上称量，再用干净的药匙把样品慢慢地加入至称量容器内，称得样品加容器的总质量，则样品的质量等于总质量减去空的容器质量。电子分析天平有除皮功能，使用电子天平称量时可用"除皮键"简化称量过程。先将称量容器除皮，则加在容器中的样品质量可以直接从天平显示屏读出，不需要额外的计算。用固定重量称量法准确称取0.2g NaCl样品三份，要求称样量在±10%范围内。

（4）减重称量　减重称量法特别适合于在空气中不稳定的样品，如易吸潮、易氧化、易与CO_2起反应的样品，基准物质也常用这种方法称量。将样品装入干燥洁净的称量瓶中，先称出样品和称量瓶的总质量。然后，从称量瓶中定量倾出部分样品（如敲击的方法：取出称量瓶至容器上方，使瓶倾斜，打开称量瓶瓶盖，用瓶盖轻敲称量瓶瓶口上缘，使称量瓶中的样品慢慢倾出至容器中，估计达需称质量时，边敲边慢慢竖起称量瓶，使瓶口的样品落回到称量瓶中，盖上瓶盖）至容器中，再次称量出称量瓶和剩余样品的质量，则被倾出样品的质量等于这两个质量之差。用减重称量法准确称取 0.2g NaCl 样品三份，要求称样量在 ±10% 范围内。

2. 溶液的配制

（1）配制 100mL 浓度为 0.1mol/L 的 NaCl 溶液 A　用托盘天平称取 NaCl 固体 0.6g，置于 250mL 烧杯中，用量筒量取 100mL 蒸馏水加入，搅拌至溶解。

（2）配制 100mL 浓度为 0.1000mol/L 的 NaCl 溶液 B　用分析天平准确称取 NaCl 固体 0.5845g，置于 50mL 小烧杯中，加约 20mL 蒸馏水搅拌至溶解，将溶液完全、定量地转移至 100mL 容量瓶中，最后用蒸馏水稀释定容至容量瓶刻度线，塞紧瓶塞，摇匀。

实验数据记录与处理示例 •••

1. 直接称量，小烧杯的质量 $m_{烧杯}$ =
2. 固定重量称量法称取三份 0.2g NaCl 样品

	I	II	III
器皿质量（m_1/g）			
（NaCl + 器皿）质量（m_2/g）			
NaCl 的质量（g）			

3. 减重称量法称取三份 0.2g NaCl 样品

	I	II	III
（称量瓶 + 样品）质量（倾出前）m_1/g			
（称量瓶 + 样品）质量（倾出后）m_2/g			
NaCl 的质量（$m_1 - m_2$）/g			

4. 溶液的配制

	NaCl 溶液 A	NaCl 溶液 B
NaCl 的质量（m/g）		
溶液的体积（V/mL）		
溶液的浓度（mol/L）		

注意事项 •••

1. 阅读材料：本教材第一部分第二章的相关内容，包括天平及称量、有效数字、量筒、容量瓶、实验记录。

2. 当天平调水平后，不要挪动天平，也不要用重物撞击天平台。

3. 称量时，不得将需称量的药品、试剂直接放在天平秤盘上。用于称量的器皿一定要清洁、干燥。要用称量纸或表面皿称量粉末样品，不要用手直接接触被称量物品，要使用镊子、戴手套或用纸，防止手上污物沾到称量物上。

4. 热的样品要放冷后才能在天平上进行称量。

5. 读取称量数据时一定要关上天平门。

6. 称取易吸水的样品时要快速，尽量使用减重称量法称量。

7. 平行称量时，要采用相同的称量过程。

8. 称量时不要将药品撒在桌面，更不要将药品洒到天平里或天平周围！如果不慎将药品洒在天平里，要及时报告老师处理。

9. 称量和配制溶液时，要根据实验的要求选择不同精密度的仪器（分析天平或托盘天平、量筒或容量瓶）。

思考题 •••

1. 为什么在称量时，天平箱内必须保持清洁、干燥？称量用的器皿外壁为什么也必须保持清洁、干燥？

2. 为什么在读数时天平箱门必须是关闭的？

3. 减重称量时，如果样品洒在小烧杯外面时，对称量结果有什么影响？用药匙将称量瓶中的样品取到小烧杯里进行减重称量，对称量结果有什么影响？

4. 在称量记录和计算中，如何正确运用有效数字？

Experiment 1　Weighing Exercise and Preparing Solutions

◆ Purpose

1. To grasp the weighing methods: direct weighing, weighing by difference, weighing substance of fixed weight, and the methods of preparing solutions.

2. To know the precision of measuring apparatus and the significant digits.

3. To learn how to use the electronic analytical balance, and how to record the experimental raw data correctly and normatively.

◆ Principle

1. The analytical balance is an instrument for measuring weight, which can carry and weigh objects of a specified size. Due to the different requirements for quality accuracy in different experiments, different precision balances will be used for weighing. The commonly used balances in general chemistry experiments include table balances and analytical balances. Table balances can accurately weigh to 0.1g and are typically used for rough weighing. The mass weighing result of an analytical balance is more accurate and precise than that of a table balance. A regular analytical balance can accurately weigh to 0.0001g. Electronic analytical balances are widely used in practice due to their advantages of simple operation, accurate and reliable weighing. They are essential experimental instruments in many laboratories. When the weighed object is placed on the weighing plate of the electronic analytical balance, the mass value can be displayed almost immediately. It can also be used in conjunction with printers, computers, recorders, etc.

2. In many experiments, there are usually different requirements for the concentration precision of different solutions. For example, the precision requirements are different for the sulfuric acid solution used in adjusting

acidity and the sulfuric acid standard solution used in quantitative analysis, so different methods used to prepare the solutions should satisfy the concentration precision requirements. When preparing a solution with low concentration accuracy, low accuracy instruments such as a table balance, measuring cylinder, and beaker can meet the experimental requirements. However, when preparing solutions with high concentration accuracy, the mass and volume measured during the experiment must be accurate, which requires the use of high precision instruments such as analytical balances, volumetric flasks, and pipettes.

◆ Equipment and Chemicals

1. Equipment   electronic analytical balance; table balance; weighing bottle; beaker(50mL or 100mL); voumeri fask(100mL).

2. Chemicals   NaCl (M = 58.45g/mol).

◆ Procedure

1. Weighing exercise

(1)Weighing preparation   Before weighing, check whether the balance is placed horizontally and whether the balance pan is clean.

(2)Direct weighing   Place the objects, such as clean and dry small beakers, weighing bottles, crucibles, etc., directly on the balance pan, read its weight directly.

(3)Fixed weight weighing   The fixed weight weighing method is suitable for weighing samples that are stable in the air. Firstly, place an empty weighing container (such as weighing paper, weighing bottle, or small beaker) on the balance pan for weighing, and then slowly add the sample with a clean spatula to the weighing container. Weigh the total mass of the sample plus the container, and the mass of the sample is equal to the total mass minus the mass of the empty container. The weighing is simplied by using the key of "tare" when the electronic balance is applied. The weight of the sample can be read directly in the container when you previously tare the contalner. Accurately weigh three parts of 0.2g NaCl sample using fixed weight weighing method with a requied weight range of ±10%.

(4)Weighting by difference   The method of weighing by difference is particularly suitable for samples that are unstable in the air, such as samples that are prone to moisture absorption, oxidation, or reaction with CO_2. Reference materials are also commonly weighed using this method. The sample is placed in a dry and clean weighing bottle and the sample in the wcighing botle is weighed. Then quantitatively pour out a portion of the sample from the weighing bottle (such as tapping method: take out the weighing bottle to the top of the container, tilt the bottle, open the weighing bottle cap, gently tap the upper edge of the weighing bottle mouth with the cap, let the sample in the weighing bottle slowly pours out into the container. When the sample tapped out is estimately reaches the required weight, slowly lift the weighing bottle while tapping, let the sample in the bottle mouth falls back into the weighing bottle, and cover the cap) into the container. The weighing bottle and sample are reweighed. From the differere between the two weights, the weight of the tapped out sample is calculated. Weigh accurately three parts of 0.2g NaC1 by difference with a required sample weight range of ±10%.

2. Preparing solutions

(1)Preparing solution A: 100mL 0.1mol/L NaCl solution

Weigh 0.6g NaCl by a table balance and place it into a 250mL beaker, then measure 100mL distilled water by a graduated cylinder and add the water into the beaker, stir the mixture until the solid of NaCl is dissolved.

(2)Preparing solution B: 100mL 0. 1000mol/L NaCl solution

Weigh accurately 0. 5845g NaCl. Pour it into a small beaker, add about 20mL water, stir the mixture until the solid is dissolved completely, then quantitatively transfer the solution to a 100mL volumetric flask, dilute it with distilled water to the scale and shake well.

◆ Demonstration of experimental data record and analysis

a. Direct weighing   weight of the small beaker $m_{beaker} =$

b. Weighting three parts of 0. 2g NaCl by fixed weight weighing

	I	II	III
the weight of the container(m_1/g)			
the weight of(NaCl + container) (m_2/g)			
the weight of NaCl (g)			

c. Weighting three parts of 0. 2g NaCl by difference

	I	II	III
the weight of(bottle + sample)before tapping(m_1/g)			
the weight of(bottle + sample)after tapping(m_2/g)			
the weight of NaCl ($m_1 - m_2$) /g			

d. Preparing solutions

	NaCl solution A	NaCl solution B
the weight of NaCl(m/g)		
the volume of solution(V/mL)		
the concentration of the solution(mol/L)		

◆ Notes

1. Reading materials: Part 1 Chapter 2 in the testbook, including digital balance and weighing, significant digit, graduated cylinder, volumetric flask, notebook.

2. Do not move the balance or hit the platform with heavy objects after leveling.

3. Do not directly place any chemical on the balance pan. The utensils used for weighing must be clean and dry. Mass powders on weighing paper or dishes. Handle the weighed objects with tongs, tweezers, gloves, or paper to prevent fingerprints.

4. Hot objects must be cooled before being weighed on a balance.

5. Be sure to close the chamber door when reading.

6. Weigh hygroscopic materials rapidly or by difference method.

7. When weighing in parallel, the same weighing process should be used.

8. Do not leave the spilled chemicals on or around the balance! Report any spill to the teacher for handling in a timely manner.

9. When weighing or preparing solutions, instruments with different precision (e. g. , analytical balance or

table balance, measuring cylinder or volumetric flask) should be used according to the experimental requirements.

◆ **Questions**

1. Why must the chamber always be clean and dry? Why must the out wall of the weighing container be kept clean and dry?

2. Why should the chamber door be close when reading?

3. What is the effect when a litle of sample is spilled outside of the weighing container? What is the effect of using a spoon to take the sample out from the weighing bottle when weighing by difference?

4. How to correctly use sigificant figures in weighing records and calculations?

实验二 氯化钡结晶水的测定

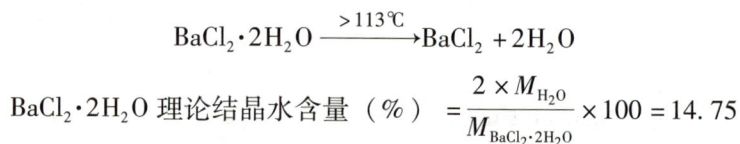
PPT

实验目的

1. 学习干燥器的使用。
2. 进一步熟悉电子天平的使用。
3. 掌握挥发重量法测定结晶水含量的原理和方法。
4. 明确恒重的意义。

实验原理

结晶水是水合结晶物质中结构内部的水，加热到一定温度可以失去。物质不同，失去结晶水的温度也不同。$BaCl_2 \cdot 2H_2O$ 中结晶水的蒸气压在20℃时为0.17 kPa，35℃时为1.57 kPa。所以氯化钡除了在特别干燥的环境中外，一般情况下，均含2分子结晶水。$BaCl_2 \cdot 2H_2O$ 失去结晶水温度为113℃，由于无水氯化钡不挥发也不易变质分解，故干燥温度可高于113℃。

$$BaCl_2 \cdot 2H_2O \xrightarrow{>113℃} BaCl_2 + 2H_2O$$

$$BaCl_2 \cdot 2H_2O \text{ 理论结晶水含量（\%）} = \frac{2 \times M_{H_2O}}{M_{BaCl_2 \cdot 2H_2O}} \times 100 = 14.75$$

本实验应用挥发重量法，根据 $BaCl_2 \cdot 2H_2O$ 干燥后所减失的重量可计算出 $BaCl_2 \cdot 2H_2O$ 样品结晶水的含量。计算公式如下：

$$\text{结晶水含量（\%）} = \frac{\text{样品失重（g）}}{\text{样品重量（g）}} \times 100$$

仪器和药品

1. 仪器 电子分析天平，称量瓶（扁平型），电热干燥箱，坩埚钳，干燥器。

2. 药品 $BaCl_2 \cdot 2H_2O$ 样品（A.R.）。

实验步骤

1. 称量瓶恒重 取直径约为3cm的扁平型称量瓶3只，洗净，于115℃电热干燥箱中干燥1h后，再置于干燥器中放冷至室温（30min）后，用电子天平称重。在上述条件下重复干燥、放冷并称重，至连续两次干燥的重量差异在0.3mg以下，即为恒重。

2. 样品结晶水测定　将 $BaCl_2 \cdot 2H_2O$ 样品放置研钵中研成粗粉，在台秤上粗称出 3 份试样，每份约 1g，置已恒重的称量瓶中，于电子天平准确称重。

将称量瓶中的样品尽量平铺于瓶底（厚度不超过 5mm），称量瓶盖斜放于瓶口以利于通气，然后置称量瓶于炉温达 115℃ 电热干燥箱中干燥 1h，移至干燥器中，放置 30min，冷至室温，盖好称量瓶盖，准确称量其重量。再重复上述操作，直至恒重。

实验数据记录与处理示例 ●●●

1. 原始数据

（1）空称量瓶恒重（g）

Ⅰ　20.0241	Ⅱ　17.2232	Ⅲ　18.8541
20.0240	17.2233	18.8540

（2）称量瓶加样重量（g）

Ⅰ　21.1088	Ⅱ　18.1947	Ⅲ　19.9071

（3）干燥恒重（g）

Ⅰ　20.9495	Ⅱ　18.0518	Ⅲ　19.7518
20.9484	18.0519	18.7516
20.9482		

2. 数据处理

		Ⅰ	Ⅱ	Ⅲ
空称量瓶恒重/g	第一次干燥	20.0241	17.2232	18.8541
	第二次干燥	20.0240	17.2233	18.8540
称量瓶加样品重/g		21.1088	18.1947	19.9071
样品重/g		1.0848	0.9715	1.0532
称量瓶加样品干燥恒重/g	第一次干燥	20.9495	18.0518	19.7518
	第二次干燥	20.9484	18.0519	18.7516
	第三次干燥	20.9482		
结晶水重/g		0.1606	0.1429	0.1556
结晶水含量/%		14.80	14.71	14.77
结晶水含量平均值/%		14.76		
相对平均偏差/%		0.2		

注意事项 ●●●

1. 要求恒重的称量，应注意平行原则，即同批次样品的实验条件，如扁形称量瓶（或加样品后）在烘箱中的干燥温度、在干燥器中冷却时间等，应保持一致。

2. 在称扁形称量瓶与样品时，要盖好称量瓶盖子，称量速度要快，以免称量过程中样品吸湿。

3. 在使用干燥器之前，应检查干燥剂是否失效。

4. 正确使用干燥器和坩埚钳。干燥器打开或盖上时应采用推开方法。搬动干燥器应用双手拿干燥器两侧底和盖子的边缘，以免干燥器的盖子滑落打破。

5. 从烘箱中取出的扁形称量瓶置于干燥器中冷却时，先不要把干燥器的盖子盖严，10min 后再盖严，以防干燥器内空气冷却后形成负压使干燥器难于打开。

6. 样品要均匀地铺在扁形称量瓶底部,以便样品中的水分得到充分挥发。

思考题 •••

1. 什么是恒重?如何才能达到恒重?

2. 空称量瓶为何要干燥至恒重?

3. 粗颗粒样品为何需要研碎?是否研得越细越好?

Experiment 2　Determination of the Water of Crystallization in Barium Chloride

◆ Purpose

1. Learn how to use a desiccators.

2. Furtherpractise the usage of analytical balance.

3. Master the principle and method of determination the content of crystalline water by the volatile weight method.

4. Know the meaning of constant weight.

◆ Principle

Crystalline water is the water within the structure of hydrated crystalline substances, which can be lost when heated to a certain temperature. Different substances have different temperatures at which crystalline water can be lost. The vapor pressure of crystalline water in $BaCl_2 \cdot 2H_2O$ is 0.17 kPa at 20℃ and 1.57 kPa at 35℃. So, except in particularly dry environments, barium chloride generally contains 2 molecules of crystalline water. $BaCl_2 \cdot 2H_2O$ loses its crystalline water at 113℃. Anhydrous barium chloride is not volatile and is not easily decomposed, so the drying temperature can be higher than 113℃.

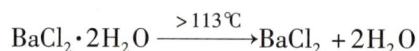

$$BaCl_2 \cdot 2H_2O \xrightarrow{>113℃} BaCl_2 + 2H_2O$$

$$\text{The theoretical content of crystallization water}(\%) = \frac{2 \times M_{H_2O}}{M_{BaCl_2 \cdot 2H_2O}} \times 100 = 14.75$$

This experiment uses the volatile weight method to calculate the content of crystalline water in the $BaCl_2 \cdot 2H_2O$ sample based on the weight loss after drying. The calculation formula is as follows:

$$\text{The practical content of crystallization water}(\%) = \frac{\text{weight lost(g)}}{\text{sample weight(g)}} \times 100$$

◆ Equipment and Chemicals

1. Equipment　electronic analytical balance; weighing bottles; dry oven; crucible clamp; desiccators.

2. Chemicals　$BaCl_2 \cdot 2H_2O$ (A.R.).

◆ Procedure

2. Dry the weighing bottles to constant weight

Take three flat weighing bottles of 3 candimeter, wash them and dry them in a drying oven at 115℃ for 1h, then place them in a desiccator to cool to room temperature (30 minutes) before weighing them using an electronic balance. Repeat drying, cooling, and weighing under the above conditions until the weight difference between two consecutive drying cycles does not exceed 0. 3mg, which is considered constant weight.

3. Determination of the sample's crystallization water

Place the $BaCl_2 \cdot 2H_2O$ sample in a mortar and grind it into powder. Roughly weigh three 1g samples on a platform scale. Place them in the weighing bottle dried to constant weight and accurately weigh them on an electronic balance.

Let the sample in the weighing bottle be a thin layer less than 5mm, place the weighing bottle cap diagonally at the bottle mouth to facilitate ventilation, then dry the weighing bottle in the drying oven at 115℃ for 1h. After the sample has been dried, allow them to cool in the desiccator for 30min, accurately weigh the weighting bottles. Repeat the above operation again until the constant weight.

◆ **Example of experimental data recording and data analysis**

1. Original data

(1)Constant weight of the empty weighing bottles(g)

Ⅰ	20. 0241	Ⅱ	17. 2232	Ⅲ	18. 8541
	20. 0240		17. 2233		18. 8540

(2)The weight of the sample and bottle(g)

Ⅰ	21. 1088	Ⅱ	18. 1947	Ⅲ	19. 9071

(3)Weight after drying to constant weight(g)

Ⅰ	20. 9495	Ⅱ	18. 0518	Ⅲ	19. 7518
	20. 9484		18. 0519		18. 7516
	20. 9482				

2. Data analysis

		Ⅰ	Ⅱ	Ⅲ
constant weight of the empty weighing bottles/g	first time	20. 0241	17. 2232	18. 8541
	second time	20. 0240	17. 2233	18. 8540
weight of the sample and bottle/g		21. 1088	18. 1947	19. 9071
weight of sample/g		1. 0848	0. 9715	1. 0532
weight of the sample and bottle after dry/g	first time	20. 9495	18. 0518	19. 7518
	second time	20. 9484	18. 0519	18. 7516
	third time	20. 9482		
weight of the crystallization water/g		0. 1606	0. 1429	0. 1556
the content of the crystallization water /%		14. 80	14. 71	14. 77
means/%		14. 76		
RAD/%		0. 2		

◆ **Notes**

1. When weighing with constant weight, pay attention to the parallel principle including the same drying temperature, cooling time and so on during the course of experiment.

2. Weigh sample quickly to avoid moisture absorption during weighing.

3. Use the desiccator and crucible clamp correctly. When open or close the desiccator, slide the lid sideways with a steady pace. When carry the desiccator, hold both edge of the bottom and the lid with both hands to prevent the lid from sliding off.

4. Before using the desiccator, check if the desiccant needs to be replaced.

5. When cooling the weighing bottle in the desiccator, do not cover the lid of the desiccator tightly first. After 10min, cover it tightly to prevent negative pressure caused by air cooling inside the desiccator.

6. In order to evaporate thoroughly, the sample should be put as a thin layer at the bottom of the weighing bottle.

◆ Questions

1. What does constant weight mean? How to get it?

2. Why must the empty weighing bottles be dried to constant weight?

3. Why should the sample be ground into powder? Is it better to grind it as finely as possible?

实验三　容量仪器的校正

PPT

实验目的 ●●●

1. 熟悉滴定管、容量瓶、移液管的操作方法。
2. 了解容量器皿校正的意义及学会滴定管、移液管的校正方法。
3. 学会容量瓶与移液管的相对校正。

实验原理 ●●●

容量瓶、移液管及滴定管是化学定量分析中常用的三种精密容量仪器。它们的容积在生产过程中已经检定，其所刻容积准确性可满足一般分析的要求。但在准确度要求较高的分析工作中，需对以上三种容器的容积进行进一步校正。对分析仪器常见的校正方法有直接（绝对）校正法和相对校正法两类。

1. 直接校正法　将滴定管、移液管或容量瓶所盛或移取的蒸馏水称重，通过水的密度和称量水的质量直接得到器皿的容积。校正一件玻璃器，需要称量它所容纳或放出的蒸馏水的质量，然后按其质量与密度的关系换算成20℃时的标准容积。

$$V_t = W_t / d_t$$

式中，V_t 为在温度 t 时水的容积，mL；W_t 为在空气中温度 t 时水的质量，g；d_t 为温度 t 时在空气中用黄铜砝码称量1mL 水（在玻璃容器中）的校正质量，g/mL。

校正时必须考虑下述三种因素：①水的密度随温度而改变；②空气浮力使重力改变；③玻璃容器本身的容积随温度而改变。

通过上述三项校正，可根据在温度 t 时容器所装入水的质量计算20℃时该容器的容积，即 t℃时容器所装水的计算容积恰好等于20℃时该容器的容积。为了便于计算，将20℃容量为1mL的玻璃容器在不同温度时所水的质量盛列于表6-1。

表 6-1 在不同温度下 1mL 的玻璃量各所量得的水在空气中的校正质量（用黄铜砝码）

温度 t/℃	d_t/（g/mL）	温度 t/℃	d_t/（g/mL）	温度 t/℃	d_t/（g/mL）
5	0.99853	14	0.99804	23	0.99655
6	0.99853	15	0.99792	24	0.99634
7	0.99852	16	0.99778	25	0.99612
8	0.99849	17	0.99764	26	0.99588
9	0.99845	18	0.99749	227	0.99566
10	0.99839	19	0.99733	28	0.99539
11	0.99833	20	0.99715	29	0.99512
12	0.99824	21	0.99695	30	0.99485
13	0.99815	22	0.99676		

2. 相对校正法 在有些实验中，我们常常只需要知道两种仪器间的相对体积比例关系，而不需要知道仪器的绝对体积，这时就会用到相对校正。例如，配套使用的容量瓶和移液管取一定比例的样品时，样品会溶解并稀释在一定体积的容量瓶中，然后用一支移液管移取溶液。在这个分析测定中，并不需要知道容量瓶和移液管的确切体积，而是需要移液管所放出的体积为容量瓶容积的一定比例。进行相对校正时，使用移液管移取比例数次的蒸馏水至干净且倒立晾干的容量瓶中，观察容量瓶中液体凹液面下缘，若跟刻度线不相切，在容量瓶的液体凹面的切线处做上记号线，此后该容量瓶和移液管配套使用时就用校正的记号线。

仪器和药品 •••

1. 仪器 酸式滴定管（25mL）、容量瓶（100mL）、移液管（20mL）、温度计、磨口塞锥形瓶（50mL）、洗耳球。

2. 药品 蒸馏水。

实验步骤 •••

1. 20mL 移液管的校正

（1）取一洁净且干燥的 50mL 磨口塞锥形瓶，在分析天平上称重，称准至 1mg。

（2）洗净一支移液管，练习移液管的使用方法。

（3）用待校正的移液管，移取蒸馏水至上述磨口锥形瓶中，再次称重，称准至 1mg。

（4）算出水的质量。测量实验时水温，按公式算出移波管的真实容积。

2. 25mL 滴定管的校正

（1）洗涤、润洗 25mL 滴定管，装入已测过温度的蒸馏水。

（2）取一干燥洁净的 50mL 具塞磨口锥形瓶，在分析天平上称重，称准至 1mg。

（3）将滴定管的液面调节至 0.00mL 刻度处。按滴定时常用速度（如每秒 3 滴）将水放入已称重的锥形瓶中，使其体积至 5.00mL 刻度处时，再次称重，得瓶加水的质量，称准至 1mg。

（4）算出水的质量，并查出 d_t，即可算出滴定管 0.00～5.00mL 刻度之间的真实容积。

（5）用上述方法继续校正滴定管 0.00～10.00mL，0.00～15.00mL，0.00～20.00mL，0.00～25.00mL 段的真实容积。

（6）重复校正一次，两次校正所得同一刻度的体积相差不应大于 0.02mL。

（7）算出各个体积处的校正值（两次平均值）。以读数值为横坐标，校正值为纵坐标作出校正曲线，以备滴定时查取。

（8）如校正 50mL 滴定管，方法同上，只是每隔 10mL 测一个校正值。现将水温 21℃时校正 25mL

滴定管的实验数据列于表 6 – 2 供参考。

表 6 – 2 25mL 滴定管的校正水温为 21℃ （1mL 水的质量 = 0.99695g）

滴定管计数/mL	瓶 + 水重/g	空瓶重/g	水重/g	真实体积/mL	校正值/mL
0.00 ~ 5.00	34.14	29.20	4.94	4.96	− 0.04
0.00 ~ 10.00	39.31	29.31	10.00	10.03	+ 0.03
0.00 ~ 15.00	44.30	29.35	14.95	15.00	0.00
0.00 ~ 20.00	49.39	29.43	19.96	20.02	+ 0.02
0.00 ~ 25.00	54.28	29.38	24.90	24.98	− 0.02

注：如果使用电子天平有除皮键，可以先将空瓶除皮，天平上可以直接读出水的质量。

3. 容量瓶与移液管的相对校正

用洗净的 20mL 移液管吸取蒸馏水，放入干燥洁净的 100mL 容量瓶中，共放入 5 次，观察容量瓶中弯月面下缘是否与体积刻度线相切。若不相切，记录弯月面下缘的位置。再重复上述实验一次。连续两次实验结果符合后，做出新标记。校正后，这对容量瓶与移液管即可在其他实验中配套使用。

注意事项

1. 永远不要想当然地认为仪器是干净的，除非是经过自己认真洗净的仪器。

2. 实验用的玻璃仪器要用洗涤剂洗涤，再用自来水冲洗，最后用蒸馏水润洗 2 ~ 3 次。

3. 用洗涤剂洗不干净的玻璃仪器，可用 1mol/L HNO₃ 溶液或 1mol/L KOH 的乙醇溶液浸泡，然后用洗涤剂洗，用自来水冲洗，最后用蒸馏水润洗 2 ~ 3 次。

4. 容量仪器绝不能加热！否则其准确度会受到影响。

5. 所有与质量相关的实验数据均需使用分析天平称量，分析天平可精确到 0.1mg，故所有原始记录的质量也必须精密到 0.1mg。

6. 每个学生都必须按照学号在固定的位置上使用天平。

思考题

1. 怎样检查滴定管是否洗净？使用未洗净的滴定管滴定，对测定结果有什么影响？

2. 校正 50mL 或 25mL 滴定管及 20mL 移液管时为什么要称准至 1mg？

3. 使用移液管的操作要领是什么？放完液体后为什么要停留 15s？最后留在管尖的液体如何处理？为什么？

4. 容量仪器的体积为什么规定为 20℃时的体积？

Experiment 3 Calibration of Volumetric Glassware

◆ Purpose

1. To mask the operation of burettes, volumetric flasks, and pipettes.

2. To understand the significance of glassware calibration and to learn the calibration methods.

3. To learn the relative calibration of volumetric flasks and pipettes.

◆ Principle

The volumetric flask, pipette, and burette are three precision volumetric glassware commonly used in chemical quantitative analysis. Their volume has been verified during the manufacturing process, and the accuracy of their engraved volume is enough for general analysis. However, further calibration is recommended for extremely accurate analysis. There are two common calibration methods for analytical glassware: direct (absolute) calibration and relative calibration.

1. Direct calibration Weigh the distilled water contained or taken from a burette, pipette, or volumetric flask, directly obtained the volume of the glassware by measuring the density and mass of the water. To calibrate a glassware, it is necessary to weigh the mass of distilled water it contains or releases, and then convert it into a standard volume at 20℃ based on its relationship between mass and density.

$$V_t = W_t / d_t$$

Where V_t (mL) is the volume of water at t℃, W_t (g) is the mass of water at t℃ in the air, d_t (g/mL) is the weight of 1mL water (in glassware) weighed by poise at t℃ in the air.

Three factors must be consideredwhen calibration: ①The density of water changes with temperature; ②The gravity changes with air buoyancy; ③The volume of the glassware itself changes with temperature.

After the above three corrections, the volume of the container at 20℃ can be calculated based on the mass of water filled in the container at temperature t. That is, the calculated volume of water filled in the container at t℃ is exactly equal to the volume of the container at 20℃. For the convenience of calculation, the mass of water in a glass container with a capacity of 1mL at different temperatures is listed in Table 6 - 1.

Table 6 - 1 The weight of 1mL water at different temperatures

t/℃	d_t/(g/mL)	t/℃	d_t/(g/mL)	t/℃	d_t/(g/mL)
5	0.99853	14	0.99804	23	0.99655
6	0.99853	15	0.99792	24	0.99634
7	0.99852	16	0.99778	25	0.99612
8	0.99849	17	0.99764	26	0.99588
9	0.99845	18	0.99749	227	0.99566
10	0.99839	19	0.99733	28	0.99539
11	0.99833	20	0.99715	29	0.99512
12	0.99824	21	0.99695	30	0.99485
13	0.99815	22	0.99676		

2. Relative calibration In some experiments, we often only need to know the relative volume ratio relationship between the two types of glassware without knowing its absolute volume. The relative calibration can be used. For example, when a volumetric flask and pipette were used as a pair to take a certain proportion of the sample, the sample might be dissolved and diluted in a certain volume of volumetric flask, and a pipette would then be used to transfer the solution. In this analysis, it is not necessary to know the exact volumes of the volumetric flask or the pipette, but rather to require the volume transfered by the pipette to be a certain proportion of the volumetric flask. When performing relative correction, a pipette is used to transfer a proportional amount of distilled water certain times to a clean and inverted dry volumetric flask. Observe the position of down - edge of meniscus. If it is not tangent to the scale line, make a new mark on the flask. When the volumetric flask and pipette is used as a pair, use the new mark.

◆　Equipment and Chemicals

1. Equipment　acidic burette(25mL), volumetric flask(100mL), pipette(20mL), thermometer, cone flask (50mL), rubber suction bulb.

2. Chemicals　distilled water.

◆　Procedure

1. Calibration of pipette（20mL）

（1）Weigh a clean and dry 50mL flask on analytical balance to the nearest milligram (1mg).

（2）Clean a 20 ml pipette and exercise its operation.

（3）Pipette with distilled water using rubber suction bulb, adjust the meniscus to the etched line, and allow the water to drain into the flask above, reweigh it to the nearest milligram (1mg).

（4）Measure the temperature of water and calculate the true volume of the pipette according to the equation.

2. Calibration of burette（25mL）

（1）Clean and rinse the burette(25mL), fill it with distilled water of known temperature.

（2）Obtain a 50mL clean and dry flask, weigh it on analytical balance to the nearest milligram (1mg).

（3）Adjust the meniscus of water to the etched line of 0.00mL at eye level. Drain the water slowly into the flask till the meniscus touched 5.00mL etched line. Reweigh the bottle to the nearest milligram (1mg).

（4）Calculate the true volume between 0.00mL and 5.00mL according to the equation.

（5）Repeat step (2) to (4) to calibrate the true volume of burette between 0.00 ~ 10.00mL, 0.00 ~ 15.00mL, 0.00 ~ 20.00mL and 0.00 ~ 25.00mL.

（6）Repeat the calibration and the volumetric difference should be within 0.02mL.

（7）Calculate the calibration value of each volume (average of the two times). The calibration line could be drawn with the volume as x – coordinate and calibration value as y – coordinate.

（8）The method is the same as above to calibrate a 50mL burette at a volume interval of 10mL. Table 6 – 2 shows experiment data of a 25mL burette calibration at 21℃.

Table 6 – 2　Experiment data of a 25mL burette calibration at 21℃

Volume from berette/mL	Weight of bottle + waer/g	Weight of empty bottle/g	Weight of water/g	True volume/mL	Correction/mL
0.00 ~ 5.00	34.14	29.20	4.94	4.96	− 0.04
0.00 ~ 10.00	39.31	29.31	10.00	10.03	+ 0.03
0.00 ~ 15.00	44.30	29.35	14.95	15.00	0.00
0.00 ~ 20.00	49.39	29.43	19.96	20.02	+ 0.02
0.00 ~ 25.00	54.28	29.38	24.90	24.98	− 0.02

Notes: If the electronic balance has "Tare" key, the empty bottle can be tared first, and the weight of water can be directly read.

3. Relative calibration of pipette and volumetric flask

Fill a clean 20mL pipette with distilled water and drain into a clean and dry 100mL volumetric flask for 5 times, observe weather the meniscus is tangent with the etched line. If it is not tangent, record the position of the meniscus. Repeat the step above. Make a new mark at the position if the meniscus of water is at the same place. The pipette and the flask can be used as a pair in other experiment.

◆ Notes

1. Never assume any glassware is clean unless you washed it yourself.

2. Clean all glassware with soap, tap water. Use distilled water as the final rinse of your glassware, preferably a double or triple rinse.

3. Glassware that will not come clean in soap and water should be cleaned by soaking in a solution of 1mol/L HNO₃, or 1mol/L KOH in alcohol, followed by a soap and water wash and distilled water rinse.

4. Never heat any piece of volumetric glassware. Heating will change the volume that will cause it not accurate anymore!

5. All items in this laboratory should be massed on the analytical balance. These balances read to the nearest 0.1mg. Thus, all masses recorded must be to the nearest 0.1mg.

6. You will be assigned a balance to use. You will use the same analytical balance for all measurements.

◆ Questions

1. How to check up whether buret is clean? What are the effects on the experiment results when we use an unclean buret?

2. Why we weigh it on analytical balance to the nearest milligram 1mg when calibrating 50mL or 25mL buret and 25mL pipette?

3. How to use pipette correctly? Why should we wait for 15 second after delivereing liquid from a buret? How to deal with a small amount of liquid remaining in the tip of the buret? Why?

4. Why is the volumetric glassware always calibrated at 20℃?

实验四　NaOH 标准溶液的配制与标定

PPT

实验目的 ●●●

1. 学会标准溶液的配制及用基准物质标定标准溶液浓度的方法。
2. 学会滴定管的使用和滴定终点的判断。

实验原理 ●●●

标准溶液是指已知准确浓度的试剂溶液，其配制方法通常有直接配制法和间接配制法两种。由于 NaOH 易吸收空气中的水分及 CO_2，因此，准确浓度的 NaOH 溶液不能够通过直接称量法称出计算量的 NaOH 并将其溶于定量水中制得，而只能用间接法配制，先配制成近似于所需浓度的溶液，然后再用基准物或标准溶液准确地测出它的准确浓度。

标定 NaOH 标准溶液的基准物质常用邻苯二甲酸氢钾，其具有稳定、不吸潮、易得到纯品、摩尔质量大等优点，可溶于水中产生钾离子和邻苯二甲酸氢根离子，后者含有一个可滴定的质子（$pK_a = 4.01$），其滴定反应如下：

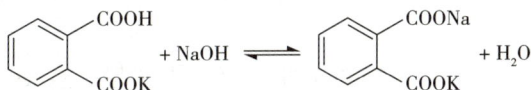

化学计量点时，溶液呈弱碱性，可选用酚酞指示剂指示滴定终点。根据滴定所消耗的 NaOH 体积，按下式计算 NaOH 标准溶液的浓度：

$$c_{\mathrm{NaOH}} = \frac{w_{\mathrm{KHC_8H_8O_4}}}{V_{\mathrm{NaOH}} \dfrac{M_{\mathrm{KHC_8H_8O_4}}}{1000}}$$

$$M_{\mathrm{KHC_8H_8O_4}} = 204.2\,\mathrm{g/mol}$$

仪器和药品

1. 仪器　锥形瓶（3×250mL），滴定管（25mL），量筒（10mL、100mL），移液管（20mL），聚乙烯或玻璃试剂瓶（2×500mL），橡皮塞，称量瓶。

2. 药品　NaOH 固体，酚酞指示剂：0.1% 乙醇溶液，邻苯二甲酸氢钾（基准物质）。

实验步骤

1. NaOH 标准溶液的配制　在台秤上称量 2.2g NaOH 固体，置于烧杯中，加入去离子水 500mL 溶解，转移至具橡皮塞的试剂瓶中，盖上瓶塞，充分摇匀，即得。

2. NaOH 标准溶液的标定　准确称取三份 0.45g 基准物质邻苯二甲酸氢钾于三个已经做了标记的 250mL 锥形瓶中，加入 50mL 去离子水或蒸馏水将其溶解，再在每份中加入 2～3 滴酚酞指示剂，用 NaOH 溶液滴定至溶液由无色到浅粉色，且 30s 内不褪色，即为滴定终点。

实验数据记录与处理示例

1. 原始记录　邻苯二甲酸氢钾的称量（g）

Ⅰ　$w_1 = 14.6758$　　　Ⅱ　$w_2 = 14.2200$　　　Ⅲ　$w_3 = 13.7639$

　　$\dfrac{w_2 = 14.2200}{0.4558}$　　　　$\dfrac{w_3 = 13.7639}{0.4561}$　　　　$\dfrac{w_4 = 13.2877}{0.4762}$

滴定所消耗的 NaOH 溶液体积（mL）

NaOH 终读数：　Ⅰ　21.78　　　Ⅱ　21.78　　　Ⅲ　22.77

NaOH 初读数：　　　0.00　　　　　0.00　　　　　0.00

2. 数据处理

实验编号	Ⅰ	Ⅱ	Ⅲ
基准物重/g	0.4558	0.4561	0.4762
V_{NaOH}/mL	21.78	21.78	22.77
C_{NaOH}/（mol/L）	0.1025	0.1026	0.1024
平均值/（mol/L）	0.1025		
相对平均偏差（RAD）/%	0.06		

注意事项

1. NaOH 具有腐蚀性！接触时要小心，避免和皮肤及衣服接触，如果不小心溅出，要立刻冲洗，实验结束后要认真洗手。

2. 固体 NaOH 应在表面皿或小烧杯中称量，不能在称量纸上称量。

3. 配制好的 NaOH 溶液，应立即在储液瓶上贴好标签，注明试剂名称、配制日期、使用者姓名等信息。

4. 滴定近终点时，须用少量蒸馏水吹洗锥形瓶内壁，将溅起的溶液淋下。

5. 滴定管在使用前应检漏，装满标准溶液前应用该溶液润洗 3 次，滴定之前应排尽气泡后再调节至零点。

6. 滴定管读数要估读到 0.01mL。

思考题 •••

1. 为什么 NaOH 不能作为基准物质？

2. 为什么要用沸腾蒸馏水配制 NaOH 溶液？

3. 为什么用 50% NaOH 溶液配制 0.1mol/L NaOH 溶液比用固体 NaOH 配制 0.1mol/L NaOH 溶液好？

4. 滴定时，若 NaOH 溶液不慎滴到实验台上而没有滴到锥形瓶中，对计算 NaOH 溶液浓度有影响吗？为什么？

5. 什么是滴定终点（或化学计量点）？滴定时如何得知已经到达滴定终点？

6. 请说明，如果滴定终点滴过头（粉色太深），对计算 NaOH 溶液的浓度有什么影响？

Experiment 4   Preparation and Standardization of Standard Sodium Hydroxide Solution

◆ Purpose

1. To learn how to prepare and standardize NaOH solution with reference substance.

2. To learn how to use a burette and determine the titration end point.

◆ Principle

Standard solutionis a solution with known accurate concentration. There are usually two methods for preparing standard solutions: direct preparation and indirect preparation. Due to the easy absorption of water and CO_2 in the air by solid NaOH, an accurate concentration of NaOH solution cannot be prepared by directly weighing the calculated amount of NaOH and dissolving it in quantitative water, and can only be prepared indirectly. Usually, sodium hydroxide solutions can only be prepared by indirect method. First a solution of approximately the desired concentration is prepared, and then its accurate concentration is determind using a reference substance or standard solution.

The reference substance for standardization of NaOH standard solutions is commonly potassium hydrogen phthalate, which has the advantages of stability, non hygroscopicity, easy to obtain pure products, and high molar mass. It is soluble in water to produce potassium ions and hydrogen phthalate ions, which contain a titratable proton ($pK_a = 4.01$). The titration reaction is as follows:

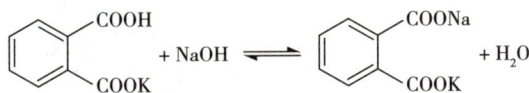

At the stoichiometric point, the solution is weakly alkaline, and a phenolphthalein indicator can be used to indicate the titration endpoint. According to the volume of NaOH consumed by titration, the concentration of NaOH standard solution can be calculated using the following equation:

$$c_{NaOH} = \frac{w_{KHC_8H_8O_4}}{V_{NaOH}\dfrac{M_{KHC_8H_8O_4}}{1000}}$$

$$M_{KHC_8H_8O_4} = 204.2 \text{g/mol}$$

◆ Equipment and Chemicals

1. Equipment Conical flasks（3 × 250mL）, buret（25mL）, graduated cylinder（10mL, 100mL）, pipet, （20mL）, polyethylene bottles or glass bottles,（2 × 500mL）, weighing bottle, Rubber stopper.

2. Chemicals sodium hydroxide（solid）; phenolphthalein solution（0.1% in ethanol）; potassium hydrogen phthalate solid, primary standard.

◆ Procedure

1. Preparation of a standard NaOH solution

Weigh 2.2g of NaOH solid into a beaker, add 500mL of deionized water to dissolve, transfer it to a reagent bottle with a rubber stopper, and shake well.

2. Standardization of NaOH standard solution

Accurately weigh three portions of 0.45g reference substance potassium hydrogen phthalate to the nearest 0.1mg into three labeled 250mL conical flasks. Add 50mL of deionized or distilled water to dissolve it, and then add 2 ~ 3 drops of phenolphthalein indicator to each portion. Titrate with NaOH solution until the solution changes from colorless to light pink and does not fade within 30 seconds.

◆ Demonstration of experimental data record and analysis

1. Original data

Weighing of potassium hydrogen phthalate（g）

Ⅰ $w_1 = 14.6758$ Ⅱ $w_2 = 14.2200$ Ⅲ $w_3 = 13.7639$

$\underline{w_2 = 14.2200}$ $\underline{w_3 = 13.7639}$ $\underline{w_4 = 13.2877}$

0.4558   0.4561   0.4762

Volume of NaOH solution consumed for titration（mL）

NaOH final reading：Ⅰ 21.78   Ⅱ 21.78   Ⅲ 22.77

NaOH initial reading：   0.00   0.00   0.00

2. Data analysis

No.	Ⅰ	Ⅱ	Ⅲ
Reference weight/g	0.4558	0.4561	0.4762
V_{NaOH}/mL	21.78	21.78	22.77
C_{NaOH}/（mol/L）	0.1025	0.1026	0.1024
Average of C_{NaOH}/（mol/L）	0.1025		
Relative Average Deviation（RAD）/%	0.06		

◆ Notes

1. NaOH is very corrosive! Handle this solution very carefully, and avoid contact with skin and clothing. If accidentally splashed, rinse immediately and wash hands carefully after the experiment.

2. Solid NaOH should be weighed on a watch glass or small beaker, not on a weighing paper.

3. The prepared NaOH solution should be immediately labeled on the storage bottle, indicating informations such as the reagent name, preparati on date, and user name.

4. When the titration is approaching the end point, a small amount of distilled water must be used to rinse the splashed solution on the inside wall of the conical flask off.

5. The burette should be leak tested before use. Before filling it with a standard solution, it should be rinsed three times. Before titration, all bubbles should be drained before adjusting to zero.

6. The burette reading should be read to 0. 01 mL.

◆ Questions

1. Why cannot NaOH be used as the reference substance?

2. Why use the boiling distilled water to prepare NaOH solution?

3. Why is it better to use a 50% NaOH solution, instead of solid NaOH, to prepare the 0. 1 mol/L NaOH solution?

4. If some NaOH solution accidentally drops onto the table instead of into the flask during titration, how would this effect the final concentration of NaOH? Why?

5. What is the definition of titration end point (or equivalence point)? How do you know this point has been reached?

6. If the endpoint in the titration is surpassed (too pink), what effect does this have on the calculated molarity of the NaOH solution? Why?

实验五　药用阿司匹林的含量测定

PPT

实验目的 ●●●

1. 掌握用酸碱滴定法测定阿司匹林含量的原理和操作。
2. 掌握酚酞指示剂的滴定终点。

实验原理 ●●●

阿司匹林（乙酰水杨酸）属于芳酸酯类药物，在水中微溶，在乙醇中易溶。阿司匹林分子结构中含有一个羧基，呈酸性，在25℃时 $K_a = 3.27 \times 10^4$，可用 NaOH 标准溶液在乙醇溶液中直接滴定，其滴定反应为：

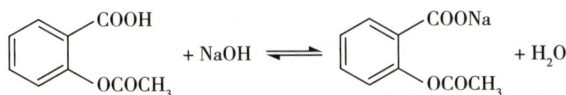

计量点时，溶液呈弱碱性，可选用酚酞指示滴定终点。根据试样量和 NaOH 标准溶液的浓度及用量，按下式计算出阿司匹林的含量：

$$\text{Aspirin}\% = \frac{(cV)_{\text{NaOH}} \times \dfrac{M_{\text{Aspirin}}}{1000}}{W_{\text{sample}}} \times 100 \qquad M_{\text{Aspirin}} = 180.2 \, \text{g/mol}$$

仪器和药品 ●●●

1. 仪器 锥形瓶（250mL×3），滴定管（25mL），量筒（10mL、100mL），烧杯（100mL），称量瓶。

2. 药品 阿司匹林原料药，NaOH 标准溶液（0.1mol/L），酚酞指示剂（0.1%乙醇溶液），中性乙醇（40mL 95%乙醇，加酚酞指示 8 滴，用 0.1mol/L NaOH 标准溶液滴定至粉红色即得）。

实验步骤 ●●●

精密称取阿司匹林样品 0.4g 于锥形瓶中，加中性乙醇 10mL 溶解后，用 NaOH 标准溶液（0.1mol/L）滴定至淡红色。平行测定三次。

实验数据记录与处理示例 ●●●

1. 原始记录

阿司匹林的称量（g）

$$
\begin{array}{lll}
Ⅰ \quad w_1 = 16.8845 & Ⅱ \quad w_2 = 16.4630 & Ⅲ \quad w_3 = 16.0588 \\
\underline{w_2 = 16.4630} & \underline{w_3 = 16.0588} & \underline{w_4 = 15.6646} \\
\quad\quad 0.4215 & \quad\quad 0.4042 & \quad\quad 0.3942
\end{array}
$$

滴定所消耗的 NaOH 溶液体积（mL）

NaOH 终读数：Ⅰ 22.80   Ⅱ 21.82   Ⅲ 21.25

NaOH 初读数：   0.00   0.00   0.00

2. 数据处理

实验编号	Ⅰ	Ⅱ	Ⅲ
阿司匹林重/g	0.4215	0.4042	0.3942
V_{NaOH}/mL	22.80	21.82	21.25
C_{NaOH}/(mol/L)	0.1025		
Aspirin/%	99.91	99.71	99.57
平均值/%	99.73		
相对平均偏差（RAD）/%	0.1		

注意事项 ●●●

1. 为避免样品水解，实验中应尽可能少用水，滴定终点应用中性乙醇代替水冲洗锥形瓶内壁。
2. 滴定终点溶液颜色应能持续 30s。

思考题 ●●●

1. 在 NaOH 滴定阿司匹林时，可能选择甲基橙指示滴定终点吗？
2. 为什么在滴定中所用的乙醇需对酚酞显中性？

Experiment 5   Determination of purity of aspirin

◆ Purpose

1. Master the principle and method of determining aspirin using acid – base titration.

2. Master the titration ending point of phenolphthalein.

◆ Principle

Aspirin (salicyclic acetate) is an aromatic ester drug that is slightly soluble in water and easily soluble in ethanol. The molecular structure of aspirin contains a carboxyl group, which is acidic. At 25℃, K_a of aspirin is 3.27×10^4, which can be titrated in alcoholic media by sodium hydroxide standard solution. The titration reaction is:

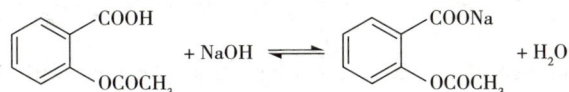

The ending point is weakly alkaline, so phenophthalein indicator is applied to indicate the titration endpoint. According to the sample weight, concentration and volume of NaOH standard solution, the content of aspirin can be calculated according to the following formula:

$$\text{Aspirin\%} = \frac{(cV)_{\text{NaOH}} \times \dfrac{M_{\text{Aspirin}}}{1000}}{W_{\text{sample}}} \times 100 \qquad M_{\text{Aspirin}} = 180.2 \text{g/mol}$$

◆ Equipment and Chemicals

1. Equipment　conical flasks (3 × 250mL), buret (25mL), graduated cylinder (10mL, 100mL), beaker (100mL), weighing bottle.

2. Chemicals　aspirin API, sodium hydroxide standard solution (0.1mol/L), phenolphthalein solution (0.1% in ethanol), neutral alcohol (40mL of 95% ethanol, add 8 drops of phenolphthalein indicator, and titrate with 0.1mol/L NaOH standard solution to pink).

◆ Procedure

Accurately weigh 0.4g of aspirin sample into a conical flask, dissolve with 10mL of neutral ethanol, and titrate with NaOH standard solution (0.1mol/L) until light pink. Perform three times in parallel.

◆ Demonstration of experimental data record and analysis

1. Original data

Weighing of aspirin (g)

I　$w_1 = 16.8845$　　　　Ⅱ　$w_2 = 16.4630$　　　　Ⅲ　$w_3 = 16.0588$

$\underline{w_2 = 16.4630}$　　　　　$\underline{w_3 = 16.0588}$　　　　　$\underline{w_4 = 15.6646}$

　0.4215　　　　　　　　　0.4042　　　　　　　　　0.3942

The volume of NaOH solution (mL)

NaOH final reading: I　22.80　　　　Ⅱ　21.82　　　　Ⅲ　21.25

NaOH initial reading:　0.00　　　　　　0.00　　　　　　0.00

2. Data analysis

Number	I	Ⅱ	Ⅲ
The weight of aspirin/g	0.4215	0.4042	0.3942
V_{NaOH}/mL	22.80	21.82	21.25
C_{NaOH}/(mol/L)	0.1025		
Aspirin/%	99.91	99.71	99.57
Average/%	99.73		
RAD /%	0.1		

◆ Notes

1. To avoid the hydrolysis of aspirin, water should be used as little as possible in the experiment. So neutral alcohol should be used instead of water to rinse the inner wall of the conical flask near the titration end point.

2. The color of the titration endpoint solution should last not less than 30s.

◆ Questions

1. Can we choose methyl orange as the indicator when titrating aspirin with NaOH?

2. Why does ethanol used in the titration need to be neutral to phenolphthalein?

实验六　HCl 标准溶液的配制与标定

PPT

实验目的 ●●●

1. 掌握 HCl 标准溶液的配制方法。

2. 掌握用无水碳酸钠作为基准物质标定盐酸标准溶液的原理及方法。

3. 掌握用比较法标定盐酸标准溶液的方法。

4. 正确判断甲基红 – 溴甲酚绿混合指示剂的滴定终点。

实验原理 ●●●

浓盐酸易挥发，无法直接配制准确浓度的标准溶液，因此配制盐酸标准溶液只能采用间接配制法，先配制大概浓度的溶液，再用基准物质标定或碱标准溶液比较的方法来确定其准确浓度。

1. 标定法 标定盐酸标准溶液的基准物质有无水碳酸钠、硼砂等。本实验采用无水碳酸钠为基准物质，滴定反应如下：

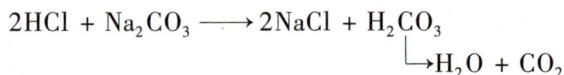

$$2HCl + Na_2CO_3 \longrightarrow 2NaCl + H_2CO_3$$
$$\quad\quad\quad\quad\quad\quad\quad\quad\quad\quad \rightarrow H_2O + CO_2$$

滴定终点以甲基红 – 溴甲酚绿混合指示剂指示，终点颜色由绿色转变为暗紫色。根据基准物质的质量和终点时消耗 HCl 标准溶液的体积，按下式计算 HCl 标准溶液的浓度：

$$c_{HCl} = \frac{2w_{Na_2CO_3}}{V_{HCl} \dfrac{M_{Na_2CO_3}}{1000}}$$

$$M_{Na_2CO_3} = 105.99 \, \text{g/mol}$$

2. 比较法 酸标准溶液的浓度可以用已知准确浓度的碱标准溶液滴定而测得。本实验采用 NaOH 标准溶液来滴定盐酸溶液，滴定反应如下：

$$NaOH + HCl \longrightarrow NaCl + H_2O$$

反应终点可以用酚酞指示，终点颜色由无色变为粉红色，按下式计算 HCl 标准溶液的浓度：

$$c_{HCl} = \frac{(cV)_{NaOH}}{V_{HCl}}$$

仪器和药品 ●●●

1. 仪器 滴定管（25mL），移液管（10mL），锥形瓶（250mL），量筒（10mL，100mL），试剂瓶（500mL）。

2. 药品　浓盐酸（36.5%～38%），无水碳酸钠（固体，基准物质），NaOH 标准溶液（0.1mol/L），酚酞指示剂（0.1% 乙醇溶液），甲基红 – 溴甲酚绿混合指示剂（3 份 0.1% 溴甲酚绿乙醇溶液和 1 份 0.2% 甲基红乙醇溶液）。

实验步骤 •••

1. 0.2mol/L HCl 溶液的配制　用小量筒取浓盐酸 9mL，倒入洁净具塞的 500mL 试剂瓶中，用蒸馏水稀释至 500mL，混匀。

2. 比较法标定 HCl 溶液浓度　移取 10.00mL HCl 溶液至锥形瓶中，加入 2 滴酚酞指示剂。用 0.1mol/L NaOH 溶液进行滴定，终点为溶液颜色由无色恰好变为粉红色，记录滴定剂的体积。平行滴定 3 份。

3. 标定法标定 HCl 溶液浓度　准确称取在 270～290℃ 干燥至恒重的无水 Na_2CO_3 基准物质 0.24g，加 50mL 蒸馏水使其完全溶解后，加甲基红 – 溴甲酚绿混合指示剂 10 滴，用 HCl 标准溶液滴定至溶液由绿色变为紫红色。煮沸 2min，冷却至室温（或旋摇 2min），继续滴定至溶液由绿色变为暗紫色终点。平行滴定 3 份。

注意事项 •••

1. Na_2CO_3 易吸水，称量要快速。

2. 在标定实验中，近终点时由于形成 H_2CO_3–$NaHCO_3$ 缓冲体系，pH 变化不大，终点不敏锐，为此需旋摇或煮沸 2min 后再滴定。

3. 旋摇或煮沸溶液时，需防止溶液溅出。

4. 混合指示剂颜色变化由绿色 – 青灰色 – 暗紫色 – 紫红色。在滴定时，首先滴定至紫红色，随着旋摇的进行，碳酸分解，溶液逐渐变回绿色，继续用 HCl 缓慢滴定至暗紫色终点。

5. 比较法平行滴定体积间的差距应该在 0.05mL 之内。

思考题 •••

1. 为什么称取无水碳酸钠基准物质时要用减重称量法？

2. 用基准物质碳酸钠标定 HCl 溶液中，近终点时为何需要将溶液煮沸或旋摇 2min？煮沸后为何又要冷却至室温后再滴定至终点？

3. 在比较法用 NaOH 标准溶液标定 HCl 溶液的实验中，是否可以选择甲基橙指示剂指示终点？

4. 实验中使用的锥形瓶是否需要干燥？加入蒸馏水的量是否需要精确？

Experiment 6　Preparation and Standardization of Standard Hydrochloric Acid Solution

◆ Purpose

1. To learn how to prepare hydrochloric acid standard solution.

2. To learn how to standardize hydrochloric acid standard solution with primary standard anhydrous sodium carbonate.

3. To learn how to standardize hydrochloric acid standard solution by comparative method.

4. Properly judge the end point of methyl red – bromocresol green mixed indicator.

◆ Principle

Concentrated hydrochloric acid is volatile and cannot be directly prepared as a standard solution with accurate concentration. Therefore, indirect preparation method can only be used to prepare a hydrochloric acid standard solution. First, a solution with approximate concentration can be prepared, and then the accurate concentration can be determined using the standardization method or the comparison method.

1. The standardization method

Theprimary standard substances for standardizing hydrochloric acid solutions are anhydrous sodium carbonate, borate, etc. In this experiment, we use anhydrous sodium carbonate as the primary standard substance. The titration reaction is as follows:

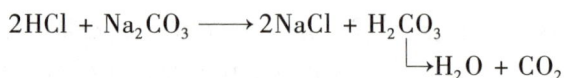

$$2HCl + Na_2CO_3 \longrightarrow 2NaCl + H_2CO_3$$
$$\downarrow H_2O + CO_2$$

The titration endpoint is indicated by a mixed indicator of methyl red – bromocresol green. The color of the end point changes from green to dark purple. Based on the mass of the reference substance and the volume of HCl standard solution consumed at the end point, the concentration of HCl standard solution can be calculated as follows:

$$c_{HCl} = \frac{2w_{Na_2CO_3}}{V_{HCl}\dfrac{M_{Na_2CO_3}}{1000}}$$

$$M_{Na_2CO_3} = 105.99 g/mol$$

2. The comparison method

The concentration of an acid standard solution can be measured by titration with an alkali standard solution of known accurate concentration. In this experiment, we use NaOH standard solution to titrate hydrochloric acid solution, and the titration reaction is as follows:

$$NaOH + HCl \longrightarrow NaCl + H_2O$$

The endpoint can be indicated by phenolphthalein, and the endpoint color changes from colorless to pink. the concentration of HCl standard solution can be calculated as follows:

$$c_{HCl} = \frac{(cV)_{NaOH}}{V_{HCl}}$$

◆ Equipment and Chemicals

1. Equipment   buret(25mL), pipet(10mL), conical flask(250mL), cylinder(10mL, 100mL), reagent bottle(500mL).

2. Chemicals   hydrochloric acid (the concentration is 36.5% ~ 38%), anhydrous sodium carbonate (solid, primary standard substance), NaOH standard solution(0.1mol/L), phenolphthalein indicator(0.1% ethanol solution), methyl red – bromocresol green mixed indicator(3 portions of 0.1% bromocresol green's ethanol solution and 1 portion of 0.2% methyl red's ethanol solution).

◆ Procedure

1. Prepatation of 0.2mol/L hydrochloride acid solution

Transter 9mL concentrated hydrochloric acid to a 500mL reagent bottle with a clean stopper, dilute with distilled water to 500mL, and mix well.

2. Standardization of HCl solution by comparative method

Transfer 10.00mL HCl solution to a conical flask and add 2 drops of phenolphthalein indicator. Titrate with a 0.1mol/L NaOH standard solution until the color of the solution changes from colorless to pink. Record the volume of the titrant. Perform 3 parallel measurements.

3. Standardization of HCl solution by standardization method

Weigh accurately about 0.24g anhydrous sodium carbonate which has previously been dried to constant weight at $270 \sim 290$℃ in a conical flask. Dissolve it with 50mL distilled water and add 10 drops of methyl red – bromocresol green mixed indicator. Titrate it with 0.2mol/L HCl solution till the color change from green to violet red. Boil it for 2 minutes, then cool it down to room temperature (or shake it by swirling for 2 minutes), Then titrate it from green to purple, that's the end – point. Perform 3 parallel measurements.

◆ Notes

1. Sodium carbonate should be weighted rapidly because it's easy to absorb water.

2. In the standardization experiment, due to the formation of $H_2CO_3 - NaHCO_3$ buffer system near the end-point, the pH change is not significant and the endpoint is not sensitive. Therefore, it is necessary to shake or boil for 2 minutes near the endpoint.

3. Be careful not to spill out the solution when shaking or boiling.

4. The color of the mixed indicator changes from green to blue – grey to dark purple to purplish red. We first titrate to a purplish red color. As the shaking or boiling proceeds, the carbonic acid decomposes and the solution gradually turns back green. Then continue to titrate slowly to dark purple endpoint.

5. The volume difference between parallel titrations in comparative method should be within 0.05mL.

◆ Questions

1. Why is the weight by difference method used when weighing anhydrous sodium carbonate as a primary substance?

2. Why is boiling or shaking necessary near the endpoint when standardizing HCl with primary substance sodium carbonate? Why do we need to cool down to the room temperature after boiling before titrating it to the endpoint?

3. Can methyl orange indicator be chosen to indicate the endpoint in the experiment of calibrating HCl solution with NaOH standard solution?

4. Do the conicalflask used in the experiment need to be dried? Does the amount of distilled water added need to be precise?

实验七　药用氧化锌的含量测定

PPT

实验目的 ●●●

1. 掌握用酸碱滴定法中返滴定法测定氧化锌含量的原理和操作。

2. 进一步练习移液管操作。

实验原理 •••

氧化锌是一种两性氧化物，不溶于水，但如果过量 HCl 溶解后，剩余量的 HCl 可用标准碱溶液进行滴定，以甲基橙为指示剂，其滴定反应式如下：

$$ZnO + 2HCl \longrightarrow ZnCl_2 + H_2O$$
$$（定量过量）$$
$$HCl + NaOH \longrightarrow NaCl + H_2O$$
$$（剩余量）$$

氧化锌的质量分数按下式计算：

$$w_{ZnO} = \frac{\frac{1}{2}\left[(cV)_{HCl} - (cV)_{NaOH}\right] \times \frac{M_{ZnO}}{1000}}{W_{sample}} \times 100\%$$

$$M_{ZnO} = 81.38 \text{g/mol}$$

仪器和药品 •••

1. 仪器 滴定管（25mL）；锥形瓶（250mL）；移液管（25mL）；洗耳球。

2. 药品 药用氧化锌；HCl 标准溶液（0.2mol/L）；NaOH 标准溶液（0.1mol/L）；甲基橙指示液：0.1% 水溶液。

实验步骤 •••

取氧化锌样品约 0.11g 精密称定，置锥形瓶中。用移液管移取盐酸标准溶液（0.2mol/L）25.00mL 于锥形瓶中，微微加热并旋摇使之溶解。待冷却后加甲基橙指示液 3 滴，用 NaOH 标准溶液（0.1mol/L）滴定至黄色，记下体积 V。

注意事项 •••

1. 返滴定法所用两种标准溶液的浓度要准确，否则将严重影响测定结果。

2. 加热使 ZnO 溶解时，只能微微加热，不能温度过高，否则 HCl 挥发，测定结果偏高。

3. 终点黄色中不能有橙色的成分，否则滴定不完全，也使结果偏高。

思考题 •••

1. 什么情况下可采用返滴定法？

2. 为什么采用返滴定法测定结果误差较大？

3. 采用返滴定法测定物质的含量，应如何估计样品的质量？

4. 本实验是否可用甲基红作指示剂？

Experiment 7   Determination of Pharmaceutical Zinc Oxide

◆ Purpose

1. Master the principle and operation of determining zinc oxide by back titration in acid – base titration.

2. Further practise the pipette operation.

◆ Principle

Zinc oxide is an amphoteric oxide that is insoluble in water. However it can be dissolved in excess hydrochoride, and the rest acid can be titrated by standard base solution with methyl orange as indicator. The tritration reaction equation is as follows：

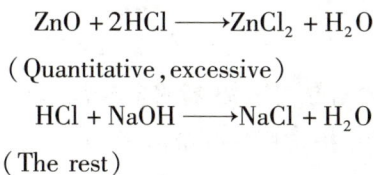

$$ZnO + 2HCl \longrightarrow ZnCl_2 + H_2O$$
$$(\text{Quantitative}, \text{excessive})$$
$$HCl + NaOH \longrightarrow NaCl + H_2O$$
$$(\text{The rest})$$

The percent of zinc oxide can be calculated as follows：

$$w_{ZnO} = \frac{\frac{1}{2}[(cV)_{HCl} - (cV)_{NaOH}] \times \frac{M_{ZnO}}{1000}}{W_{sample}} \times 100\%$$

$$M_{ZnO} = 81.38 \text{g/mol}$$

◆ Equipment and Chemicals

1. Equipment　brettet (25mL), conical flask (250mL), pipette (25mL), rubber suction bulb.

2. Chemicals　zine oxide (for pharmaceutical use); HCl standard solution(0.2mol/L); NaOH standard solution(0.1mol/L); methyl orange indicator：0.1% aqueous solution.

◆ Procedure

Weigh accurately about 0.11g ZnO, put them in 250mL conical flask, pipette 25.00mL HCl(0.2mol/L) standard solution into the flask to dissolve it, heat the flask if necessary. When the solution is cooled down to room temperature, add three drops of methyl orange indicator. Titrate with standard NaOH solution (0.1mol/L) until the color changes to yellow, record the volume consumed.

◆ Notes

1. The concentration of the two standard solutions used in the back titration method must be accurate, otherwise it will seriously affect the results.

2. Gentle heating is preferred when it is necessary, otherwise HCl will volatilize at higher-temperature and therefore the result will be higher.

3. At the end point, yellow color of indicator should not include any orange, or the titration is incomplete; the result will be higher, too.

◆ Questions

1. Under which conditions can the back titration be applied?

2. Why is the error of back titration often high?

3. How to estimate the quantity of the sample when using back titration method to determine its content?

4. Could methyl red be used as the indicator in this experiment?

实验八　EDTA 标准溶液的配制与标定

PPT

实验目的 •••

1. 掌握 EDTA 标准溶液的配制和标定。

2. 掌握配位滴定的原理。

3. 掌握铬黑 T 指示剂的终点判断。

实验原理 •••

乙二胺四乙酸（简称 EDTA）是含有羧基和氨基的螯合剂，在一定的条件下，能与大多数金属离子形成 1∶1 的稳定螯合物，是络合滴定中常用的标准溶液。

完全质子化的 EDTA（H_4Y）难溶于水，在分析中常常使用其二钠盐 $Na_2H_2Y \cdot 2H_2O$（$M = 372.24 g/mol$）配制标准溶液。乙二胺四乙酸二钠盐不易得到纯品，常用间接法配制标准溶液。本实验以 ZnO 为基准物质，以铬黑 T（In）为指示剂，在 pH≈10 条件下对 EDTA 溶液进行标定，终点溶液颜色由紫红色变为纯蓝色，滴定反应为：

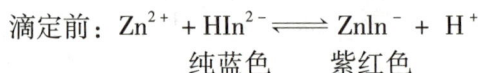

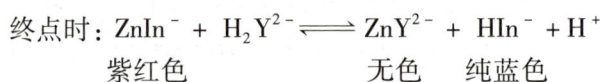

滴定前：$Zn^{2+} + HIn^{2-} \rightleftharpoons ZnIn^- + H^+$
　　　　　　　纯蓝色　　紫红色

终点前：$Zn^{2+} + H_2Y^{2-} \rightleftharpoons ZnY^{2-} + 2H^+$

终点时：$ZnIn^- + H_2Y^{2-} \rightleftharpoons ZnY^{2-} + HIn^- + H^+$
　　　　　紫红色　　　　　　　无色　　纯蓝色

EDTA 标准溶液的浓度计算公式如下：

$$c_{EDTA} = \frac{W_{ZnO} \times \frac{20}{100}}{V_{EDTA} \times \frac{M_{ZnO}}{1000}} \qquad M_{ZnO} = 81.38 g/mol$$

仪器和药品 •••

1. 仪器 滴定管（25mL），锥形瓶（250mL），烧杯（50mL），容量瓶（100mL），移液管（20mL），量筒（10mL，100mL）。

2. 药品 乙二胺四乙酸二钠（$Na_2H_2Y \cdot 2H_2O$，分析纯），ZnO（基准物质，800℃下灼烧至恒重），稀盐酸（3mol/L），氨试液（40mL 浓氨水加水稀释至100mL），甲基红指示剂（0.1% 的 60% 乙醇液），NH_3–NH_4Cl 缓冲液（pH≈10，54g NH_4Cl 溶于水，加氨水 350mL，用水稀释至1000mL），铬黑 T 指示剂（0.2g 铬黑 T 溶于15mL 三乙醇胺中后，加入5mL 无水乙醇，混匀即得）。

实验步骤 •••

1. EDTA 标准溶液（0.05mol/L）的配制 称取 $Na_2H_2Y \cdot 2H_2O$ 约9.5g，加蒸馏水 500mL 使其溶解，摇匀，贮存于硬质玻璃瓶中。

2. EDTA 标准溶液的标定 准确称取已在 800℃ 灼烧至恒重的基准物质 ZnO 约0.45g 至一小烧杯中，加稀盐酸（3mol/L）10mL，搅拌使其溶解，并定量转移至100mL 容量瓶中，加水稀释至刻度，摇匀。用移液管吸取 20.00mL 溶液至锥形瓶中，加甲基红指示剂 1 滴，用氨试液调至溶液刚呈微黄色，再加蒸馏水 25mL，加 NH_3–NH_4Cl 缓冲液（pH=10）10mL，加铬黑 T 指示剂4 滴，摇匀。用 EDTA 标准液滴定至溶液由紫红色转变为纯蓝色，即为终点，记录 EDTA 标准溶液的体积，平行做三次，计算 EDTA 标准溶液的浓度。

注意事项 •••

1. $Na_2H_2Y \cdot 2H_2O$ 溶解较慢，可先于温水中溶解，再稀释至一定体积，或放置过夜。

2. EDTA 标准溶液应贮存于硬质玻璃瓶中。长期存放时要放在聚乙烯瓶中，以免 EDTA 与玻璃瓶壁中的金属离子作用而改变 EDTA 溶液浓度。

3. 用稀盐酸溶解 ZnO 时应仔细操作防止溅失，待 ZnO 完全溶解后方可定量转移至 100mL 容量瓶中。

4. 滴加氨试液后若出现 $Zn(OH)_2$ 沉淀，通常加缓冲液后即可溶解。

5. 配位反应进行的速度较慢，故滴定时加入 EDTA 标准溶液的速度不能太快，在室温低时尤其要注意。特别是近终点时，应逐滴加入，并充分振摇。

思考题 ●●●

1. 为什么在滴定前要加 $NH_3 - NH_4Cl$ 缓冲液？

2. $Na_2H_2Y \cdot 2H_2O$ 的基本性质怎样？为什么不用 EDTA 直接来配制标准溶液？

3. 酸碱滴定和配位滴定有什么不同？

4. 实验中为何需要加甲基红指示剂，用氨试液调节溶液至微黄色？

Experiment 8　Preparation and Standardization of EDTA Solution

◆ Purpose

1. To learn the preparation and standardization of EDTA standard solution.

2. To learn the principle of coordination titration.

3. To learn the endpoint determination of eriochrome black T indicator.

◆ Principle

Ethylene diamine tetraacetic acid (EDTA) is an excellent chelating agent containing carboxyl and amino groups. EDTA can form a 1 : 1 stable chelate with most metal ions, so it is a commonly used standard solution in complexometric titration.

Completely protonated EDTA is slight soluble in water. EDTA standard solution is commonly prepared by dissolving disodium dihydrogen ethylene diami etracetate ($Na_2H_2Y \cdot 2H_2O$) in water. It is not easy to get the pure product. Therefore, its standard solution is prepared by indirect method and its concentration is standardized with primary standard. In this experiment, we use ZnO as the primary substance and eriochrome black T (in) as the indicator to standardize EDTA solution under $pH \approx 10$. At the end point the color change is from violet – red to pure blue. The titration reaction is as follows:

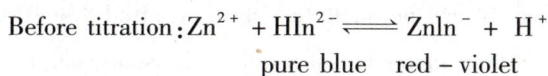

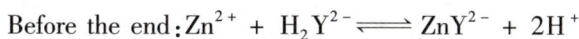

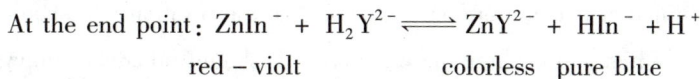

$$\text{Before titration:} Zn^{2+} + HIn^{2-} \rightleftharpoons ZnIn^- + H^+$$
$$\text{pure blue}\quad\text{red – violet}$$

$$\text{Before the end:} Zn^{2+} + H_2Y^{2-} \rightleftharpoons ZnY^{2-} + 2H^+$$

$$\text{At the end point:} ZnIn^- + H_2Y^{2-} \rightleftharpoons ZnY^{2-} + HIn^- + H^+$$
$$\text{red – violt}\qquad\qquad\text{colorless}\quad\text{pure blue}$$

The concentration calculation formula for EDTA standard solution is as follows:

$$c_{EDTA} = \frac{W_{ZnO} \times \dfrac{20}{100}}{V_{EDTA} \times \dfrac{M_{ZnO}}{1000}} \qquad M_{ZnO} = 81.38 g/mol$$

◆ Equipment and Chemicals

1. Equipment　buret(25mL), conical flask(250mL), beaker(50mL), volumetric flask(100mL), pipet(20mL), measuring cylinder(10mL, 100mL).

2. Chemicals　Disodium dihydrogen ethlene diamine tetraacetate ($Na_2H_2Y \cdot 2H_2O$, AR), Zinc oxide(primay standard, ignited to constant weight at about 800℃), diluted hydrochloric acid(3mol/L), ammonia test solution(dilute 40mL of cocentrated ammonia solution with water to make 100mL), methyl red indicator(dissolve 0.1g of methyl red in 100mL 60% ethanol solution), ammonia – ammonium chloride buffer solution (pH = 10, dissolve 54g ammonium chloride in water, add 350mL cncentrated ammonia solution, dilute with water to 1000mL), eriochrome black T indicator(dissolve 0.2g of eriochrome black T in 15mL of triethanolamine, add 5mL of anhydrous ethanol, and mix well).

◆ Procedure

1. Preparation of 0.05mol/L EDTA solution

Dissolve about 9.5g of $Na_2H_2Y \cdot 2H_2O$ with water to make 500mL and mix well. Preserve it in a hard glass bottle.

2. Standardization of standard EDTA solution

Accurately weigh 0.45g of zinc oxide primary standard, previously ignited to constant weight at about 800℃, in a small beaker, add 10mL of dilute hydochloric acid (3mol/L). After it is dissolved, quantity transfer to 100mL volumetric flask, dilute it with water to make 100mL and mix well. Transfer 20.00mL of the solution to a conical flask, add 1 drop of methyl red and add ammonia test solution dropwise until a slight yellow color is obtained. Add 25mL of water, 10mL of NH_3–NH_4Cl buffer solution (pH = 10.0) and 4 drops of eriochrome black T indicator. Titrate with standard EDTA solution until the color changes from red – violet to pure blue. Record the volume of standard EDTA solution. Repeat it twice and calculate the concentration of standard EDTA solution.

◆ Notes

1. $Na_2H_2Y \cdot 2H_2O$ dissolves slowly, so it can be dissolved in warm water first, and then diluted to a certain volume, or left overnight.

2. EDTA standard solution should be stored in a hard glass bottle. When stored for a long time, it should be placed in a polyethylene bottle to prevent the concentration of EDTA solution from changing due to the interaction between EDTA and metal ions of the glass bottle wall.

3. When dissolving ZnO with dilute hydrochloric acid, careful operation should be taken to prevent splashing. After ZnO is completely dissolved, it can be quantitatively transferred to a 100mL volumetric flask.

4. If $Zn(OH)_2$ precipitates after adding ammonia test solution, it usually can be dissolved after adding buffer solution.

5. The rate of coordination reaction is relatively slow, so the titration rate should not be too fast, especially at low room temperature. When approaching the end point, it should be added dropwise and thoroughly shaken.

◆ Questions

1. Why should NH_3–NH_4Cl buffer be added before the titration?

2. What are the basic properties of $Na_2H_2Y \cdot 2H_2O$? Why not use EDTA to prepare standard solutions directly?

3. What is the difference between acid – base titration and complexometric titration?

4. Why do we add methyl red indicator and adjust the solution to a slightly yellow color with ammonia test solution in the experiment?

实验九　药用葡萄糖酸锌中锌含量测定

PPT

实验目的 •••

1. 掌握络合滴定测定锌盐含量的原理及方法。
2. 进一步熟悉铬黑 T 指示剂的终点判断。

实验原理 •••

锌是人体必需的微量元素，葡萄糖酸锌作为补锌药物，主要用于预防或治疗由于缺锌所导致的相关疾病。因 EDTA 可以和大部分金属离子生成稳定的 1∶1 的配合物，很多金属离子的浓度可以利用配位滴定来测定。本实验以 EDTA 为滴定剂，铬黑 T 为指示剂，在 pH≈10 条件下对葡萄糖酸锌进行滴定，终点溶液颜色由紫红色变为纯蓝色，滴定反应为：

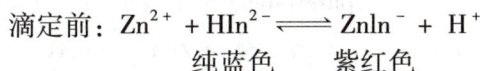

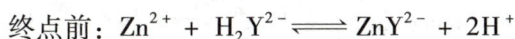

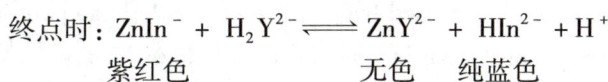

$$\text{滴定前：} Zn^{2+} + HIn^{2-} \Longrightarrow ZnIn^- + H^+$$
$$\text{纯蓝色} \qquad \text{紫红色}$$

$$\text{终点前：} Zn^{2+} + H_2Y^{2-} \Longrightarrow ZnY^{2-} + 2H^+$$

$$\text{终点时：} ZnIn^- + H_2Y^{2-} \Longrightarrow ZnY^{2-} + HIn^{2-} + H^+$$
$$\text{紫红色} \qquad\qquad \text{无色} \quad \text{纯蓝色}$$

葡萄糖酸锌中锌的含量计算公式如下：

$$Zn\% = \frac{(cV)_{EDTA} \times M_{Zn}}{W_{sample} \times 1000} \times 100\% \qquad M_{Zn} = 65.41\,g/mol$$

仪器和药品 •••

1. 仪器　滴定管（25mL），锥形瓶（250mL），量筒（10mL，100mL）。

2. 药品　药用葡萄糖酸锌；EDTA 标准溶液（0.05mol/L），氨 – 氯化铵缓冲溶液（pH = 10.0），铬黑 T 指示剂。

实验步骤 •••

准确称取 0.48g 药用葡萄糖酸锌于锥形瓶中，加 20mL 蒸馏水溶解（必要时可适当加热），加 10mL 氨 – 氯化铵缓冲溶液和 4 滴铬黑 T 指示剂，用 0.05mol/L EDTA 标准溶液滴定至溶液颜色由紫红变为纯蓝色。平行测定三次。

注意事项 •••

1. 葡萄糖酸锌在冷水中溶解速度较慢，可微温溶解，待冷却后再测定。
2. 为了能够准确地判断终点的颜色变化，可先进行预滴定。
3. 络合反应的速率较慢，滴定速度不宜太快。终点时溶液颜色的变化并非突然由紫红色变到纯蓝色，而是从紫红色 – 紫色 – 蓝紫色 – 纯蓝色。

思考题 •••

用 EDTA 测定锌离子含量时，能用二甲酚橙为指示剂吗？

Experiment 9　Determination of Zinc Content in Pharmaceutical Zinc Gluconate

◆ Purpose

　　1. To learn the principle and method of complexometric titration of zinc salt.

　　2. Further familiar the endpoint determination of eriochrome black T indicator.

◆ Principle

　　Zinc is an essential trace element in the human body. Zinc gluconate, as a zinc supplement drug, is mainly used to prevent or treat related diseases caused by zinc deficiency. Because EDTA can form stable 1 : 1 complexes with most metal ions, the concentration of many metal ions can be determined by coordination titration. This experiment used EDTA as the titrant and eriochrome black T as the indicator to titrate zinc gluconate under pH ≈ 10. The color change at the end point is from purple red to pure blue. The titration reaction is as follows:

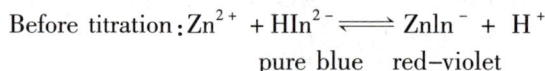

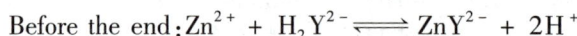

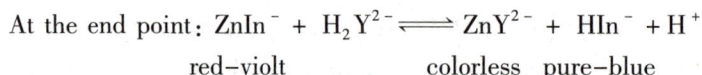

$$\text{Before titration:} Zn^{2+} + HIn^{2-} \Longrightarrow ZnIn^- + H^+$$
$$\quad\quad\quad\quad\quad \text{pure blue}\quad \text{red-violet}$$

$$\text{Before the end:} Zn^{2+} + H_2Y^{2-} \Longrightarrow ZnY^{2-} + 2H^+$$

$$\text{At the end point:} ZnIn^- + H_2Y^{2-} \Longrightarrow ZnY^{2-} + HIn^- + H^+$$
$$\quad\quad \text{red-violt}\quad\quad\quad\quad \text{colorless}\quad \text{pure-blue}$$

　　The zinc content in zinc gluconate can be calculated as follows:

$$Zn\% = \frac{(cV)_{EDTA} \times M_{Zn}}{W_{sample} \times 1000} \times 100\% \quad\quad M_{Zn} = 65.41 \text{g/mol}$$

◆ Apparatus and Chemicals

　　Apparatus: buret(25mL), conical flask(250mL), measuring cylinder(10mL, 100mL).

　　Chemicals: pharmaceutical zinc gluconate, EDTA standard solution(0.05mol/L), ammonia – ammonium chloride buffer solution (pH = 10), eriochrome black T indicator.

◆ Procedure

　　Accurately weigh 0.48g of pharmaceutical zinc gluconate in a conical flask, dissolve it in 20mL of distilled water (heated appropriately if necessary), add 10mL of ammonium chloride buffer solution and 4 drops of chromium black T indicator, and titrate with 0.05mol/L EDTA standard solution until the color of the solution changes from purple red to pure blue. Measure three times in parallel.

◆ Notes

　　1. Zinc gluconate dissolves slowly in cold water and can be slightly heated for dissolution.

　　2. In order to accurately determine the color changeat the end point, pre titration can be performed first.

　　3. The rate of complex reaction is slow, and the titration rate should not be too fast. The color change of the solution at the end point is not a sudden change from purplish red to pure blue, but rather from purplish red to purple to blue purple to pure blue.

◆ Question

　　Can xylenol orange be used as the indicator when measuring zinc ion content with EDTA?

实验十 碘标准溶液的配制与标定

PPT

实验目的 ●••

1. 掌握碘标准溶液的配制与保存方法。
2. 掌握碘标准溶液的标定原理和方法。
3. 了解碘量法的操作过程。

实验原理 ●••

碘在水中的溶解度较小，25℃时100mL水仅能溶解0.035g碘。除了在水溶液中较小的溶解度，水溶液中碘还具有可观的蒸气压，因此操作时由于碘挥发会引起溶液浓度的变化。将碘溶解于碘化钾的水溶液可以克服上述两个问题。碘溶解度增加是由于三碘负离子的形成反应：

$$I_2 + I^- \rightleftharpoons I_3^-$$

这样形成的溶液与碘的纯水溶液相比，碘的蒸气压要低得多，因而由于碘挥发引起的损失大大降低。

直接法制备标准碘溶液需要使用重新升华的碘和不含碘酸盐的碘化钾，因此在实际中通常不用直接法而采用间接法来制备碘标准溶液。为避免碘化钾中碘酸钾杂质的干扰，须向碘溶液中加一些盐酸。在酸性条件下，少量存在的碘酸钾会和碘化钾反应生成碘。

碘标准溶液可用三氧化二砷（砒霜）基准物质标定或用标准硫代硫酸钠溶液比较标定。

在用三氧化二砷标定碘溶液时，三氧化二砷首先溶解于氢氧化钠溶液中：

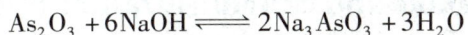

$$As_2O_3 + 6NaOH \rightleftharpoons 2Na_3AsO_3 + 3H_2O$$

但是标定过程中强碱的存在是不容许的，因此可加硫酸中和过量的碱。此外，由于三氧化二砷和碘的反应是可逆的：

$$AsO_3^{3-} + I_2 + H_2O \rightleftharpoons AsO_4^{3-} + 2I^- + 2H^+$$

若滴定反应中的碘化氢一生成就被中和，上述反应就能向右定量进行。碘化氢可以通过加碳酸氢钠中和。

在用硫代硫酸钠标准溶液比较法标定碘溶液浓度时，标定反应为：

$$2Na_2S_2O_3 + I_2 \rightleftharpoons Na_2S_4O_6 + 2NaI$$

仪器和药品 ●••

1. 仪器 滴定管，量筒（10mL×2），研钵，棕色瓶（500mL），锥形瓶（250mL×3）。

2. 药品 碘（分析纯），碘化钾（分析纯），浓盐酸，淀粉指示剂，标准硫代硫酸钠（0.1mol/L），三氧化二砷（基准物质级），氢氧化钠溶液（1mol/L），酚酞指示剂，硫酸（1mol/L），碳酸氢钠。

实验步骤 ●••

1. 碘溶液制备（0.05mol/L） 称取18g碘化钾、6.5g碘于研钵中，加入10mL蒸馏水，研磨至所有碘溶解后，加一滴盐酸，然后用蒸馏水稀释至500mL，转移至带玻璃塞的棕色瓶中，置阴凉处保存。

2. 碘溶液的标定

（1）用三氧化二砷标定 准确称取0.11g三氧化二砷（预先于105℃干燥2h），溶解于4mL的氢氧化钠（1mol/L）后，加20mL蒸馏水、一滴酚酞指示剂，滴加硫酸（0.5mol/L）调节溶液至微酸性。然后加2g碳酸氢钠、30mL蒸馏水，再加2mL淀粉指示剂，慢慢用碘溶液滴定至溶液恰好呈蓝色，按下式计算碘溶液的物质的量浓度：

$$c_{I_2} = \frac{\dfrac{W_{As_2O_3}}{M_{As_2O_3}}}{V_{I_2} \times \dfrac{1}{1000}} \qquad M_{As_2O_3} = 197.84\text{g/mol}$$

（2）用硫代硫酸钠标准溶液标定　移取20.00mL硫代硫酸钠溶液至250mL锥形瓶，用25mL水稀释，加1mL淀粉指示剂，然后用碘标准溶液滴定至溶液恰好呈蓝色，按下式计算碘溶液的物质的量浓度：

$$c_{I_2} = \frac{(cV)_{Na_2S_2O_3}}{2V_{I_2}}$$

注意事项 •••

1. 碘须首先溶解在浓碘化钾溶液后再稀释。
2. 碘溶液腐蚀橡胶，因此应使用带玻璃活塞的酸式滴定管盛碘溶液。

思考题 •••

1. 配制碘溶液时，为什么要在溶液非常浓的情况下将碘和碘化钾一起研磨，当碘和碘化钾溶解后才能用水稀释？
2. 当用三氧化二砷标定碘溶液时，为什么加氢氧化钠、硫酸和碳酸氢钠？

Experiment 10　Preparation and Standardization of Iodine Standard Solution

◆ Purpose

1. To master the preparation and preservation methods of iodine standard solutions.

2. To master the standardization principle and method of iodine standard solution.

3. To learn the operation process of iodine measurement method.

◆ Principle

Iodine has a low solubility in water, and only 0.035g of iodine can be dissolved in 100mL of water at 25℃. In addition to its relatively small solubility in aqueous solutions, iodine also has a considerable vapor pressure in aqueous solutions. The evaporation of iodine during titration can cause concentration change of the solution. Dissolving iodine in an aqueous solution of potassium iodide can overcome the above two problems. The increase in iodine solubility is due to the formation of triiodide negative ions:

$$I_2 + I^- \rightleftharpoons I_3^-$$

The iodine solution in potassium iodide solution has a much lower vapor pressure than the iodine solution in a pure water. Thus the loss caused by iodine volatilization is greatly reduced.

The direct preparation of standard iodine solutions requires the use of re sublimated iodine and potassium iodide without iodate salts. Therefore, indirect methods are usually used instead of direct methods to prepare iodine standard solutions. To avoid the interference from potassium iodate impurities in potassium iodide, it is necessary to add some hydrochloric acid into the iodine solution. Under acidic conditions, a small amount of potassium iodate will react with potassium iodide to produce iodine.

Iodine standard solution can be standardized with arsenic trioxide (arsenic) primary substance or compared with standard sodium thiosulfate solution.

When standardizing iodine solution with arsenic trioxide, arsenic trioxide is first dissolved in sodium hydroxide solution:

$$As_2O_3 + 6NaOH \rightleftharpoons 2Na_3AsO_3 + 3H_2O$$

However, the presence of strong bases during the standardization process is not allowed, so sulfuric acid can be added to neutralize the excess bases. Furthermore, due to the reversible reaction between arsenic trioxide and iodine:

$$AsO_3^{3-} + I_2 + H_2O \rightleftharpoons AsO_4^{3-} + 2I^- + 2H^+$$

If the hydrogen iodide in the titration reaction is neutralized as fast as it is formed, the above reaction can proceed quantitatively to the right. Hydrogen iodide can be neutralized by sodium bicarbonate.

When using the sodium thiosulfate standard solution to standardize the concentration of iodine solution, the standardization reaction is as follows:

$$2Na_2S_2O_3 + I_2 \rightleftharpoons Na_2S_4O_6 + 2NaI$$

◆ Equipment and Chemicals

1. Equipment buret(25mL), measuring cylinder(10mL), mortar; conical flask(250mL), brown reagent bottle(500mL).

2. Chemicals iodine(A. R.), potassium iodide(A. R.), concentrated hydrochloric acid, starch indicator; sodium thiosulphate standard solution(0.1mol/L), phenolphthalein indicator, arsenic trioxide(primary substance), sulfuric acid(1mol/L); sodium bicarbonate.

◆ Procedure

1. Preparation of 0.05mol/L iodine solution

Weigh 18g of potassium iodide and 6.5g of iodine in a mortar, add 10mL of distilled water, grind until all the iodine is dissolved, add a drop of hydrochloric acid, and then dilute with distilled water to 500mL, transfer to a brown bottle with a glass stopper, and store in a cool place.

2. Standardization of iodine solution

(1) Standardization with arsenic trioxide Accurately weigh 0.11g of arsenic trioxide (pre dried at 105℃ for 2 hours), and dissolve it in 4mL of sodium hydroxide (1mol/L). Dilute with 20mL of distilled water, add one drop of phenolphthalein indicator and sulfuric acid (0.5mol/L) dropwise to adjust the solution to slightly acidic. Then add 2g of sodium bicarbonate, 30mL of distilled water, and 2mL of starch indicator. Slowly titrate with iodine solution until the solution turns blue. Calculate the concentration of the iodine solution using the following formula:

$$c_{I_2} = \frac{W_{As_2O_3}}{V_{I_2} \times \frac{M_{As_2O_3}}{1000}} \qquad M_{As_2O_3} = 197.84 \text{g/mol}$$

(2) Standardization with sodium thiosulfate standard solution Transfer 20.00mL of sodium thiosulfate solution to a 250mL conical flask, dilute with 25mL of water, add 1mL of starch indicator, and then titrate with i-

odine standard solution until the solution turns blue. Calculate the concentration of iodine solution using the following formula:

$$c_{I_2} = \frac{(cV)_{Na_2S_2O_3}}{2V_{I_2}}$$

◆ Notes

1. Iodine must first be dissolved in a concentrated potassium iodide solution before dilution.

2. Iodine solution corrodes rubber, therefore an acid burette with a glass piston should be used.

◆ Question

1. Why grind iodine and potassium iodide together when preparing iodine solution in a very concentrated solution, and only dilute them with water after they dissolve?

2. Why add sodium hydroxide, sulfuric acid, and sodium bicarbonate when titrating iodine with arsenic trioxide?

实验十一　维生素 C 含量测定

PPT

实验目的 •••

掌握碘量法测定维生素 C 含量的原理及操作。

实验原理 •••

维生素 C 也称抗坏血酸，是中等强度的还原剂，分子结构中还原性的烯二醇基能被碘单质定量的氧化为二酮基，因此可以用碘标准溶液滴定，滴定反应为：

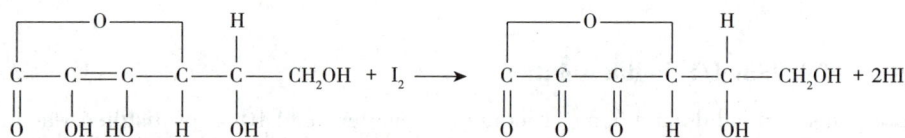

由于维生素 C 还原性很强，在空气中容易被氧化，特别是在碱性溶液中更容易被空气氧化，因此滴定须在稀醋酸溶液中（pH 3~4）进行以避免副反应。以淀粉为指示剂，滴定终点时溶液由无色变为蓝色。

仪器和药品 •••

1. 仪器　滴定管（25mL），量筒（10mL），锥形瓶（250mL×3）。

2. 药品　碘标准溶液（0.05mol/L），维生素 C，稀醋酸，淀粉指示剂（5% 水溶液）。

实验步骤 •••

准确称取 0.2g 维生素 C，加入 100mL 新煮沸放冷的蒸馏水和 10mL 稀醋酸，待样品溶解后加入 1mL 淀粉指示剂，立即用 0.05mol/L 的碘标准溶液滴定至溶液呈现蓝色。

按如下公式计算维生素 C 的含量：

$$维生素\ C\% = \frac{(cV)_{I_2} \times \dfrac{M_{Vc}}{1000}}{W_{sample}} \qquad M_{Vc} = 176.12g/mol$$

1. 维生素 C 不稳定，溶解后应立即滴定。

2. 量取稀醋酸和淀粉指示剂的量筒不得混用。

3. 由于淀粉容易变质，因此淀粉指示剂应新鲜配制或加入防腐剂。

思考题 •••

1. 为什么溶解维生素 C 需用新煮沸放冷的蒸馏水？

2. 在滴定过程中，如果溶解维生素 C 后没有立即滴定，对测定结果有什么影响？

Experiment 11　Determination of Vitamin C

◆ Purpose

To master the principle and operation of determining vitamin C using iodimetry.

◆ Principle

Vitamin C, also known as ascorbic acid, is a medium strength reducing agent. The reducing ethylene glycol group in the molecular structure can be quantitatively oxidized to a diketone group by iodine. Therefore, it can be titrated with iodine standard solution. The titration reaction is as follows:

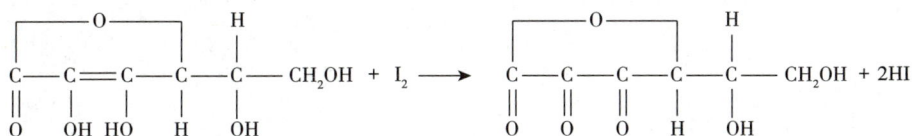

Due to the strong reducibility of vitamin C, it is easily oxidized in air, especially in alkaline solutions. Therefore, titration must be carried out in a dilute acetic acid solution (pH 3~4) to avoid side reactions. Using starch as the indicator, the solution changes from colorless to blue at the endpoint.

◆ Equipment and Chemicals

1. Equipment　buret(25mL), conical flask(250mL), measuring cylinder(10mL).

2. Chemicals　iodine standard solution(0.05mol/L), vitamin C, dilute acetic acid solution (pH 3~4), starch indicator(5% water solution).

◆ Procedure

Accurately weigh 0.2g of vitamin C, add 100mL of freshly boiled and cooled distilled water, 10mL of dilute acetic acid. After the sample is dissolved, add 1mL of starch indicator and immediately titrate with 0.05mol/L iodine standard solution until the solution turns blue.

Calculate the content of vitamin C as follows:

$$\text{Vitamin C\%} = \frac{(cV)_{I_2} \times \frac{M_{Vc}}{1000}}{W_{sample}} \qquad M_{Vc} = 176.12 \text{g/mol}$$

◆ Notes

1. Vitamin C is unstable and should be titrated immediately after dissolution.

2. The measuring cylinder for measuring dilute acetic acid and starch indicator should not be mixed.

3. As starch is prone to spoilage, starch indicators should be freshly prepared or add preservatives.

◆ Questions

1. Why does it take freshly boiled and cooled distilled water to dissolve vitamin C?

2. What is the impact on the results if vitamin C is not immediately titrated after dissolution during the titration process?

书网融合……

习题

附 录

附表一　UPAC元素周期表

Key:
- atomic number
- **Symbol**
- name
- abridged standard atomic weight

1	2	3	4	5	6	7	8	9	10	11	12	13	14	15	16	17	18
1 **H** hydrogen 1.0080 ±0.0002																	2 **He** helium 4.0026 ±0.0001
3 **Li** lithium 6.94 ±0.06	4 **Be** beryllium 9.0122 ±0.0001											5 **B** boron 10.81 ±0.02	6 **C** carbon 12.011 ±0.002	7 **N** nitrogen 14.007 ±0.001	8 **O** oxygen 15.999 ±0.001	9 **F** fluorine 18.998 ±0.001	10 **Ne** neon 20.180 ±0.001
11 **Na** sodium 22.990 ±0.001	12 **Mg** magnesium 24.305 ±0.002											13 **Al** aluminium 26.982 ±0.001	14 **Si** silicon 28.085 ±0.001	15 **P** phosphorus 30.974 ±0.001	16 **S** sulfur 32.06 ±0.02	17 **Cl** chlorine 35.45 ±0.01	18 **Ar** argon 39.95 ±0.16
19 **K** potassium 39.098 ±0.001	20 **Ca** calcium 40.078 ±0.004	21 **Sc** scandium 44.956 ±0.001	22 **Ti** titanium 47.867 ±0.001	23 **V** vanadium 50.942 ±0.001	24 **Cr** chromium 51.996 ±0.001	25 **Mn** manganese 54.938 ±0.001	26 **Fe** iron 55.845 ±0.002	27 **Co** cobalt 58.933 ±0.001	28 **Ni** nickel 58.693 ±0.001	29 **Cu** copper 63.546 ±0.003	30 **Zn** zinc 65.38 ±0.02	31 **Ga** gallium 69.723 ±0.001	32 **Ge** germanium 72.630 ±0.008	33 **As** arsenic 74.922 ±0.001	34 **Se** selenium 78.971 ±0.008	35 **Br** bromine 79.904 ±0.003	36 **Kr** krypton 83.798 ±0.002
37 **Rb** rubidium 85.468 ±0.001	38 **Sr** strontium 87.62 ±0.01	39 **Y** yttrium 88.906 ±0.001	40 **Zr** zirconium 91.224 ±0.002	41 **Nb** niobium 92.906 ±0.001	42 **Mo** molybdenum 95.95 ±0.01	43 **Tc** technetium [97]	44 **Ru** ruthenium 101.07 ±0.02	45 **Rh** rhodium 102.91 ±0.01	46 **Pd** palladium 106.42 ±0.01	47 **Ag** silver 107.87 ±0.01	48 **Cd** cadmium 112.41 ±0.01	49 **In** indium 114.82 ±0.01	50 **Sn** tin 118.71 ±0.01	51 **Sb** antimony 121.76 ±0.01	52 **Te** tellurium 127.60 ±0.03	53 **I** iodine 126.90 ±0.01	54 **Xe** xenon 131.29 ±0.01
55 **Cs** caesium 132.91 ±0.01	56 **Ba** barium 137.33 ±0.01	57-71 lanthanoids	72 **Hf** hafnium 178.49 ±0.01	73 **Ta** tantalum 180.95 ±0.01	74 **W** tungsten 183.84 ±0.01	75 **Re** rhenium 186.21 ±0.01	76 **Os** osmium 190.23 ±0.03	77 **Ir** iridium 192.22 ±0.01	78 **Pt** platinum 195.08 ±0.02	79 **Au** gold 196.97 ±0.01	80 **Hg** mercury 200.59 ±0.01	81 **Tl** thallium 204.38 ±0.01	82 **Pb** lead 207.2 ±1.1	83 **Bi** bismuth 208.98 ±0.01	84 **Po** polonium [209]	85 **At** astatine [210]	86 **Rn** radon [222]
87 **Fr** francium [223]	88 **Ra** radium [226]	89-103 actinoids	104 **Rf** rutherfordium [267]	105 **Db** dubnium [268]	106 **Sg** seaborgium [269]	107 **Bh** bohrium [270]	108 **Hs** hassium [269]	109 **Mt** meitnerium [277]	110 **Ds** darmstadtium [281]	111 **Rg** roentgenium [282]	112 **Cn** copernicium [285]	113 **Nh** nihonium [286]	114 **Fl** flerovium [290]	115 **Mc** moscovium [290]	116 **Lv** livermorium [293]	117 **Ts** tennessine [294]	118 **Og** oganesson [294]

57 **La** lanthanum 138.91 ±0.01	58 **Ce** cerium 140.12 ±0.01	59 **Pr** praseodymium 140.91 ±0.01	60 **Nd** neodymium 144.24 ±0.01	61 **Pm** promethium [145]	62 **Sm** samarium 150.36 ±0.02	63 **Eu** europium 151.96 ±0.01	64 **Gd** gadolinium 157.25 ±0.03	65 **Tb** terbium 158.93 ±0.01	66 **Dy** dysprosium 162.50 ±0.01	67 **Ho** holmium 164.93 ±0.01	68 **Er** erbium 167.26 ±0.01	69 **Tm** thulium 168.93 ±0.01	70 **Yb** ytterbium 173.05 ±0.02	71 **Lu** lutetium 174.97 ±0.01
89 **Ac** actinium [227]	90 **Th** thorium 232.04 ±0.01	91 **Pa** protactinium 231.04 ±0.01	92 **U** uranium 238.03 ±0.01	93 **Np** neptunium [237]	94 **Pu** plutonium [244]	95 **Am** americium [243]	96 **Cm** curium [247]	97 **Bk** berkelium [247]	98 **Cf** californium [251]	99 **Es** einsteinium [252]	100 **Fm** fermium [257]	101 **Md** mendelevium [258]	102 **No** nobelium [259]	103 **Lr** lawrencium [262]

附表二　一些物质的摩尔质量

化学式	M_B (g/mol)	化学式	M_B (g/mol)	化学式	M_B (g/mol)
Ag	107.87	$Ba(OH)_2 \cdot 8H_2O$	315.46	$CdCO_3$	172.42
$AgBr$	187.77	$BaSO_4$	233.39	CdS	144.48
$AgBrO_3$	235.77	$Ba_3(AsO_4)_2$	689.82	Ce	140.12
$AgCN$	133.89	Be	9.012	CeO_2	172.11
$AgCl$	143.32	BeO	25.01	$Ce(SO_4)_2$	332.24
AgI	234.77	Bi	208.98	$Ce(SO_4)_2 \cdot 4H_2O$	404.30
$AgNO_3$	169.87	$BiCl_3$	315.34	$Ce(SO_4)_2$	
$AgSCN$	165.95	$Bi(NO_3)_3 \cdot 5H_2O$	485.07	$\cdot 2(NH_4)_2SO_4 \cdot 2H_2O$	632.55
Ag_2CrO_4	331.73	$BiOCl$	260.43	Cl	35.45
Ag_2SO_4	311.80	$BiOHCO_3$	286.00	Cl_2	70.91
Ag_3AsO_4	462.52	$BiONO_3$	286.98	Co	58.93
Ag_3PO_4	418.58	Bi_2O_3	465.96	$CoCl_2$	129.84
Al	26.98	Bi_2S_3	514.16	$CoCl_2 \cdot 6H_2O$	237.93
$AlBr_3$	266.69	Br	79.90	$Co(NO_3)_2$	182.94
$AlCl_3$	133.34	BrO_3^-	127.90	$Co(NO_3)_2 \cdot 6H_2O$	291.03
$AlCl_3 \cdot 6H_2O$	241.43	Br_2	159.81	CoS	91.00
$Al(NO_3)_3$	213.00	C	12.01	$CoSO_4$	155.00
$Al(NO_3)_3 \cdot 9H_2O$	375.13	CH_3COOH （醋酸）	60.05	$CoSO_4 \cdot 7H_2O$	281.10
Al_2O_3	101.96	$(CH_3CO)_2O$ （醋酐）	102.09	Co_2O_3	165.86
$Al(OH)_3$	78.00	CN^-	26.01	Co_3O_4	240.80
$Al_2(SO_4)_3$	342.15	CO	28.01	Cr	52.00
$Al_2(SO_4)_3 \cdot 18H_2O$	666.43	$CO(NH_2)_2$ （尿素）	60.05	$CrCl_3$	158.35
As	74.92	CO_2	44.01	$CrCl_3 \cdot 6H_2O$	266.44
AsO_4^{3-}	138.92	CO_3^{2-}	60.01	CrO_4^{2-}	115.99
As_2O_3	197.84	$C_2O_4^{2-}$	88.02	Cr_2O_3	151.99
As_2O_5	229.84	Ca	40.08	$Cr_2(SO_4)_3$	392.18
As_2S_3	246.04	$CaCl_2$	110.98	Cu	63.55
B	10.81	$CaCl_2 \cdot 2H_2O$	147.01	$CuCl$	99.00
B_2O_3	69.62	$CaCl_2 \cdot 6H_2O$	219.08	$CuCl_2$	134.45
Ba	137.33	$CaCO_3$	100.09	$CuCl_2 \cdot 2H_2O$	170.48
$BaBr_2$	297.14	CaC_2O_4	128.10	CuI	190.45
$BaCO_3$	197.34	CaO	56.08	$Cu(NO_3)_2$	187.55
$BaCl_2$	208.23	$Ca(OH)_2$	74.09	$Cu(NO_3)_2 \cdot 3H_2O$	241.60
$BaCl_2 \cdot 2H_2O$	244.26	$CaSO_4$	136.14	CuO	79.55
$BaCrO_4$	253.32	$Ca_3(PO_4)_2$	310.18	CuS	95.61

化学式	M_B (g/mol)	化学式	M_B (g/mol)	化学式	M_B (g/mol)
BaO	153.33	$CdCl_2$	183.32	$CuSCN$	121.63
$Ba(OH)_2$	171.34	HF	20.01	$CuSO_4$	159.61
$CuSO_4 \cdot 5H_2O$	249.69	HI	127.91	K	39.10
Cu_2O	143.09	HIO_3	175.91	$KAl(SO_4)_2 \cdot 12H_2O$	474.38
$Cu_2(OH)_2CO_3$	221.12	HNO_2	47.01	KBr	119.00
Cu_2S	159.16	HNO_3	63.01	$KBrO_3$	167.00
F	19.00	H_2	2.016	KCN	65.12
F_2	38.00	H_2CO_3	62.02	KCl	74.55
Fe	55.85	$H_2C_2O_4$	90.04	$KClO_3$	122.55
$FeCO_3$	115.86	$H_2C_2O_4 \cdot 2H_2O$	126.07	$KClO_4$	138.55
$FeCl_2$	126.75	H_2O	18.01	$KFe(SO_4)_2 \cdot 12H_2O$	503.25
$FeCl_2 \cdot 4H_2O$	198.81	H_2O_2	34.01	$KHC_2O_4 \cdot H_2O$	146.14
$FeCl_3$	162.21	H_2S	34.08	$KHC_2O_4 \cdot H_2C_2O_4 \cdot 2H_2O$	254.19
$FeCl_3 \cdot 6H_2O$	270.30	H_2SO_3	82.08	$KHC_4H_4O_6$	188.18
$FeNH_4(SO_4)_2 \cdot 12H_2O$	482.20	$H_2SO_3 \cdot NH_2$	98.10	（酒石酸氢钾）	
$Fe(NO_3)_3$	241.86	（氨基磺酸）		$KHC_8H_4O_4$	204.22
$Fe(NO_3)_3 \cdot 9H_2O$	404.00	H_2SO_4	98.08	（邻苯二甲酸氢钾）	
FeO	71.85	H_3AsO_3	125.94	$KHSO_4$	136.17
$Fe(OH)_3$	106.87	H_3AsO_4	141.94	KI	166.00
FeS	87.91	H_3BO_3	61.83	KIO_3	214.00
FeS_2	119.98	H_3PO_3	82.00	$KIO_3 \cdot HIO_3$	389.91
$FeSO_4$	151.91	H_3PO_4	98.00	$KMnO_4$	158.03
$FeSO_4 \cdot 7H_2O$	278.02	Hg	200.59	KNO_2	85.10
$FeSO_4 \cdot (NH_4)_2SO_4$	392.14	$Hg(CN)_2$	252.63	KNO_3	101.10
$\cdot 6H_2O$		$HgCl_2$	271.50	$KNaC_4H_4O_6 \cdot 4H_2O$	282.22
Fe_2O_3	159.69	HgI_2	454.40	（酒石酸钾钠）	
$Fe_2(SO_4)_3$	399.88	$Hg(NO_3)_2$	324.60	KOH	56.10
$Fe_2(SO_4)_3 \cdot 9H_2O$	562.02	HgO	216.59	K_2CO_3	138.21
Fe_3O_4	231.54	HgS	232.66	K_2CrO_4	194.19
H	1.008	$HgSO_4$	296.65	$K_2Cr_2O_7$	294.18
HBr	80.91	Hg_2Br_2	560.99	K_2O	94.20
HCN	27.02	Hg_2Cl_2	472.09	K_2PtCl_6	485.99
$HCOOH$ （甲酸）	46.02	Hg_2I_2	654.99	K_2SO_4	174.26
$HC_2H_3O_2$ （醋酸）	60.05	$Hg_2(NO_3)_2$	525.19	$K_2SO_4 \cdot Al_2(SO_4)_3$	948.78
$HC_7H_5O_2$ （苯甲酸）	122.12	$Hg_2(NO_2)_2 \cdot 2H_2O$	561.22	$\cdot 24H_2O$	
HCl	36.46	Hg_2SO_4	497.24	$K_2S_2O_7$	254.32
Cd	112.41	I	126.90	K_3AsO_4	256.22

化学式	M_B (g/mol)	化学式	M_B (g/mol)	化学式	M_B (g/mol)
$HClO_4$	100.46	I_2	253.81	$K_3Fe(CN)_6$	329.25
K_3PO_4	212.27	$NH_4C_2H_3O_2$（醋酸铵）	77.08	$Na_2C_2O_4$	134.00
$K_4Fe(CN)_6$	368.35	NH_4Cl	53.49	Na_2HAsO_3	169.91
Li	6.941	NH_4HCO_3	79.06	Na_2HPO_4	141.96
$LiCl$	42.39	$NH_4H_2PO_4$	115.03	$Na_2HPO_4 \cdot 12H_2O$	358.14
$LiOH$	23.95	NH_4NO_3	80.04	Na_2H_2Y（EDTA 钠）	336.21
Li_2CO_3	73.89	NH_4VO_3	116.98	$Na_2H_2Y \cdot 2H_2O$	372.24
Li_2O	29.88	$(NH_4)_2CO_3$	96.09	Na_2O	61.98
Mg	24.30	$(NH_4)_2C_2O_4$	124.10	Na_2O_2	77.98
$MgCO_3$	84.31	$(NH_4)_2C_2O_4 \cdot H_2O$	142.11	Na_2S	78.05
MgC_2O_4	112.32	$(NH_4)_2HPO_4$	132.06	$Na_2S \cdot 9H_2O$	240.18
$MgCl_2$	95.21	$(NH_4)_2MoO_4$	196.01	Na_2SO_3	126.04
$MgCl_2 \cdot 6H_2O$	203.30	$(NH_4)_2PtCl_6$	443.87	Na_2SO_4	142.04
$MgNH_4AsO_4$	181.26	$(NH_4)_2S$	68.14	$Na_2S_2O_3$	158.11
$MgNH_4PO_4$	137.31	$(NH_4)_2SO_4$	132.14	$Na_2S_2O_3 \cdot 5H_2O$	248.19
$Mg(NO_3)_2 \cdot 6H_2O$	256.41	$(NH_4)_3PO_4 \cdot 12MoO_3$	1 876.32	Na_3AsO_3	191.89
MgO	40.30	NO_3^-	62.00	Na_3AsO_4	207.89
$Mg(OH)_2$	58.32	Na	22.99	Na_3PO_4	163.94
$MgSO_4$	120.37	$NaBiO_3$	279.97	$Na_3PO_4 \cdot 12H_2O$	380.12
$MgSO_4 \cdot 7H_2O$	246.48	$NaBr$	102.89	Ni	58.34
$Mg_2P_2O_7$	222.55	$NaBrO_3$	150.89	$NiC_8H_{14}O_4N_4$	288.56
Mn	54.94	$NaCHO_2$（甲酸钠）	68.01	（丁二酮肟镍）	
$MnCO_3$	114.95	$NaCN$	49.01	$NiCl_2 \cdot 6H_2O$	237.34
$MnCl_2 \cdot 4H_2O$	197.90	$NaC_2H_3O_2$（醋酸钠）	82.03	$Ni(NO_3)_2 \cdot 6H_2O$	290.44
$Mn(NO_3)_2 \cdot 6H_2O$	287.04	$NC_2H_3O_2 \cdot 3H_2O$	136.08	NiO	74.34
MnO	70.94	$NaCl$	58.44	NiS	90.41
MnO_2	86.94	$NaClO$	74.44	$NiSO_4 \cdot 7H_2O$	280.51
MnS	87.00	$NaHOC_3$	84.01	O	16.00
$MnSO_4$	151.00	NaH_2PO_4	119.98	OH^-	17.01
$MnSO_4 \cdot 4H_2O$	223.06	$NaH_2PO_4 \cdot H_2O$	137.99	O_2	32.00
Mn_2O_3	157.87	NaI	149.89	P	30.97
$Mn_2P_2O_7$	283.82	$NaNO_2$	69.00	PO_4^{3-}	94.97
Mn_3O_4	228.81	$NaNO_3$	84.99	P_2O_5	141.94
N	14.01	$NaOH$	40.00	Pb	207.20
N_2	28.01	$Na_2B_4O_7$	201.22	$PbCO_3$	267.21
NH_3	17.03	$Na_2B_4O_7 \cdot 10H_2O$	381.37	PbC_2O_4	295.22
NH_4^+	18.04	Na_2CO_3	105.99	$Pb(C_2H_3O_2)_2$	325.29
$NH_4C_2H_3O_2$	77.08	$Na_2CO_3 \cdot 10H_2O$	286.14	$Pb(C_2H_3O_2)_2 \cdot 3H_2O$	379.34

化学式	M_B(g/mol)	化学式	M_B(g/mol)	化学式	M_B(g/mol)
$PbCl_2$	278.11	SnO_2	150.71	$UO_2(C_2H_3O_2)_2$	
$PbCrO_4$	323.19	SnS	150.78	$\cdot 2H_2O$	424.15
PbI_2	461.01	SnS_2	182.84	UO_3	286.03
$Pb(IO_3)_2$	557.00	Sr	87.62	U_3O_8	842.08
$Pb(NO_3)_2$	331.21	$SrCO_3$	147.63	V	50.94
PbO	223.20	SrC_2O_4	175.64	VO_2	82.94
PbO_2	239.20	$SrCl_2 \cdot 6H_2O$	266.62	V_2O_5	181.88
PbS	239.27	$Sr(NO_3)_2$	211.63	W	183.84
$PbSO_4$	303.26	$Sr(NO_3)_2 \cdot 4H_2O$	283.69	WO_3	231.85
Pb_2O_3	462.40	SrO	103.62	Zn	65.39
Pb_3O_4	685.60	$SrSO_4$	183.68	$ZnCO_3$	125.40
$Pb_3(PO_4)_2$	811.54	$Sr_3(PO_4)_2$	452.80	ZnC_2O_4	153.41
S	32.07	Th	232.04	$Zn(C_2H_3O_2)_2$	183.48
SO_2	64.06	$Th(C_2O_4)_2 \cdot 6H_2O$	516.17	$Zn(C_2H_3O_2)_2 \cdot 2H_2O$	219.51
SO_3	80.06	$ThCl_4$	373.85	$ZnCl_2$	136.30
SO_4^{2-}	96.06	$Th(NO_3)_4$	480.06	$Zn(NO_3)_2$	189.40
Sb	121.78	$Th(NO_3)_4 \cdot 4H_2O$	552.11	$Zn(NO_3)_2 \cdot 6H_2O$	297.49
$SbCl_3$	228.12	$Th(SO_4)_2$	424.16	ZnO	81.39
$SbCl_5$	299.02	$Th(SO_4)_2 \cdot 9H_2O$	586.30	ZnS	97.46
Sb_2O_3	291.52	Ti	47.88	$ZnSO_4$	161.45
Sb_2O_5	323.52	$TiCl_3$	154.24	$ZnSO_4 \cdot 7H_2O$	287.56
Si	28.09	$TiCl_4$	189.69	$Zn_2P_2O_7$	304.72
$SiCl_4$	169.90	TiO_2	79.88	Zr	91.22
SiF_4	104.08	$TiOSO_4$	159.94	$Zr(NO_3)_4$	339.24
SiO_2	60.08	U	238.03	$Zr(NO_3)_4 \cdot 5H_2O$	429.32
Sn	118.71	UCl_4	379.84	$ZrOCl_2 \cdot 8H_2O$	322.25
$SnCl_2$	189.62	UF_4	314.02	ZrO_2	123.22
$SnCl_2 \cdot 2H_2O$	225.65	$UO_2(C_2H_3O_2)_2$	388.12	$Zr(SO_4)_2$	283.35

附表三　实验室常用酸、碱溶液的浓度

溶液名称	密度（g/mL,20℃）	质量分数/%	物质的量浓度（mol/L）
浓 H_2SO_4	1.84	98	18
稀 H_2SO_4	1.18	25	3
	1.06	9	1
浓 HNO_3	1.42	69	16
稀 HNO_3	1.20	33	6
	1.07	12	2
浓 HCl	1.19	28	12
稀 HCl	1.10	20	6
	1.03	7	2
H_3PO_4	1.7	85	15
浓高氯酸（$HClO_4$）	1.7～1.75	70～72	12
稀 $HClO_4$	1.12	19	2
冰醋酸（HAc）	1.05	99	17
稀 HAc	1.02	12	2
氢氟酸（HF）	1.13	40	23
浓氨水（$NH_3 \cdot H_2O$）	0.88	28	15
稀氨水	0.98	4	2
浓 NaOH	1.43	40	14
	1.33	30	13
稀 NaOH	1.09	8	2
$Ba(OH)_2$（饱和）	–	2	0.1
$Ca(OH)_2$（饱和）	–	0.15	–

附表四　实验室中一些试剂的配制方法

试 剂 名 称	浓度(mol/L)	配 制 方 法
硫化钠 Na_2S	1	称取 240g $Na_2S \cdot 9H_2O$、40g NaOH 溶于适量水中，稀释至 1L，混匀
硫化铵 $(NH_4)_2S$	3	通 H_2S 于 200mL 浓 $NH_3 \cdot H_2O$ 中直至饱和，然后再加 200mL 浓 $NH_3 \cdot H_2O$，最后加水稀释至 1L，混匀
氯化亚锡 $SnCl_2$	0.25	称取 56.4g $SnCl_2 \cdot 2H_2O$ 溶于 100mL 浓 HCl 中，加水稀释至 1L，在溶液中放几颗纯锡粒
氯化铁 $FeCl_3$	0.5	称取 135.2g $FeCl_3 \cdot 6H_2O$ 溶于 100mL 6mol/L HCl 中，加水稀释至 1L
三氯化铬 $CrCl_3$	0.1	称取 26.7g $CrCl_3 \cdot 6H_2O$ 溶于 30mL 6mol/L HCl 中，加水稀释至 1L
硝酸亚汞 $Hg_2(NO_3)_2$	0.1	称取 56g $Hg_2(NO_3)_2 \cdot 2H_2O$ 溶于 250mL 6mol/L HNO_3 中，加水稀释至 1L，并加入少许金属汞
硝酸铅 $Pb(NO_3)_2$	0.25	称取 83g $Pb(NO_3)_2$ 溶于少量水中，加入 15mL 6mol/L HNO_3，用水稀释至 1L
硝酸铋 $Bi(NO_3)_3$	0.1	称取 48.5g $Bi(NO_3)_3 \cdot 5H_2O$ 溶于 250mL 1mol/L HNO_3 中，加水稀释至 1L
硫酸亚铁 $FeSO_4$	0.25	称取 69.5g$FeSO_4 \cdot 7H_2O$ 溶于适量水中，加入 5mL 18mol/L H_2SO_4，再加水稀释至 1L，并置入小铁钉数枚
Cl_2 水	Cl_2 的饱和水溶液	将 Cl_2 通入水中至饱和为止（用时临时配制）
Br_2 水	Br_2 的饱和水溶液	在带有良好磨口塞的玻璃瓶内，将市售的 Br_2 约 50g（16mL）注入 1L 水中，在 2h 内经常剧烈振荡，每次振荡之后微开塞子，使积聚的 Br_2 蒸气放出，在储存瓶底产总有过量的溴。将 Br_2 倒入试剂瓶时，剩余的 Br_2 应留于储存瓶中，而不倒入试剂瓶（倾倒 Br_2 或 Br_2 水时，应在通风橱中进行，将凡士林涂在手上或带橡皮手套操作，以防 Br_2 蒸气灼伤）
I_2 水	~0.005	将 1.3g I_2 和 5g KI 溶解在尽可能少量的水中，待 I_2 完全溶解后（充分搅动）再加水稀释至 1L
亚硝酰铁氰化钠	3	称取 3g $Na_2[Fe(CN_5)NO] \cdot 2H_2O$ 溶于 100mL 水中
淀粉溶液	0.5	称取易溶淀粉 1g 和 $HgCl_2$5mg（作防腐剂）置于烧杯中，加水少许调成薄浆，然后倾入 200mL 沸水中
奈斯勒试剂		称取 115g HgI_2 和 80g KI 溶于足量的水中，稀释至 500mL，然后加入 500mL 6mol/L NaOH 溶液，静置后取其清液保存于棕色瓶中
对氨基苯磺酸	0.34	0.5g 对氨基苯磺酸溶于 150mL 2mol/L HAc 溶液中
α-萘胺	0.12	0.3g α-萘胺加 20mL 水，加热煮沸，在所得溶液中加入 150mL 2mol/L HOAc
钼酸铵		5g 钼酸铵溶于 100mL 水中，加入 35mL HNO_3（密度 1.2g/mL）
硫代乙酰胺	5	5g 硫代乙酰胺溶于 100mL 水中
钙指示剂	0.2	0.2g 钙指示剂溶于 100mL 水中
镁试剂	0.001	0.001g 对硝基偶氮间苯二酚溶于 100mL 2mol/L NaOH 中
铝试剂	1	1g 铝试剂溶于 1L 水中
二苯硫腙	0.01	10mg 二苯硫腙溶于 100mL CCl_4 中
丁二酮肟	1	1g 丁二酮肟溶于 100mL 95% 乙醇中

续表

试 剂 名 称	浓度(mol/L)	配 制 方 法
醋酸铀酰锌		(1) 10g UO₂(OAc)₂·2H₂O 和 6mL 6mol/L HOAc 溶于 50mL 水中 (2) 30g Zn(OAc)₂·2H₂O 和 3mL 6mol/L HCl 溶于 50mL 水中将 (1)、(2) 两种溶液混合，24h 后取清液使用
二苯碳酰二肼 （二苯偕肼）	0.04	0.04g 二苯碳酰二肼溶于 20mL 95% 乙醇中，边搅拌，边加入 80mL(1:9)H₂SO₄(存于冰箱中可用一个月)
六亚硝酸合钴 （Ⅲ）钠盐		Na₃[Co(NO₂)₆] 和 NaOAc 各 20g，溶解于 20mL 冰醋酸和 80mL 水的混合溶液中，贮于棕色瓶中备用（久置溶液，颜色由棕变红即失效）
NH₃·H₂O-NH₄Cl 缓冲溶液	pH = 10.0	称取 20.00gNH₄Cl(s)溶于适量水中，加入 100.00mL 浓氨水（密度 0.9g/mL）混合后稀释至 1L，即为 pH = 10.0 的缓冲溶液
邻苯二甲酸氢钾-氢氧 化钠缓冲溶液	pH = 4.00	量取 0.200mol/L 邻苯二甲酸氢钾溶液 250.00mL、0.100mol/L 氢氧化钠溶液 4.00mL，混合后稀释至 1L，即为 pH = 4.00 的缓冲溶液

附表五　常见阳离子、阴离子的主要鉴定反应

离子	试剂	鉴定反应	介质条件	主要干扰离子
NH_4^+	NaOH	$NH_4^+ + OH^- \xrightarrow{\text{加热}} NH_3 \uparrow + H_2O$ NH_3 气使湿润的红色石蕊试纸变蓝或 pH 试纸呈碱性反应	强碱性	CN^- $CH^- + 2H_2O \xrightarrow{(OH^-\text{并加热})} HCOO^- + NH_3 \uparrow$
	奈氏试剂〔四碘合汞（Ⅱ）酸钾的碱性溶液〕	$NH_4^+ + 2[HgI_4]^{2-} + 4OH^- \longrightarrow Hg_2NI \downarrow$ （棕色）$+ 7I^- + 4H_2O$	碱性介质	Fe^{3+}、Cr^{3+}、Co^{2+}、Ni^{2+}、Ag^+、Hg^{2+} 等离子能与奈斯勒试剂生成有色沉淀，妨碍 NH_4^+ 的鉴定
Na^+	醋酸铀酰锌	$Na^+ + Zn^{2+} + 3UO_2^{2+} + 9Ac^- + 9H_2O$ $\longrightarrow ZaZn(UO_2)_3(OAc)_9 \cdot 9H_2O \downarrow$ （淡黄绿色）	中性或乙酸溶液中	大量 K^+ 存在有干扰〔生成 $KOAc \cdot UO_2 \cdot (OAc)_2$ 针状结晶〕，Ag^+、Hg_2^{2+}、$Sb(Ⅲ)$ 存在亦有干扰
	焰色反应	挥发性钠盐在煤气灯的无色火焰（氧化焰）中灼烧时，火焰呈黄色		
K^+	$Na_3[Co(NO_2)_6]$	$2K^+ + Na^+ + [Co(NO_2)_6]^{3-} \longrightarrow K_2Na$ $[CO(NO_2)_6] \downarrow$（亮黄色）	中性或弱酸性	Rb^+、Cs^+、NH_4^+ 能与试剂形成相似的化合物，妨碍鉴定
	焰色反应	挥发性钾盐在煤气灯的无色火焰（氧化焰）中灼烧时，火焰呈紫色		Na^+ 存在时，K^+ 所显示的紫色被黄色遮盖，可透过蓝色玻璃去观察
Mg^{2+}	镁试剂	镁试剂被氧化镁吸附后呈天蓝色沉淀	强碱性介质	1. 除碱金属外，在强碱性介质中能形成有色沉淀的离子如 Ag^+、Hg^{2+}、Ni^{2+}、Co^{2+}、Cr^{3+}、Cu^{2+}、Mn^{2+}、Fe^{3+} 等离子对反应均有干扰 2. 大量 NH_4^+ 存在会降低 OH^- 的浓度，从而降低 Mg^{2+} 鉴定反应的灵敏度
Ba^{2+}	K_2CrO_4	$Ba^{2+} + CrO_4^{2-} \longrightarrow BaCrO_4 \downarrow$ （黄色）	中性或弱酸性介质	Sr^{2+}、Pb^{2+}、Ag^+、Ni^{2+}、Hg_2^{2+} 等离子与 CrO_4^{2-} 能生成有色沉淀，影响 Ba^{2+} 的检出
	焰色反应	挥发性钡盐使火焰呈黄绿色		
Ca^{2+}	$(NH_4)_2C_2O_4$	$Ca^{2+} + C_2O_4^{2-} \longrightarrow CaC_2O_4 \downarrow$ （白色）	中性或弱酸性介质	Ag^+、Pb^{2+}、Cd^{2+}、Hg^{2+}、Hg_2^{2+} 等金属离子均能与 $C_2O_4^{2-}$ 作用生成沉淀对反应有干扰，可在氨性试液中加入 Zn 粉，将它们还原而除去
	焰色反应	挥发性钙盐使火焰呈砖红色		
Al^{3+}	铝试剂	形成红色絮状沉淀	$pH = 4 \sim 5$ 左右	Fe^{3+}、Cr^{3+}、Co^{2+}、Mn^{2+} 等离子也能生成与铝相类似的红色沉淀而有干扰
Sb^{3+}	Sn 片	$2Sb^{3+} + 3Sn \longrightarrow 2Sb \downarrow$（黑色）$+ 3Sn^{2+}$	酸性介质	Ag^+、AsO_2^-、Bi^{3+} 等离子也能与 Sn 发生氧化还原反应，析出黑色金属，妨碍 Sb^{3+} 鉴定
Bi^{3+}	$Na_2[Sn(OH)_4]$	$2Bi^{3+} + 3[Sn(OH)_4]^{2-} + 6OH^- \longrightarrow$ $2Bi \downarrow$（黑色）$+ 3[Sn(OH)_6]^{2-}$ 注意：试剂必须临时配制	强碱性介质	Pb^{2+} 存在时，也会慢慢地被 $[Sn(OH)_4]^{2-}$ 还原而析出黑色金属 Pb，干扰 Bi^{3+} 的鉴定

续表

离子	试剂	鉴定反应	介质条件	主要干扰离子
Sn^{2+}	$HgCl_2$	$Sn^{2+} + 2HgCl_2 + 4Cl^- \longrightarrow Hg_2Cl_2\downarrow$（白色）$+ [SnCl_6]^{2-}$ $Sn^{2+} + Hg_2Cl_2 + 4Cl^- \longrightarrow 2Hg\downarrow$（黑色）$+ [SnCl_6]^{2-}$	酸性介质	
Pb^{2+}	K_2CrO_4	$Pb^{2+} + CrO_4^{2-} \longrightarrow PbCrO_4\downarrow$ （黄色）	中性或弱酸性介质	Ba^{2+}、Sr^{2+}、Ag^+、Ni^{2+}、Zn^{2+} 等离子与 CrO_4^{2-} 作用生成有色沉淀，影响 Pb^{2+} 的检出
Cr^{3+}（或 CrO_4^{2-}）	用 H_2O_2 氧化后加可溶性 Pb^{2+} 盐（或 Ag^+ 盐或 Ba^{2+} 盐）	$Cr^{3+} + 4OH^- \longrightarrow [Cr(OH)_4]^-$ $2[Cr(OH)_4]^- + 3H_2O_2 + 2OH^- \longrightarrow 2CrO_4^{2-} + 8H_2O$	碱性介质	凡与 CrO_4^{2-} 生成有色沉淀的金属离子均有干扰
		$CrO_4^{2-} + Pb^{2+} \longrightarrow PbCrO_4\downarrow$ （黄色） $CrO_4^{2-} + 2Ag^+ \longrightarrow Ag_2CrO_4\downarrow$ （砖红色） $CrO_4^{2-} + Ba^{2+} \longrightarrow BaCrO_4\downarrow$ （黄色）	用 HOAc 酸化成弱酸性介质	
Cr^{3+}（或 CrO_4^{2-}）	在 NaOH 条件下用 H_2O_2 氧化后再酸化并用戊醇（或乙醇）萃取	$Cr^{3+} + 4OH^- \longrightarrow [Cr(OH)_4]^-$ $2[Cr(OH)_4]^- + 3H_2O_2 + 2OH^- \longrightarrow 2CrO_4^{2-} + 8H_2O$	碱性介质	
		$2CrO_4^{2-} + 2H^+ \longrightarrow Cr_2O_7^{2-} + H_2O$ $Cr_2O_7^{2-} + 4H_2O_2 + 2H^+ \longrightarrow 3H_2O + 2H_2CrO_6$（蓝色） 反应要求在较低温度下进行	酸性介质	
Mn^{2+}	$NaBiO_3$	$2Mn^{2+} + 5NaBiO_3 + 14H^+ \longrightarrow 2MnO_4^-$（紫红色）$+ 5Na^+ + 5Bi^{3+} + 7H_2O$	HNO_3 或 H_2SO_4 介质	
Fe^{2+}	$K_3[Fe(CN)_6]$	$K^+ + Fe^{2+} + [Fe(CN)_6]^{3-} \longrightarrow KFe[Fe(CN)_6]\downarrow$（深蓝色）	酸性介质	
Fe^{3+}	$K_4[Fe(CN)_6]$	$K^+ + Fe^{3+} + [Fe(CN)_6]^{4-} \longrightarrow KFe[Fe(CN)_6]\downarrow$（深蓝色）	酸性介质	
	NH_4SCN	$Fe^{3+} + SCN^- \longrightarrow [Fe(NCS)]^{2+}$（血红色）	酸性介质	氟化物、磷酸、草酸、酒石酸、柠檬酸，含有 α-或 β-OH 基的有机酸能与 Fe^{3+} 生成稳定的配合物，妨碍 Fe^{3+} 的检出；大量 Cu^{2+} 存在能与 SCN^- 生成黑绿色的 $Cu(SCN)_2$ 沉淀，干扰 Fe^{3+} 的检出
Co^{2+}	NH_4SCN（饱和或固体），并用丙酮或戊醇萃取	$Co^{2+} + 4SCN^- \longrightarrow [Co(NCS)_4]^{2-}$（艳蓝绿色）	酸性介质	Fe^{3+} 干扰 Co^{2+} 的检出

续表

离子	试剂	鉴定反应	介质条件	主要干扰离子
Ni^{2+}	丁二酮肟 CH_3CNOH 丨 CH_3CNOH	Ni^{2+}能与丁二酮肟生成玫瑰红色的螯合物沉淀 	在氨性或醋酸钠溶液中进行，合适的酸度为 pH = 5~10	Co^{2+}、Fe^{2+}、Bi^{3+}分别与本试剂反应生成棕色、红色可溶物和黄色沉淀，Fe^{3+}、Mn^{2+}与氨水生成有色沉淀，均干扰Ni^{2+}的检出
Cu^{2+}	$K_4[Fe(CN)_6]$	$2Cu^{2+}+[Fe(CN)_6]^{4-}\longrightarrow$ $Cu_2[Fe(CN_6)]\downarrow$（红褐色）	中性或酸性介质	Fe^{3+}、Bi^{3+}、Co^{2+}等离子能与本试剂反应生成深色沉淀，均有干扰
Ag^+	HCl	$Ag^++Cl^-\longrightarrow AgCl\downarrow$（白色） 沉淀溶于过量氨水，用$HNO_3$酸化后沉淀又重新析出 $AgCl+2NH_3\cdot H_2O\longrightarrow$ $[Ag(NH_3)_2]^++Cl^-+2H_2O$ $[Ag(NH_3)_2]^++Cl^-+2H^+$ $\longrightarrow 2NH_4^++AgCl\downarrow$	酸性介质	Pb^{2+}、Hg_2^{2+}（与Cl^-形成$PbCl_2$、Hg_2Cl_2白色沉淀）干扰Ag^+的鉴定，但$PbCl_2$、Hg_2Cl_2难溶于氨水，可与$AgCl$分离
	K_2CrO_4	$CrO_4^{2-}+2Ag^+\longrightarrow 2Ag_2CrO_4\downarrow$ （砖红色）	中性或微酸性介质	Pb^{2+}、Hg_2^{2+}、Ba^{2+}有干扰（生成有色沉淀）
Zn^{2+}	$(NH_4)_2S$	$Zn^{2+}+S^{2-}\longrightarrow ZnS\downarrow$（白色）	$c(H^+)<$ 0.3mol/L	凡能与S^{2-}生成有色硫化物的金属离子均有干扰
	二苯硫腙（打萨宗）	加入二苯硫腙振荡后水层呈粉红色	强碱性	在中性或弱酸性条件下，许多重金属离子都能与二苯硫腙生成有色的配合物，因而应注意鉴定的介质条件
Cd^{2+}	H_2S或Na_2S	$Cd^{2+}+H_2S\longrightarrow CdS\downarrow$（黄色）$+2H^+$	碱性介质	凡能与S^{2-}生成有色硫化物沉淀的金属离子均有干扰
Hg^{2+}	$SnCl_2$	见Sn^{2+}的鉴定	酸性介质	
	KI 和 $NH_3\cdot H_2O$	（1）先加入过量KI： $Hg^{2+}+2I^-\longrightarrow HgI_2\downarrow$ $HgI_2+2I^-\longrightarrow [HgI_4]^{2-}$ （2）再加入$NH_3\cdot H_2O$或NH_4^+盐溶液并加入浓碱溶液，则生成红棕色沉淀： $NH_4^++2[HgI_4]^{2-}+4OH^-\longrightarrow Hg_2N\downarrow$ （棕色）$+7I^-+4H_2O$		凡能与I^-、OH^-生成深色沉淀的金属离子均有干扰
Cl^-	$AgNO_3$	$Cl^-+Ag^+\longrightarrow AgCl\downarrow$（白色） $AgCl$溶于过量氨水或$(NH_4)_2CO_3$，用HNO_3酸化后沉淀重新析出	酸性介质	

续表

离子	试剂	鉴定反应	介质条件	主要干扰离子
Br^-	氯水，CCl_4（或苯）	$2Br^- + Cl_2 \longrightarrow Br_2 + 2Cl^-$ 析出的溴溶于 CCl_4（或苯）溶剂中呈橙黄色或橙红色	中性或酸性介质	
SO_4^{2-}	$BaCl_2$	$SO_4^{2-} + Ba^{2+} \longrightarrow BaSO_4 \downarrow$（白色）	酸性介质	
SO_3^{2-}	稀 HCl	$SO_3^{2-} + 2H^+ \longrightarrow SO_2 \uparrow + H_2O$ SO_2 的检验 （1）SO_2 可使 MnO_4^- 还原而褪色 （2）SO_2 可将 I_2 还原为 I^-，使淀粉-I_2 溶液褪色 （3）SO_2 可使品红溶液褪色 因此，可用蘸有 MnO_4^- 溶液，或淀粉-I_2 溶液，或品红溶液的试纸检验	酸性介质	$S_2O_3^{2-}$、S^{2-} 对 SO_3^{2-} 的鉴定有干扰
	$Na_2[Fe(CN)_5NO]$ $ZnSO_4$ $K_4[Fe(CN)_6]$	生成红色沉淀	中性介质	S^{2-} 与 $Na_2[Fe(CN)_5NO]$ 生成紫红色配合物，干扰 SO_3^{2-} 的鉴定
$S_2O_3^{2-}$	稀 HCl	$S_2O_3^{2-} + 2H^+ \longrightarrow SO_2 \uparrow + S \downarrow + H_2O$ 反应中有硫析出使溶液变浑浊	酸性介质	SO_3^{2-}、S^{2-} 同时存在时，干扰 $S_2O_3^{2-}$ 的鉴定
	$AgNO_3$	$2Ag^+ + S_2O_3^{2-} \longrightarrow Ag_2S_2O_3 \downarrow$（白色） $Ag_2S_2O_3$ 沉淀不稳定，生成后立即发生水解反应，并且伴随明显的颜色变化，由白→黄→棕，最后变成黑色的 Ag_2S $Ag_2S_2O_3 + H_2O \longrightarrow Ag_2S \downarrow$（黑色）$+ 2H^+ + SO_4^{2-}$	中性介质	S^{2-}
S^{2-}	稀 HCl	$S^{2-} + 2H^+ \longrightarrow H_2S \uparrow$ H_2S 气体可使蘸有 $Pb(NO_3)_2$ 或 $Pb(OAc)_2$ 的试纸变黑	酸性介质	SO_3^{2-}、$S_2O_3^{2-}$ 存在有干扰
	$Na_2[Fe(CN)_5NO]$	$S^{2-} + [Fe(CN)_5NO]^{2-} \longrightarrow [Fe(CN)_5NOS]^{4-}$（紫红色）	碱性介质	
NO_2^-	对氨基苯磺酸 α-萘胺	溶液呈现红色	中性或乙酸介质	MnO_4^- 等强氧化剂有干扰
NO_3^-	$FeSO_4$ 浓 H_2SO_4	$NO_3^- + 3Fe^{2+} + 4H^+ \longrightarrow 3Fe^{3+} + NO + 2H_2O$ $Fe^{2+} + NO \longrightarrow [FeNO]^{2+}$（棕色） 在混合液与浓硫酸分层处形成棕色环	酸性介质	NO_2^- 有同样反应，妨碍鉴定
CO_3^{2-}	稀 HCl （稀 H_2SO_4）	$CO_3^{2-} + 2H^+ \longrightarrow CO_2 \uparrow + H_2O$ CO_2 气体使饱和 $Ba(OH)_2$ 溶液变混浊 $CO_2 + 2OH^- + Ba^{2+} \longrightarrow BaCO_3 \downarrow$（白色）$+ H_2O$	酸性介质	

离子	试剂	鉴定反应	介质条件	主要干扰离子
PO_4^{3-}	$AgNO_3$	$3Ag^+ + PO_4^{3-} \longrightarrow Ag_3PO_4 \downarrow$（黄色）	中性或弱酸性介质	CrO_4^{2-}、S^{2-}、AsO_4^{3-}、AsO_3^{3-}、I^-、$S_2O_3^{2-}$ 等离子能与 Ag^+ 生成有色沉淀，妨碍鉴定
	$(NH_4)_2MoO_4$（过量试剂）	$PO_4^{3-} + 3NH_4^+ + 12MoO_4^{2-} + 24H^+ \longrightarrow$ $(NH_4)_3PO_4 \cdot 12MoO_3 \cdot 6H_2O \downarrow$（黄色） $+6H_2O$ （1）无干扰离子时不必加 HNO_3 （2）磷钼酸铵能溶于过量磷酸盐溶液生成配合物，因此需要加入过量钼酸铵试剂	HNO_3 介质	（1）SO_3^{2-}、$S_2O_3^{2-}$、S^{2-}、I^-、Sn^{2+} 等还原性离子易将钼酸铵还原为低价钼的化合物—钼蓝，严重干扰 PO_4^{3-} 的检出 （2）SiO_3^{2-}、AsO_4^{3-} 与钼酸铵试剂也能形成相似的黄色沉淀，妨碍鉴定 （3）大量 Cl^- 存在时，可与 $Mo(\text{VI})$ 形成配合物而降低反应的灵敏度
SiO_3^{2-}	NH_4Cl（饱和）（加热）	$SiO_3^{2-} + 2NH_4^+ \longrightarrow H_2SiO_3 \downarrow$（白色胶状）$+ 2NH_3 \uparrow$	碱性介质	
F^-	H_2SO_4	$CaF_2 + H_2SO_4 \rightarrow 2HF \uparrow + CaSO_4$ HF 与硅酸盐或 SiO_2 作用，生成 SiF_4 气体。当 SiF_4 与水作用时，立即转化为不溶于水的硅酸沉淀而使水变混 $SiO_2 + 4HF \longrightarrow SiF_4 + 2H_2O$ $SiF_4 + 4H_2O \longrightarrow H_4SiO_4 \downarrow + 4HF$ 用上述方法鉴定溶液中的 F^- 时，应先将溶液蒸发至干或在乙酸存在下用 $CaCl_2$ 沉淀 F^-，将 CaF_2 离心分离后，小心烘干，然后进行试验		

附表六 常见阳离子与常用试剂的反应

离子	HCl	H_2SO_4	NaOH 适量	NaOH 过量	$NH_3 \cdot H_2O$ 适量	$NH_3 \cdot H_2O$ 过量	$c(H^+)=$ 0.3mol/L 下通 H_2S	$(NH_4)_2S$，或硫化物沉淀后加入过量 $(NH_4)_2S$
Na^+								
NH_4^+								
K^+								
Mg^{2+}			$Mg(OH)_2\downarrow$ （白色）	（不溶）	$Mg(OH)_2\downarrow$	（不溶）		
Ba^{2+}		$BaSO_4$ （白色）	1) $Ba(OH)_2\downarrow$ （白色）	（不溶）				
Sr^{2+}		$SrSO_4$ （白色）	1) $Sr(OH)_2\downarrow$ （白色）	（不溶）				
Ca^{2+}		1) $CaSO_4$ （白色）	1) $Ca(OH)_2\downarrow$ （白色）	（不溶）				
Al^{3+}			$Al(OH)_3\downarrow$ （白色）	$[Al(OH)_4]^-$ （无色）	$Al(OH)_3\downarrow$	（微溶）		$Al(OH)_3\downarrow$
Sn^{2+}			$Sn(OH)_2\downarrow$ （白色）	$[Sn(OH)_4]^{2-}$ （无色）	$Sn(OH)_2\downarrow$	（不溶）	$SnS\downarrow$ （褐色）	（不溶）
Sn^{4+}			$Sn(OH)_4\downarrow$ （白色）	$[Sn(OH)_6]^{2-}$ （无色）	$Sn(OH)_4\downarrow$	（不溶）	$SnS_2\downarrow$ （黄色）	SnS_3^{2-}
Pb^{2+}	1) $PbCl_2\downarrow$ （白色）	$PbSO_4$ （白色）	$Pb(OH)_2\downarrow$ （白色）	$[Pb(OH)_4]^{2-}$ （无色）	$Pb(OH)_2\downarrow$ 或碱式盐↓	（不溶）	$PbS\downarrow$ （黑色）	（不溶）
Sb^{3+}			$Sb(OH)_3\downarrow$ （白色）	$[Sb(OH)_4]^-$ （无色）	$Sb(OH)_3\downarrow$	（不溶）	$Sb_2S_3\downarrow$	SbS_3^{3-}
Sb^{5+}			$H_3SbO_4\downarrow$ （白色）	SbO_4^{3-}	$H_3SbO_4\downarrow$	（不溶）	$Sb_2S_5\downarrow$ （橙红）	SbS_4^{3-}
Bi^{3+}			$Bi(OH)_3\downarrow$ （白色）	（不溶）	$Bi(OH)_3\downarrow$	（不溶）	$Bi_2S_3\downarrow$ （黑褐）	（不溶）
Cu^{2+}			$Cu(OH)_2\downarrow$ （浅蓝色）	$[Cu(OH)_4]^{2-}$ （亮蓝）	碱式盐↓ （浅蓝色）	$[Cu(NH_3)_4]^{2+}$ （深蓝色）	$CuS\downarrow$ （黑色）	（不溶）
Ag^+	$AgCl\downarrow$ （白色）	1) Ag_2SO_4 （白色）	$Ag_2O\downarrow$ （棕褐）	（不溶）	$Ag_2O\downarrow$	$[Ag(NH_3)_2]^+$ （无色）	$Ag_2S\downarrow$ （黑色）	（不溶）

续表

离子	HCl	H_2SO_4	NaOH		$NH_3 \cdot H_2O$		$c(H^+) =$ 0.3mol/L 下通 H_2S	$(NH_4)_2S$，或 硫化物沉淀后 加 入 过 量 $(NH_4)_2S$
			适量	过量	适量	过量		
Zn^{2+}			$Zn(OH)_2 \downarrow$ （白色）	$[Zn(OH)_4]^{2-}$ （无色）	$Zn(OH)_2 \downarrow$	$[Zn(NH_3)_4]^{2+}$ （无色）		ZnS （白色）
Cd^{2+}			$Cd(OH)_2 \downarrow$ （白色）	（不溶）	$Cd(OH)_2 \downarrow$	$[Cd(NH_3)_4]^{2+}$ （无色）	$CdS \downarrow$ （黄色）	（不溶）
Hg^{2+}			$HgO \downarrow$ （黄色）	（不溶）	$^{2)} HgNH_2Cl \downarrow$ （白色）	（不溶）	$HgS \downarrow$ （黑色）	$[HgS_2]^{2-}$ （浓 Na_2S）
Hg_2^{2+}	$Hg_2Cl_2 \downarrow$ （白色）	$^{1)} Hg_2SO_4$ （白色）	$Hg_2O \downarrow \rightarrow$ $HgO \downarrow + Hg \downarrow$ （黑色）	（不溶）	$^{2)} HgNH_2Cl \downarrow$ $+ Hg \downarrow$ （黑色）	（不溶）	$HgS \downarrow$ $+ Hg \downarrow$	（不溶）
Cr^{3+}			$Cr(OH)_3 \downarrow$ （灰绿色）	$[Cr(OH)_4]^{-}$ （亮绿）	$Cr(OH)_3 \downarrow$	（不溶）		$Cr(OH)_3$
Mn^{2+}			$Mn(OH)_2 \downarrow$ （肉色） $\longrightarrow$ $MnO(OH)_2$ （棕色）$\downarrow$	（不溶）	$Mn(OH)_2 \downarrow$ $\longrightarrow$ $MnO(OH)_2$	（不溶）		MnS （肉色）
Fe^{2+}			$Fe(OH)_2 \downarrow$ （白色） $\longrightarrow$ $Fe(OH)_3 \downarrow$ （红棕色）	（不溶）	$Fe(OH)_2 \downarrow$ $\longrightarrow$ $Fe(OH)_3 \downarrow$	（不溶）		FeS （黑色）
Fe^{3+}			$Fe(OH)_3 \downarrow$ （红棕色）	（不溶）	$Fe(OH)_3 \downarrow$	（不溶）		Fe_2S_3 （黑色）
Co^{2+}			$Co(OH)_2 \downarrow$ （粉红） $\longrightarrow$ $CoO(OH) \downarrow$ （褐色）	（不溶）	碱式盐$\downarrow$ （蓝色）	$[Co(NH_3)_6]^{2+}$ （土黄色）$\rightarrow$ $[Co(NH_3)_6]^{3+}$ （棕红）		CoS （黑色）
Ni^{2+}			$Ni(OH)_2 \downarrow$ （绿色）	（不溶）	碱式盐$\downarrow$ （浅绿色）	$[Ni(NH_3)_6]^{2+}$ （蓝色）		NiS （黑色）

1）浓度大时才会出现沉淀。

2）$Hg(NO_3)_2$ 与 $NH_3 \cdot H_2O$ 反应则生成 $HgO \cdot NH_2HgNO_3$ 白色沉淀，$Hg_2(NO_3)_2$ 与 $NH_3 \cdot H_2O$ 反应则生成$HgO \cdot NH_2HgNO_3 + Hg$ 黑色沉淀。

附表七　常见阴离子与常用试剂的反应

离子＼试剂	稀 H_2SO_4（或稀 HCl）	$BaCl_2$ 中性或弱碱性溶液	$BaCl_2$ 酸性溶液或沉淀后加酸	$AgNO_3$ 中性或微酸性溶液	$AgNO_3$ 稀 HNO_3 溶液中	I_2－淀粉	$KMnO_4$	KI－淀粉
SO_4^{2-}		$BaSO_4\downarrow$（白色）	$BaSO_4\downarrow$（白色）	$Ag_2SO_4\downarrow$（白色）只在浓溶液中析出				
SO_3^{2-}	$SO_2\uparrow$	$BaSO_3\downarrow$（白色）	溶解	$Ag_2SO_3\downarrow$（白色）		SO_4^{2-}（蓝色褪去）	SO_4^{2-}（紫色褪去）	
$S_2O_3^{2-}$	$SO_2\uparrow+S\downarrow$	$BaS_2O_3\downarrow$（白色）	溶解	$Ag_2S_2O_3\downarrow\rightarrow Ag_2S\downarrow$ 颜色由白→黄→棕→黑	$S\downarrow$	$S_4O_6^{2-}$（蓝色褪去）	SO_4^{2-}（紫色褪去）	
CO_3^{2-}	$CO_2\uparrow$	$BaCO_3\downarrow$（白色）	溶解	Ag_2CO_3（白色）	$CO_2\uparrow$			
PO_4^{3-}		$Ba_3(PO_4)_2\downarrow$（白色）	溶解	$Ag_3PO_4\downarrow$（黄色）				
SiO_3^{2-}	H_2SiO_3（胶状）	$BaSiO_3\downarrow$（白色）	$H_2SiO_3\downarrow$	$Ag_2SiO_3\downarrow$（黄色）	$H_2SiO_3\downarrow$			
AsO_3^{3-}		$Ba_3(AsO_3)_2\downarrow$（白色）	溶解	$Ag_3AsO_3\downarrow$（黄色）		AsO_4^{3-}（蓝色褪去）（碱性介质）	AsO_4^{3-}（紫色褪去）	
AsO_4^{3-}		$Ba_3(AsO_4)_2\downarrow$（白色）	溶解	$Ag_3AsO_4\downarrow$（黄色）				AsO_3^{3-}（变蓝）（酸性介质）
F^-	浓 H_2SO_4 分解氟化物生成 HF	$BaF_2\downarrow$（白色）浓溶液中析出	溶解					
Cl^-				$AgCl\downarrow$（白色）	$AgCl\downarrow$		$Cl_2\uparrow$（紫色褪去）	
Br^-				$AgBr\downarrow$（淡黄色）	$AgBr\downarrow$		Br_2（紫色褪去）	
I^-				$AgI\downarrow$（黄色）	$AgI\downarrow$		I_2（紫色褪去）	
S^{2-}	$H_2S\downarrow$			$Ag_2S\downarrow$（黑色）	$Ag_2S\downarrow$	$S\downarrow$（蓝色褪去）	$S\downarrow$（紫色褪去）	
NO_3^-								
NO_2^-	$NO_2\uparrow+NO\uparrow$			$AgNO_2\downarrow$（淡黄色）			NO_3^-（紫色褪去）	$NO\uparrow$（变蓝）
OAc^-	HOAc							

附表八　常见离子和化合物的颜色

一、离子

序号	物　质	颜　色	序号	物　质	颜　色
1	$[Ti(H_2O)_6]^{3+}$	紫色	5	$[Fe(NCS)_n]^{3-n}$	血红色 ($n \leqslant 6$)
	$[TiO(H_2O_2)]^{2+}$	橙色		$[Fe(CN)_6]^{4-}$	黄色
	TiO^{2+}	无色		$[Fe(CN)_6]^{3-}$	红棕色
2	$[V(H_2O)_6]^{2+}$	蓝紫色		$[FeCl_6]^{3-}$	黄色
	$[V(H_2O)_6]^{3+}$	绿色		$[FeF_6]^{3-}$	无色
	VO^{2+}	蓝色		$[Fe(C_2O_4)_3]^{3-}$	黄色
	VO_2^+	黄色	6	$[Co(H_2O)_6]^{2+}$	粉红色
3	$[Cr(H_2O)_6]^{2+}$	天蓝色		$[Co(NH_3)_6]^{2+}$	土黄色
	$[Cr(H_2O)_6]^{3+}$	蓝紫色		$[Co(NH_3)_6]^{3+}$	红棕色
	$[Cr(NH_3)_6]^{3+}$	黄色		$[Co(NCS)_4]^{2-}$	蓝色
	$[CrCl(H_2O)_5]^{2+}$	蓝绿色	7	$[Ni(H_2O)_6]^{2+}$	亮绿色
	$[CrCl_2(H_2O)_4]^+$	绿色		$[Ni(NH_3)_6]^{2+}$	蓝色
	$[Cr(OH)_4]^-$	亮绿色		$[Ni(NH_3)_6]^{3+}$	蓝紫色
	$[CrO_4^{2-}]$	黄色	8	$[Cu(H_2O)_4]^{2+}$	蓝色
	$[Cr_2O_7^{2-}]$	橙色		$[Cu(NH_3)_4]^{2+}$	深蓝色
4	$[Mn(H_2O)_6]^{2+}$	肉色		$[Cu(OH)_4]^{2-}$	亮蓝色
	MnO_4^{2-}	绿色		$[CuCl_2]^-$	无色
	MnO_4^-	紫红色		$[Cu(NH_3)_2]^+$	无色
5	$[Fe(H_2O)_6]^{2+}$	浅绿色		$[CuCl_4]^{2-}$	黄色
	$[Fe(H_2O)_6]^{3+}$	淡紫色	9	I_3^-	浅棕黄色

二、化合物

序号	物　质	颜　色	序号	物　质	颜　色
1. 氧化物	PbO_2	棕褐色	1. 氧化物	CrO_3	橙红色
	Pb_3O_4	红色		MoO_2	紫色
	Pb_2O_3	橙色		WO_2	棕红色
	Sb_2O_3	白色		MnO_2	棕色
	Bi_2O_3	黄色		FeO	黑色
	TiO_2	白色		Fe_2O_3	棕红色
	V_2O_5	橙或黄色		Fe_3O_4	红色
	VO_2	深蓝色		CoO	灰绿色
	V_2O_3	黑色		Co_2O_3	黑色
	VO	黑色		NiO	暗绿色
	Cr_2O_3	绿色		Ni_2O_3	黑色

序号	物　质	颜　色	序号	物　质	颜　色
1. 氧化物	CuO	黑色	3. 氯化物	$CoCl_2 \cdot 6H_2O$	粉红色
	Cu_2O	暗红色		$Co(OH)Cl$	蓝色
	Ag_2O	褐色		$CuCl$	白色
	ZnO	白色		$AgCl$	白色
	CdO	棕黄色		Hg_2Cl_2	白色
	Hg_2O	黑色		$Hg(NH_2)Cl$	白色
	HgO	红色或黄色	4. 溴化物	$PbBr_2$	白色
2. 氢氧化物	$Mg(OH)_2$	白色		$AgBr$	淡黄色
	$Al(OH)_3$	白色	5. 碘化物	PbI_2	黄色
	$Sn(OH)_2$	白色		SbI_3	黄色
	$Sn(OH)_4$	白色		BiI_3	褐色
	$Pb(OH)_2$	白色		CuI	白色
	$Sb(OH)_3$	白色		AgI	黄色
	$Bi(OH)_3$	白色		Hg_2I_2	黄绿色
	$BiO(OH)$	灰黄色		HgI_2	红色
	$Cr(OH)_3$	灰绿色	6. 硫化物	SnS	褐色
	$Mn(OH)_2$	白色		SnS_2	黄色
	$MnO(OH)_2$	棕黑色		PbS	黑色
	$Fe(OH)_2$	白色		As_2S_3	黄色
	$Fe(OH)_3$	红棕色		As_2S_5	黄色
	$Co(OH)_2$	粉红色		Sb_2S_3	橙色
	$CoO(OH)$	褐色		Sb_2S_5	橙色
	$Ni(OH)_2$	绿色		Bi_2S_3	黑色
	$NiO(OH)$	黑色		Bi_2S_5	黑褐色
	$Cu(OH)$	黄色		MnS	肉色
	$Cu(OH)_2$	浅蓝色		FeS	黑色
	$Zn(OH)_2$	白色		Fe_2S_3	黑色
	$Cd(OH)_2$	白色		CoS	黑色
3. 氯化物	$Sn(OH)Cl$	白色		NiS	黑色
	$PbCl_2$	白色		Cu_2S	黑色
	$SbOCl$	白色		CuS	黑色
	$BiOCl$	白色		Ag_2S	黑色
	$TiCl_2 \cdot 6H_2O$	紫色或绿色		ZnS	白色
	$CrCl_3 \cdot 6H_2O$	绿色		CdS	黄色
	$FeCl_3 \cdot 6H_2O$	棕黄色		HgS	红色或黑色
	$CoCl_2$	蓝色	7. 硫酸盐	$CaSO_4$	白色
	$CoCl_2 \cdot H_2O$	蓝紫色		$SrSO_4$	白色
	$CoCl_2 \cdot 2H_2O$	紫红色		$BaSO_4$	白色

续表

序号	物　质	颜　色	序号	物　质	颜　色
	$PbSO_4$	白色		$BaSiO_3$	白色
	$Cr_2(SO_4)_3$	桃红色		$MnSiO_3$	肉色
	$Cr_2(SO_4)_3 \cdot 18H_2O$	紫色		$Fe_2(SiO_3)_3$	棕红色
	$Cr_2(SO_4)_3 \cdot 6H_2O$	绿色	10. 硅酸物	$CoSiO_3$	紫色
	$[Fe(NO)]SO_4$	深棕色		$NiSiO_3$	翠绿色
	$(NH_4)_2Fe(SO_4)_2 \cdot 6H_2O$	浅绿色		$CuSiO_3$	蓝色
7. 硫酸盐	$NH_4Fe(SO_4)_2 \cdot 12H_2O$	浅紫色		$ZnSiO_3$	白色
	$CoSO_4 \cdot 7H_2O$	红色		Ag_2SiO_3	黄色
	$Cu_2(OH)_2SO_4$	浅蓝色		$CaCrO_4$	黄色
	$CuSO_4 \cdot 5H_2O$	蓝色		$SrCrO_4$	浅黄色
	Ag_2SO_4	白色		$BaCrO_4$	黄色
	Hg_2SO_4	白色	11. 铬酸盐	$PbCrO_4$	黄色
	$HgSO_4 \cdot HgO$	黄色		Ag_2CrO_4	砖红色
	$CaCO_3$	白色		Hg_2CrO_4	棕色
	$Mg_2(OH)_2CO_3$	白色		$CdCrO_4$	黄色
	$SrCO_3$	白色		$HgCrO_4$	红色
	$BaCO_3$	白色		CaC_2O_4	白色
	$Pb_2(OH)_2CO_3$	白色		BaC_2O_4	白色
	$Bi(OH)CO_3$	白色	12. 草酸盐	PbC_2O_4	白色
	$MnCO_3$	白色		FeC_2O_4	浅黄色
	$FeCO_3$	白色		$Ag_2C_2O_4$	白色
8. 碳酸盐	$CdCO_3$	白色		$CuCN$	白色
	$Co_2(OH)_2CO_3$	红色		$Cu(CN)_2$	黄色
	$Ni_2(OH)_2CO_3$	浅绿色	13. 拟	$Ni(CN)_2$	浅绿色
	$Cu_2(OH)_2CO_3$	蓝色	卤化物	$AgCN$	白色
	$Zn_2(OH)_2CO_3$	白色		$AgSCN$	白色
	$Cd_2(OH)_2CO_3$	白色		$Cu(SCN)_2$	黑绿色
	$Hg_2(OH)_2CO_3$	红褐色		$BaSO_3$	白色
	Ag_2CO_3	白色	14. 其他	BaS_2O_3	白色
	Hg_2CO_3	浅黄色	含氧酸盐	$NaBiO_3$	浅黄色
	$Ca_3(PO_4)_2$	白色		$Ag_2S_2O_3$	白色
	$CaHPO_4$	白色		$Mn_2[Fe(CN)_6]$	白色
	$BaHPO_4$	白色	15. 其他	$K[Fe(CN)_6Fe]$	深蓝色
9. 磷酸盐	$MgNH_4PO_4$	白色	化合物	$Co_2[Fe(CN)_6]$	绿色
	$FePO_4$	浅黄色		$Ni_2[Fe(CN)_6]$	浅绿色
	Ag_3PO_4	黄色		$Zn_2[Fe(CN)_6]$	白色

续表

序号	物　质	颜　色	序号	物　质	颜　色
15. 其他化合物	$Cu_2[Fe(CN)_6]$	棕红色	15. 其他化合物	$Cd_2[Fe(CN)_6]$	白色
	$Ag_4[Fe(CN)_6]$	白色		$(NH_4)_3PO_4 \cdot 12MoO_3 \cdot 6H_2O$	黄色
	$K_2Ba[Fe(CN)_6]$	白色		$HgNI$	棕红色
	$Pb_2[Fe(CN)_6]$	白色		二丁二酮肟合镍(Ⅱ)	桃红色

附表九　微溶化合物的溶度积
（18 ~ 25℃，$I = 0$）

微溶化合物	K_{sp}	pK_{sp}	微溶化合物	K_{sp}	pK_{sp}
Ag_3AsO_4	1×10^{-22}	22.0	$CaCO_3$	2.9×10^{-9}	8.54
$AgBr$	4.1×10^{-13}	12.39	CaF_2	2.7×10^{-11}	10.57
Ag_2CO_3	8.1×10^{-12}	11.09	$CaC_2O_4 \cdot H_2O$	2.0×10^{-9}	8.70
$AgCl$	1.8×10^{-10}	9.75	$Ca_3(PO_4)_2$	2.0×10^{-29}	28.70
Ag_2CrO_4	2.0×10^{-12}	11.71	$CaSO_4$	9.1×10^{-6}	5.04
$AgCN$	1.2×10^{-16}	15.92	$CaWO_4$	8.7×10^{-9}	8.06
$AgOH$	2.0×10^{-8}	7.71	$CdCO_3$	5.2×10^{-12}	11.28
AgI	9.3×10^{-17}	16.03	$Cd_2[Fe(CN)_6]$	3.2×10^{-17}	16.49
$Ag_2C_2O_4$	3.5×10^{-11}	10.46	$Cd(OH)_2$ 新析出	2.5×10^{-14}	13.60
Ag_3PO_4	1.4×10^{-16}	15.84	$CdC_2O_4 \cdot 3H_2O$	9.1×10^{-8}	7.04
Ag_2SO_4	1.4×10^{-5}	4.84	CdS	8×10^{-27}	26.1
Ag_2S	2×10^{-49}	48.7	$CoCO_3$	1.4×10^{-13}	12.84
$AgSCN$	1.0×10^{-12}	12.00	$Co_2[Fe(CN)_6]$	1.8×10^{-15}	14.74
$Al(OH)_3$ 无定型	1.3×10^{-33}	32.9	$Co(OH)_2$ 新析出	2×10^{-15}	14.7
As_2S_3 [1]	2.1×10^{-22}	21.68	$Co(OH)_3$	2×10^{-44}	43.7
$BaCO_3$	5.1×10^{-9}	8.29	$Co[Hg(SCN)_4]$	1.5×10^{-8}	7.82
$BaCrO_4$	1.2×10^{-10}	9.93	$\alpha\text{-}CoS$	4×10^{-21}	20.4
BaF_2	1×10^{-6}	6.0	$\beta\text{-}CoS$	2×10^{-25}	24.7
$BaC_2O_4 \cdot H_2O$	2.3×10^{-8}	7.64	$Co_3(PO_4)_2$	2×10^{-35}	34.7
$BaSO_4$	1.1×10^{-10}	9.96	$Cr(OH)_3$	6×10^{-31}	30.2
$Bi(OH)_3$	4×10^{-31}	30.4	$CuBr$	5.2×10^{-9}	8.28
$BiOOH$ [2]	4×10^{-10}	9.4	$CuCl$	1.2×10^{-8}	7.92
BiI_3	8.1×10^{-19}	18.09	$CuCN$	3.2×10^{-20}	19.49
$BiOCl$	1.8×10^{-31}	30.75	CuI	1.1×10^{-12}	11.96
$BiPO_4$	1.3×10^{-23}	22.89	$CuOH$	1×10^{-14}	14.0
Bi_2S_3	1×10^{-97}	Cu_2S	Cu_2S	2×10^{-47}	47.7
$CuSCN$	4.8×10^{-15}	14.32	$\gamma\text{-}NiS$	2×10^{-26}	25.7
$CuCO_3$	1.4×10^{-10}	9.86	$PbClF$	2.4×10^{-9}	8.62
$Cu(OH)_2$	2.2×10^{-20}	19.66	$PbCrO_4$	2.8×10^{-13}	12.55
Cu_2S	6×10^{-36}	35.2	$PbCO_3$	7.4×10^{-14}	13.13

续表

微溶化合物		K_{sp}	pK_{sp}	微溶化合物	K_{sp}	pK_{sp}
$FeCO_3$		3.2×10^{-11}	10.50	$PbCl_2$	1.6×10^{-5}	4.79
$Fe(OH)_2$		8×10^{-16}	15.1	PbF_2	2.7×10^{-8}	7.57
FeS		6×10^{-18}	17.2	$Pb(OH)_2$	1.2×10^{-15}	14.93
$Fe(OH)_3$		4×10^{-35}	37.4	PbI_2	7.1×10^{-9}	8.15
$FePO_4$		1.3×10^{-22}	21.89	$PbMoO_4$	1×10^{-13}	13.0
Hg_2Br_2 [3]		5.8×10^{-23}	22.24	$Pb_3(PO_4)_2$	8.0×10^{-43}	42.10
Hg_2CO_3		8.9×10^{-17}	16.05	$PbSO_4$	1.6×10^{-8}	7.79
Hg_2Cl_2		1.3×10^{-18}	17.88	PbS	8×10^{-28}	27.9
$Hg_2(OH)_2$		2×10^{-24}	23.7	$Pb(OH)_4$	3×10^{-66}	65.5
Hg_2I_2		4.5×10^{-29}	28.35	$Sb(OH)_3$	4×10^{-42}	41.4
Hg_2SO_4		7.4×10^{-7}	6.13	Sb_2S_3	2×10^{-93}	92.8
Hg_2S		1×10^{-47}	47.0	$Sn(OH)_2$	1.4×10^{-23}	27.85
$Hg(OH)_2$		3.0×10^{-26}	25.52	SnS	1×10^{-25}	25.0
HgS	红色	4×10^{-53}	52.4	$Sn(OH)_4$	1×10^{-56}	56.0
	黑色	2×10^{-52}	51.7	SnS_2	2×10^{-27}	26.7
$MgNH_4PO_4$		2×10^{-13}	12.7	$SrCO_3$	1.1×10^{-10}	9.96
$MgCO_3$		3.5×10^{-8}	7.46	$SrCrO_4$	2.2×10^{-5}	4.65
MgF_2		6.4×10^{-9}	8.19	SrF_2	2.4×10^{-9}	8.61
$Mg(OH)_2$		1.8×10^{-11}	10.74	$SrC_2O_4 \cdot H_2O$	1.6×10^{-7}	6.80
$MnCO_3$		1.8×10^{-11}	10.74	$Sr_3(PO_4)_2$	4.1×10^{-28}	27.39
$Mn(OH)_2$		1.9×10^{-13}	12.72	$SrSO_4$	3.2×10^{-7}	6.49
MnS 无定形		2×10^{-10}	9.7	$Ti(OH)_3$	1×10^{-40}	40.0
MnS 晶形		2×10^{-13}	12.7	$Ti(OH)_2$ [4]	1×10^{-29}	29.0
$NiCO_3$		6.6×10^{-9}	8.18	$ZnCO_3$	1.4×10^{-11}	10.84
$Ni(OH)_2$ 新析出		2×10^{-15}	14.7	$Zn_2[Fe(CN)_6]$	4.1×10^{-16}	15.39
$Ni_3(PO_4)_2$		5×10^{-31}	30.3	$Zn(OH)_2$	1.2×10^{-17}	16.92
α-NiS		3×10^{-19}	18.5	$Zn_3(PO_4)_2$	9.1×10^{-33}	32.04
β-NiS		1×10^{-24}	24.0	ZnS	2×10^{-22}	21.7

1) 为下列平衡的平衡常数：$As_2S_3 + 4H_2O \rightleftharpoons 2HAsO_2 + 3H_2S$

2) $BiOOH$　$K_{sp} = [BiO^+][OH^-]$

3) $(Hg_2)_mX_n$　$K_{sp} = [Hg_2^{2+}]^m[X^{-2m/n}]^n$

4) $TiO(OH)_2$　$K_{sp} = [TiO^{2+}][OH^-]^2$

附表十　弱酸、弱碱在水中的解离常数
（25℃，$I=0$）

弱　酸	分　子　式	K_a	pK_a
砷酸	H_3AsO_4	$6.3 \times 10^{-3}(K_{a_1})$	2.20
		$1.0 \times 10^{-7}(K_{a_2})$	7.00
		$3.2 \times 10^{-12}(K_{a_3})$	11.50
亚砷酸	$HAsO_3$	6.0×10^{-10}	9.22
硼酸	H_3BO_3	5.8×10^{-10}	9.24
焦硼酸	$H_2B_4O_7$	$1 \times 10^{-4}(K_{a_1})$	4
		$1 \times 10^{-9}(K_{a_2})$	9
碳酸	$H_2CO_3(CO_2 + H_2O)$	$4.2 \times 10^{-7}(K_{a_1})$	6.38
		$5.6 \times 10^{-11}(K_{a_2})$	10.25
氢氰酸	HCN	6.2×10^{-10}	9.21
铬酸	H_2CrO_4	$1.8 \times 10^{-1}(K_{a_1})$	0.74
		$3.2 \times 10^{-7}(K_{a_2})$	6.50
氢氟酸	HF	6.6×10^{-4}	3.18
亚硝酸	HNO_2	5.1×10^{-4}	3.29
过氧化氢	H_2O_2	1.8×10^{-12}	11.75
磷酸	H_3PO_4	$7.6 \times 10^{-3}(K_{a_1})$	2.12
		$6.3 \times 10^{-8}(K_{a_2})$	7.20
		$4.4 \times 10^{-13}(K_{a_3})$	12.36
焦磷酸	$H_4P_2O_7$	$3.0 \times 10^{-2}(K_{a_1})$	1.52
		$4.4 \times 10^{-3}(K_{a_2})$	2.36
		$2.5 \times 10^{-7}(K_{a_3})$	6.60
		$5.6 \times 10^{-10}(K_{a_4})$	9.25
亚磷酸	HPO_3	$5.0 \times 10^{-2}(K_{a_1})$	1.30
		$2.5 \times 10^{-7}(K_{a_2})$	6.60
氢硫酸	H_2S	$1.3 \times 10^{-7}(K_{a_1})$	6.88
		$7.1 \times 10^{-15}(K_{a_2})$	14.15
硫酸	H_2SO_4	$1.0 \times 10^{-2}(K_{a_2})$	1.99
亚硫酸	$H_2SO_3(SO_2 + H_2O)$	$1.3 \times 10^{-2}(K_{a_1})$	1.90
		$6.3 \times 10^{-8}(K_{a_2})$	7.20
偏硅酸	H_2SiO_3	$1.7 \times 10^{-10}(K_{a_1})$	9.77
		$1.6 \times 10^{-12}(K_{a_2})$	11.8
甲酸	$HCOOH$	1.8×10^{-4}	3.74
乙酸	CH_3COOH	1.8×10^{-5}	4.74

弱　酸	分子式	K_a	pK_a
一氯乙酸	CH_2ClOOH	1.4×10^{-3}	2.86
二氯乙酸	CH_2Cl_2OOH	5.0×10^{-2}	1.30
三氯乙酸	CCl_3COOH	0.23	0.64
乳酸	$CH_3CHOHCOOH$	1.4×10^{-4}	3.86
苯甲酸	C_6H_5COOH	6.2×10^{-5}	4.21
草酸	$H_2C_2O_4$	$5.9 \times 10^{-2}(K_{a_1})$	1.22
		$6.4 \times 10^{-5}(K_{a_2})$	4.19
d-酒石酸	$CH(OH)COOH$ | $CH(OH)COOH$	$9.1 \times 10^{-4}(K_{a_1})$	3.04
		$4.3 \times 10^{-5}(K_{a_2})$	4.37
邻苯二甲酸	$C_6H_4(COOH)_2$	$1.1 \times 10^{-3}(K_{a_1})$	2.95
		$3.9 \times 10^{-6}(K_{a_2})$	5.41
柠檬酸	CH_2COOH | $C(OH)COOH$ | CH_2COOH	$7.4 \times 10^{-4}(K_{a_1})$	3.13
		$1.7 \times 10^{-5}(K_{a_2})$	4.76
		$4.0 \times 10^{-7}(K_{a_3})$	6.40
苯酚	C_6H_5OH	1.1×10^{-10}	9.95
乙二胺四乙酸	$H_6\text{-EDTA}^{2+}$	$0.1(K_{a_1})$	0.9
	$H_5\text{-EDTA}^{+}$	$3 \times 10^{-2}(K_{a_2})$	1.6
	$H_4\text{-EDTA}$	$1 \times 10^{-2}(K_{a_3})$	2.0
	$H_3\text{-EDTA}^{-}$	$2.1 \times 10^{-3}(K_{a_4})$	2.67
	$H_2\text{-EDTA}^{2-}$	$6.9 \times 10^{-7}(K_{a_5})$	6.16
	$H\text{-EDTA}^{3-}$	$5.5 \times 10^{-2}(K_{a_6})$	10.26
氨水	$NH_3 \cdot H_2O$	1.8×10^{-5}	4.74
联氨	H_2NHH_2	$3.0 \times 10^{-6}(K_{b_1})$	5.52
		$7.6 \times 10^{-15}(K_{b_2})$	14.12
羟氨	NH_2OH	9.1×10^{-9}	8.04
甲胺	CH_3NH_2	4.2×10^{-4}	3.38
乙胺	$C_2H_5NH_2$	5.6×10^{-4}	3.25
二甲胺	$(CH_3)_2NH$	1.2×10^{-4}	3.93
二乙胺	$(C_2H_5)_2NH$	1.3×10^{-3}	2.89
乙醇胺	$HOCH_2CH_2NH_2$	3.2×10^{-5}	4.50
三乙醇胺	$(HOCH_2CH_2)_3N$	5.8×10^{-7}	6.24
六次甲基四胺	$(CH_2)_6N_4$	1.4×10^{-9}	8.85
乙二胺	$H_2NCH_2CH_2NH_2$	$8.5 \times 10^{-5}(K_{b_1})$	4.07
		$7.1 \times 10^{-8}(K_{b_2})$	7.15
吡啶	C_5H_5N	1.7×10^{-9}	8.77